Clean Energy Production Technologies

Series Editors

Neha Srivastava, Department of Chemical Engineering and Technology, IIT (BHU) Varanasi, Varanasi, India

P. K. Mishra, Department of Chemical Engineering and Technology, IIT (BHU) Varanasi, Varanasi, India

The consumption of fossil fuels has been continuously increasing around the globe and simultaneously becoming the primary cause of global warming as well as environmental pollution. Due to limited life span of fossil fuels and limited alternate energy options, energy crises is important concern faced by the world. Amidst these complex environmental and economic scenarios, renewable energy alternates such as biodiesel, hydrogen, wind, solar and bioenergy sources, which can produce energy with zero carbon residue are emerging as excellent clean energy source. For maximizing the efficiency and productivity of clean fuels via green & renewable methods, it's crucial to understand the configuration, sustainability and techno-economic feasibility of these promising energy alternates. The book series presents a comprehensive coverage combining the domains of exploring clean sources of energy and ensuring its production in an economical as well as ecologically feasible fashion. Series involves renowned experts and academicians as volume-editors and authors, from all the regions of the world. Series brings forth latest research, approaches and perspectives on clean energy production from both developed and developing parts of world under one umbrella. It is curated and developed by authoritative institutions and experts to serves global readership on this theme.

Neha Srivastava
Editor

Sustainable Nanostrategies for Biorefineries

Environmental, Economic, and Social Perspectives

Springer

Editor
Neha Srivastava
Department of Chemical Engineering
and Technology
IIT (Banaras Hindu University)
Varanasi, Uttar Pradesh, India

ISSN 2662-6861 ISSN 2662-687X (electronic)
Clean Energy Production Technologies
ISBN 978-981-95-5874-2 ISBN 978-981-95-5875-9 (eBook)
https://doi.org/10.1007/978-981-95-5875-9

This Springer imprint is published by the registered company Springer Nature Singapore Pte Ltd.
The registered company address is: 152 Beach Road, #21-01/04 Gateway East, Singapore 189721, Singapore

Preface

This book entitled *Sustainable Nanostrategies for Biorefineries: Environmental, Economic, and Social Perspectives* provides a way forward toward sustainable bioenergy production via feasible nanotechnology inclusion. Nanotechnology and nanomaterial have made the things smoother and have also proven its potential role in the area of bioenergy. However, in spite of technical advantages, the area has restricted limitations which restrict its sustainable and huge application in this area. High production cost, emerging developing technology in this area, and toxic use are the prominent factors of nanotechnology that raise environmental, economic, and social sustainability concerns among researchers working in this area.

Therefore, focusing on these facts, this book aims to explore and tries to provide suggested boundaries as well as sustainable application between nanostrategies and biorefineries. Based on this, the current book is divided into 13 chapters, where Chaps. 1 and 2 provide current status update on bioenergy production and socio-economic balance in bioenergy and nanotechnology. Chapters 3 and 4 focus on details about case study of commercial bioprocess technology for bioenergy production and current scenario as well as status of nanotechnology area. Chapters 5 and 6 deliver information about green nanotechnology, and its properties and applications. Chapters 7 and 8 discuss on sustainable assessment of nanotechnology industries in bioenergy area for long run application and its socioeconomic and environmental impact. Chapters 9 and 10 present comparative analysis of nanotechnology and green nanotechnology along with socioeconomic and environmental debate in this area. Chapters 11 and 12 discuss on geo-environmental consequences of bioenergy and its socioeconomic and environmental impact expansion. Finally, Chap. 13 provides future insights about the nanotechnology application in bioenergy area for long run and application. This book certainly sets a promising milestone in the emerging area of nanotechnology and bioenergy for sustainable commercial scale operations and implementations.

Varanasi, Uttar Pradesh, India Neha Srivastava

Contents

Editors and Contributors

About the Editor

Neha Srivastava is currently affiliated with the Department of Chemical Engineering and Technology at IIT (BHU), Varanasi, Uttar Pradesh, India. She previously worked in the area of bioenergy at the Department of Molecular and Cellular Engineering, SHIATS, Uttar Pradesh, India, where she also earned her PhD in Biotechnology in 2016.

Her research interests include bioprocess technology and biofuels production, with a focus on microbial screening, enzyme production and enhancement, biohydrogen and bioethanol production from waste biomass, green nanocatalysts, and sustainable process development.

She has an extensive publication record in peer-reviewed SCI-indexed journals, along with several patents, a technology transfer, and numerous edited books to her credit. She serves as the Editor-in-Chief of a few international journals and as Series Editor for multiple international book series. Her accomplishments also include contributions to popular science communication and the deposition of several potential microbial strains.

Contributors

Banzeer Ahsan Abbasi Department of Botany, Rawalpindi Women University, Rawalpindi, Pakistan

Muhammad Sajjad Ahmad Department of Biotechnology, University of Sargodha, Sargodha, Pakistan

Muhammad Zubair Ali Department of Biotechnology, University of Sargodha, Sargodha, Pakistan

Jahnavi Alwala Department of Biochemistry, Chaitanya Deemed to be University, Hyderabad, Telangana, India

Muhammad Anas Department of Soil Science and Plant Nutrition, Selçuk University, Konya, Turkey

Shailendra Kumar Arya UIET, Biotechnology Branch, Panjab University, Chandigarh, India

Olga Babich Institute of Living Systems, Immanuel Kant BFU, Kaliningrad, Russia

Mahesh Chand Department of Chemistry, Guru Gobind Singh IP University Delhi, Delhi, India

Petros Chavula Department of Agricultural Economics and Extension, School of Agricultural Sciences, University of Zambia, Lusaka, Zambia

Africa Centre of Excellence for Climate-Smart Agriculture and Biodiversity Conservation, Haramaya University, Dire Dawa, Ethiopia

Pooja Devi Department of Physics, Kalindi College, University of Delhi, Delhi, India

Marcelo Franco Department of Exact Science, State University of Santa Cruz, Ilheus, Brazil

Rajni Garg Department of Applied Sciences And Humanities, Galgotias College of Engineering and Technology, Greater Noida, UP, India

Deepti Goyal Department of Chemistry, Gautam Buddha University, Greater Noida, India

Erdogan E. Hakki Department of Soil Science and Plant Nutrition, Selçuk University, Konya, Turkey

Javed Iqbal Department of Botany, Bacha Khan University, Charsadda, Khyber Pakhtunkhwa, Pakistan

Muhammad Irfan Department of Biotechnology, University of Sargodha, Sargodha, Pakistan

Muhammad Islam Department of Biotechnology, University of Sargodha, Sargodha, Pakistan

Maryam Kainat Department of Biotechnology, University of Sargodha, Sargodha, Pakistan

Fredrick Kayusi Department of Environmental Studies, Geography and Planning, Maasai Mara University, Narok, Kenya

Department of Environmental and Earth Sciences, Pwani University, Kilifi, Kenya

Muhammad Khan Institute of Zoology, University of the Punjab New Campus, Lahore, Pakistan

Waseem Ahmed Khattak Department of Plant Sciences, Faculty of Biological Sciences, Quaid-i-Azam University Islamabad, Islamabad, Pakistan

Yuliya Kulikova Institute of Living Systems, Immanuel Kant BFU, Kaliningrad, Russia

Sakshi Kabra Malpani Independent Researcher and Consultant, Issaquah, WA, USA

Maryam Maqsood Department of Biotechnology, University of Sargodha, Sargodha, Pakistan

Tehreem Maryyam Department of Biotechnology, University of Sargodha, Sargodha, Pakistan

Rajesh Kumar Meena Department of Chemistry, Kalindi College, University of Delhi, Delhi, India

Ramchander Merugu Department of Biochemistry, Chaitanya Deemed to be University, Hyderabad, Telangana, India

Department of Biochemistry, Mahatma Gandhi University, Nalgonda, Telangana, India

Akhtar Munir Department of Chemistry, Quaid-i-Azam University, Islamabad, Pakistan

Farheen Naaz Center for Gene Regulation in Health and Disease, Department of Biological, Geological, and Environmental Sciences, College of Arts and Sciences, Cleveland State University, Cleveland, OH, USA

Vagdevi Pally Department of Biochemistry, Mahatma Gandhi University, Nalgonda, Telangana, India

Anjali Pandit Dr. B Lal Institute of Biotechnology, Malviya Industrial Area, Jaipur, Rajasthan, India

Subhadra Rajpoot University Centre for Research & Development, Department of Biotechnology, Chandigarh University, Mohali, Punjab, India

Sonia Sethi Dr. B Lal Institute of Biotechnology, Malviya Industrial Area, Jaipur, Rajasthan, India

Hafiz Abdullah Shakir Institute of Zoology, University of the Punjab New Campus, Lahore, Pakistan

Gursharan Singh School of Allied Medical Sciences, Lovely Professional University, Phagwara, India

Jatinder Singh Department of Horticulture, SAGR, Lovely Professional University, Phagwara, India

Sheetal Thakur University Centre for Research & Development, Department of Biotechnology, Chandigarh University, Mohali, Punjab, India

Madhavi Tiwari Department of Applied Sciences, Shri Rawatpura Sarkar University, Raipur, Chhattisgarh, India

Ozod Sunnatalievich Turaev Research Institute of Plant Genetic Resources, Tashkent, Uzbekistan

Rani Padmini Velamakanni Department of Biochemistry, Mahatma Gandhi University, Nalgonda, Telangana, India

Annu Yadav Department of Chemistry, Kalindi College, University of Delhi, Delhi, India

Chapter 1
Bioenergy Utilization: Current Status

Tehreem Maryyam, Maryam Maqsood, Yuliya Kulikova, Olga Babich, Marcelo Franco, and Muhammad Irfan

Abstract Bioenergy is considered a renewable source of energy that is produced from organic substances such as plants and animal waste. According to the reports, bioenergy currently makes up about 10% of the global primary energy supply. The world is facing an energy crisis, which threatens its future. Bioenergy holds significant potential to ensure energy security in the future. It can also aid in climate change mitigation through the reduction of greenhouse gas emissions. The current trend of bioenergy utilization is increasing worldwide attributed to the growing demand for sustainable energy and advancements in bioenergy production. The United States, Brazil, China, and Germany are at the forefront of bioenergy production worldwide. Bioenergy is being utilized globally to generate electricity, heat, and biofuels. The top bioenergy consuming countries are reported to be India, China, the United States, and Brazil. This chapter summarizes the current state of bioenergy worldwide, its production methods, uses, and challenges that need to be overcome to promote its use.

Keywords Bioenergy · Biofuels · Renewable energy · Bioenergy utilization · Biomass · Bioenergy technologies

1.1 Introduction

Bioenergy represents a form of renewable energy derived from diverse biomass sources (Ferrari et al. 2020). Biomass is an organic matter, which is a major source of bioenergy (Saravanan et al. 2025). Food waste, animal dung, forest and agricultural

T. Maryyam · M. Maqsood · M. Irfan (✉)
Department of Biotechnology, University of Sargodha, Sargodha, Pakistan
e-mail: irfan.ashraf@uos.edu.pk

Y. Kulikova · O. Babich
Institute of Living Systems, Immanuel Kant BFU, Kaliningrad, Russia

M. Franco
Department of Exact Science, State University of Santa Cruz, Ilheus, Brazil

N. Srivastava (ed.), *Sustainable Nanostrategies for Biorefineries*, Clean Energy Production Technologies, https://doi.org/10.1007/978-981-95-5875-9_1

remnants, industrial waste materials, and municipal solid waste (MSW) are all examples of biomass applied in bioenergy production (Jamil et al. 2024; Rajvanshi et al. 2022). Through mechanical, physical, biochemical, and thermochemical processes, these biomasses can be transformed into bioenergy. The bioenergy produced through these processes comes in various forms, including liquid biofuels like biodiesel and bioethanol, gaseous biofuels like biogas and syngas, and solid biofuels like biopellets and briquettes (Suhartini et al. 2024). At present, biochemical conversion methods are the most commonly used techniques for converting biomass into biofuels. Biochemical conversion, particularly through fermentation, produces biofuels such as biodiesel and ethanol, while thermochemical conversion produces bio-oil and syngas (Osman et al. 2023b). By 2050, it is anticipated that biomass will make up 10–50% of global energy usage (Ambaye et al. 2021).

For centuries, bioenergy was the main energy source that powered human life and activities until around 1840. By the year 1905, fossil energy emerged as the dominant source. The world's energy consumption has multiplied nearly ten times over the past century. This surge was due to a rapid increase in population and industrial development. More than 80% of the world's energy demand is currently fulfilled by fossil fuels, making them the main energy source (Guo et al. 2015). Across both established and developing countries, fossil fuels are the principal source of electricity and energy generation. This is further employed for residential use, transportation, and industrialization. These fossil fuels are limited and non-renewable with limited efficiency and low power generation capabilities. Consequently, a gap forms between the global demand for energy and its supply (Kazmi et al. 2025). In 2018, it was found that the majority of global energy needs, around 85%, were fulfilled by fossil fuels. Furthermore, many studies state that fossil fuel reservoirs are depleting rapidly, and the world will run out of fossil fuels in the coming years (Hosseini 2022). Humans are exploiting these reserves at a pace 50% faster than their production in nature. It is highly concerning that the world is facing unprecedented pressure on its finite fossil fuel reserves due to population growth, climate change, and environmental degradation. Global fossil fuels demand has increased nearly ten times since the start of the twentieth century. It is expected to increase fourfold by 2030 relative to 2010. If the present global scenario continues unchecked, human demand is expected to outpace Earth's regeneration capacity by 75% by 2030. By 2050, this demand will outstrip the planet's capacity by 100% (Potrč et al. 2022).

Bioenergy is sourced from renewable feedstocks that are sustainable and will not deplete overtime. This makes bioenergy a sustainable solution (Khan et al. 2022). Bioenergy contributes significantly to the clean energy transformation, particularly important for countries with low energy self-sufficiency. Bioenergy offers several advantages. It improves energy security and lessens dependency on fossil fuels, which eventually results in lower greenhouse gas (GHG) emissions. Additionally, bioenergy helps local economies grow and provides clean energy to underserved communities. Hence, it contributes to sustainable development (Ersoy and Ugurlu 2024). Bioenergy is currently the world's most predominant renewable energy source, providing over 55% of the primary renewable energy supply (Röder and

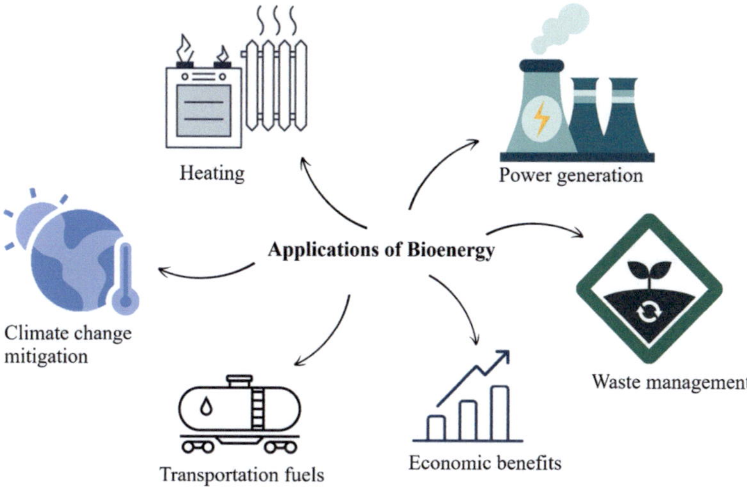

Fig. 1.1 Some key applications of the bioenergy

Welfle 2019). Moreover, it contributes about 10% to the global consumption of primary energy (Duarah et al. 2022). Further, global primary bioenergy supply could increase up to 6 times by 2050 (37% of demand). Thus, bioenergy can play a vital role in fulfilling global energy demands (Errera et al. 2023).

It is claimed that bioenergy is cost-effective and environmentally benign when it comes to managing garbage, because it facilitates the removal of garbage and other things that may lead to pollution and other negative effects on the environment. When bioenergy is produced from waste or leftovers that are effectively transformed into energy products or from sustainable biomass growth, it is a realistic way to address the effects of climate change. Figure 1.1 presents the overview of different applications of bioenergy. Because bioenergy is renewable and has a significant capacity, it is clearly superior to conventional fossil fuels in the context of environmental protection. The decarbonization of all energy sectors might be aided by sustainable bioenergy, which would quickly lessen reliance on fossil fuels and, as a result, reduce greenhouse gas emissions through replacement. Achieving environmental sustainability and reducing the adverse effects of traditional energy sources will need the efficient management and use of residues for energy (Ersoy and Ugurlu 2024).

Global output of bioenergy has increased at an unparalleled rate in the last few decades (Kang et al. 2015). According to IRENA's 2020 projections, global bioenergy production is estimated around 115.7 GW. With 38.5 GW, the European Union is the largest contributor, while Asia comes next with 36.27 GW. In 2019, the total capacity of renewable energy worldwide was reached a total of 2533 GW (Ambaye et al. 2021). Currently, 10% of the world's energy supply comes from biomass; two-thirds of this bioenergy is produced in developing nations, with the remaining portion coming from developed nations. The two top producers of bioethanol, accounting for more than 85% of the global market, are the United States and Brazil. Brazil has an abundance of biomass resources to meet its whole primary energy needs (Welfle 2017).

1.2 Global and Regional Status of Bioenergy Utilization

1.2.1 Global Bioenergy Utilization Patterns

Bioenergy is regarded as one of the most flexible and versatile forms of renewable energy with minimal carbon emissions (Aidonojie et al. 2023). The decade 2009–2019 witnessed a rapid expansion in renewable energy demand, rising by 15.1 exajoules (Duarah et al. 2022). Bioenergy has the capability to contribute significantly to the global energy mix. According to estimates, it could provide between 10% and 60% or more of the world's primary energy needs. By 2060, bioenergy production is projected to more than double, increasing from the existing level of 56 EJ to 145 EJ. This projection suggests that bioenergy can play a vital role in ensuring future energy demands. Bioenergy is now playing a more important role compared to traditional biomass uses (Duarah et al. 2022).

Globally, bioenergy is utilized across three primary sectors: electricity generation, heating/cooling, and transportation. In addition to producing power, it can be utilized for a variety of tasks like cooking, heating, and transportation (Qamar et al. 2024). At present, more than 80% of bioenergy is utilized for heating and cooking purposes in buildings and industrial settings. In terms of global heat consumption, bioenergy was responsible for around 20% in 2020. This was comprised of 8% modern bioenergy and 12% traditional biomass use. The usage of bioenergy to generate electricity is growing. It depends on using a range of resources, including forestry and agricultural wastes, biogas, pellets, and municipal solid trash (MSV). The most common biofuel for generating sustainable and clean electricity is wood pellets (Ersoy and Ugurlu 2024).

According to the FAOSTAT Bioenergy domain of the UN's Food and Agriculture Organization (FAO), world bioenergy production increased from about 28 exajoules (EJ) to 50 EJ between 1990 and 2023. During the same timeframe, world bioenergy final consumption increased from 24 EJ to 40 EJ. In 2023, solid biofuels dominated global bioenergy final consumption making up 89% (35 EJ). Additionally, liquid biofuels contributed 4.4 EJ (11%), and gaseous biofuels contributed a negligible 0.1 EJ (0.3%). Asia is the largest consumer of solid biofuels (17.5 EJ). The countries with the highest bioenergy consumption in 2023 were India (7.6 EJ), China (4.7 EJ), the United States (3.3 EJ), Brazil (2.0 EJ), and Ethiopia (1.4 EJ) (Fig. 1.2).

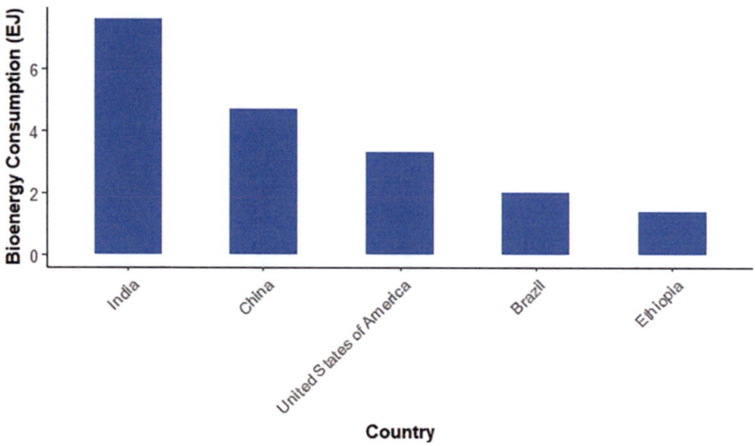

Fig. 1.2 Largest bioenergy consumers in the year 2023

1.2.2 Regional Bioenergy Utilization Patterns

Based on the comprehensive statistical data on renewable energy from 2020, the worldwide renewable energy output and capacity had reached a cumulative total of 2533 GW in 2019, out of which hydropower has the major power output (1308 GW). The next largest contributors were wind energy at 622 GW, followed by solar energy at 585, and bioenergy at 126 GW. Based on IRENA's 2020 global estimates, the overall bioenergy production was stated to be approximately 115.7 GW. Bioenergy production was highest in the European Union at 38.5 GW, and then Asia with 36.27 GW followed closely. Furthermore, the top bioenergy producer among Asian countries was China at 16.54 GW, while India followed at 10.23 GW (Ambaye et al. 2021) (Fig. 1.3).

From 2007 to 2017, global biofuels production expanded at a growth rate of 11.4% per year (Ebadian et al. 2020).

Biofuels can be categorized into several liquid forms, including bioethanol, biodiesel, biomethanol, biobutanol, bio-oil and green diesel (Fivga et al. 2019). During World War II, energy was in high demand and countries faced significant fuel shortages. At that time, bioethanol synthesis from cellulosic raw material was commenced in some countries, including United States, China, Germany, Russia, Korea, and Switzerland (Pirzadah et al. 2014). Bioethanol serves as a transportation fuel and is also utilized in pharmaceuticals, cosmetics, and various other industries. This growing consumption demands an increased production. Bioethanol has high burning limit so it can be used as an alternative to gasoline (Kazmi et al. 2025). In 2005, Brazil ranked first along with United States and produced 99% of the total ethanol production across world. Germany and France accounted for approximately 69% of the overall diesel production in world (Manikandan et al. 2022).

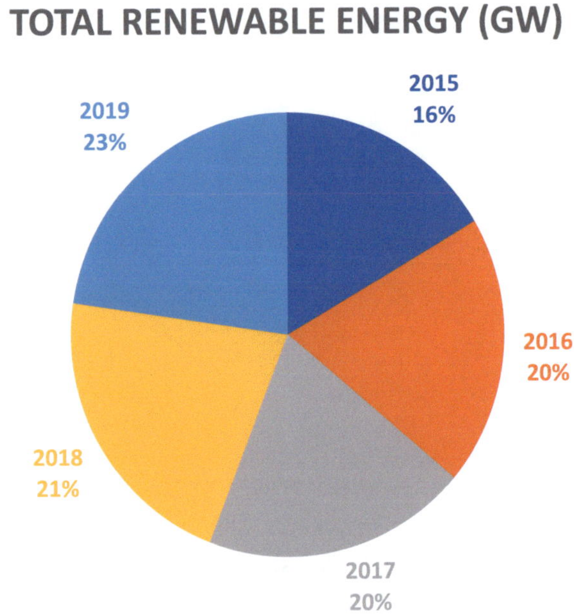

Fig. 1.3 Renewable energy generation and capacity globally (Ambaye et al. 2021)

The year 2010 witnessed global biofuel production of 105 billion liters, which was 17% more than in 2009. It is about 2.7% of the overall transportation fuels in the world. In the same year, ethanol fuel production around the world was 86,000 million liters. Out of this, 90% of ethanol fuel was made by United States and Brazil. Conversely, the European Union contributed the largest volume of biodiesel in the world. They supplied about 53% of global production in 2010 (Pereira et al. 2013).

United States and Brazil were again the dominant producers of biofuels in 2018. They have produced 69% of the world's biofuels, while the European Union made 9% (Jeswani et al. 2020). Hence, it is obvious that the United States and Brazil are at the forefront of ethanol production globally (Kazmi et al. 2025). Brazil utilizes soybeans as primary feedstock for biodiesel production. Approximately, 78% of feedstock used for producing biodiesel is derived from soybeans. The primary product from soybeans is a protein-dense meal, which is primarily used as animal feed for livestock (Evaristo et al. 2016). The United States is using corn and Brazil is using sugarcane as feedstock for ethanol production (Sarris and Papanikolaou 2016; Subramaniam and Masron 2021). Bioethanol production from sugarcane and corn has increased to almost 30,000 million gallons in 2019, in which the Brazil, Unites states, China, and EU are the leading producers (Letti et al. 2022).

In a report published by Statista's research unit, it was found that in 2024, the United states led the list of top ten biofuel production countries with 1917 petajoules, followed by Brazil with 1143 petajoules. Brazil and Indonesia took second and third place, producing approximately 1143 and 459 petajoules, respectively. In comparison, Germany produced about 149 petajoules of biofuel that year, ranking it as the top producer in Europe and one of the top five countries in the world (Fig. 1.4).

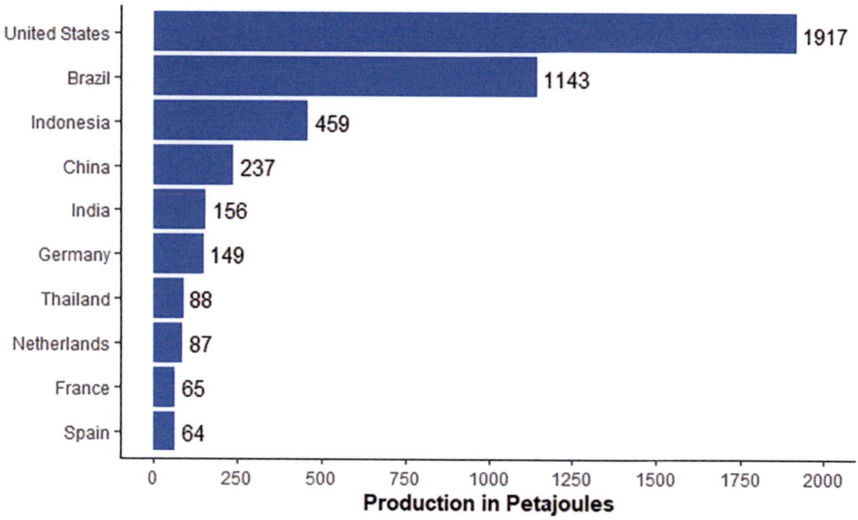

Fig. 1.4 The top global leaders in biofuel output for 2024. (Source: https://www.statista.com/statistics/274168/biofuel-production-in-leading-countries-in-oil-equivalent/)

The European Union, particularly Germany, leads in biogas capacity. The United States, United Kingdom, Brazil, Italy, and China represent more than 86% of worldwide biogas capacity as of 2020. With 37%, Germany had the largest share for global biogas capacity, followed by the United States with 11%. The United Kingdom, Italy, and China had shares of 9%, 7%, and 4%, respectively. On a global scale, of the total biogas generated, two-thirds of its output was used for power generation purposes. Half of this was dedicated to electricity generation, while the other half was used for combined heat and power (CHP) applications (Ersoy and Ugurlu 2024).

Europe has a fundamental role in biogas production and utilization. The global capacity for biogas production has seen significant growth. Global biogas production capacity has increased dramatically. Biogas production capacity increased from around 65 GW in 2010 to more than 120 GW in 2019. This growth in biogas capacity is the result of increasing demand for renewable energy and concern about climate change. With significant monetary contributions to biogas infrastructure such as biogas facilities and technologies, Europe remains the global leader in biogas generation. The region accounts for over 70% of the world's total biogas output (Abanades et al. 2022). In terms of biogas production capacity measured in megawatts (MW), Europe and the European Union have the biggest share of biogas production capacity. In contrast, African countries and South America have much lower biogas production capacities (Agdew et al. 2025) (Table 1.1).

European countries, particularly Germany, Latvia, Sweden, France, and others, are significant producers and consumers of solid biofuels like wood pellets. The EU28 utilizes solid biomass energy for several purposes, including energy

Table 1.1 Comparative overview of bioenergy utilization by leading countries

Country	Primary feedstock	Key bioenergy applications	Notable policies/ programs	Reference
Brazil	Sugarcane	Ethanol for transport, biogas for electricity	Proálcool Program, Blending Mandate	Cortez and Baldassin Jr (2016)
United States	Corn, soy, forestry residues	Ethanol, biodiesel, biomass power	Renewable Fuel Standard (RFS)	Moore et al. (2014) and Page-Dumroese et al. (2022)
Germany	Energy crops, animal waste	Biogas, district heating, co-firing	Renewable Energy Act (EEG)	Thrän et al. (2023)
China	Crop residues, MSW, forestry	Biomass power, biogas, ethanol	14th Five-Year Plan, Biogas Subsidy	Qaseem and Wu (2021)
India	Rice husk, cattle dung, molasses	CBG, electricity, ethanol blending	SATAT Scheme, Ethanol Blending Policy	Kumar et al. (2023, 2024)

production, industrial applications, and households. In 2017, 22% of solid biomass energy was used by the industrial sector (Malico et al. 2019).

The current status of bioenergy, particularly biofuel production, varies across countries. Recently in 2020, biofuel production worldwide declined by 6% in 2020. China is nominated as a significant biofuel producer, producing 1140 million liters of biodiesel and over 3000 million liters of ethanol, with ambitious goals for increasing future production. The European Union is also a major player, leading in biodiesel production and accounting for 31% of the world's total output, which totaled 14,600 million liters in 2019. These countries have implemented policies to encourage bioenergy growth, aiming for economic viability (Duarah et al. 2022).

1.3 Technologies for Bioenergy Production

Different technologies are used to produce bioenergy by converting organic materials into energy (Fig. 1.5).

1.3.1 Physical Methods

Changing biomass by preprocessing, size reduction/comminution, drying, and densification is known as physical or mechanical biomass conversion. Through this process, biomass is changed into forms that have better qualities than raw biomass, such as increased mass density, energy density, and hydrophobicity (Orisaleye et al. 2023). The three most widely utilized biomass conversion technologies are briquetting, extraction, and distillation.

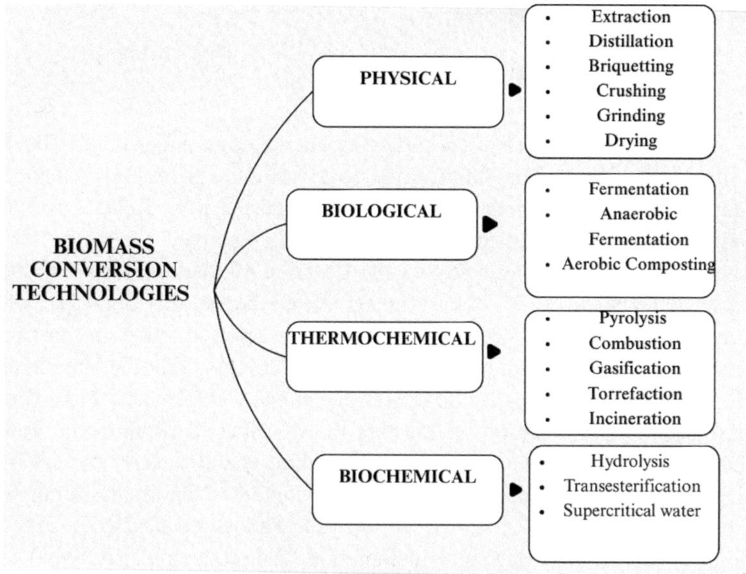

Fig. 1.5 Summary of biomass conversion technologies

1.3.1.1 Extraction Process

Extraction is a mechanical conversion process that produces oil from the seeds of various biomass crops, such as groundnuts, cotton, and so on. In addition to oil, this process produces extra solid, or "cake," which is suitable for feeding animals (Kumar et al. 2015). For full oil extraction, the cell walls must be broken down utilizing various mechanical or non-mechanical techniques (Makareviciene and Sendzikiene 2022). The extraction parameter optimization phases, which are reliant on the oil accumulation in the vegetable fractions, the physical-chemical characteristics of the oil, and its possible uses, have an impact on the efficiency of the oil recovery process (Sorita et al. 2023).

1.3.1.2 Distillation Process

Oilseeds are crushed in order to extract oil during the distillation process. It also includes the vaporization of essential oils and steam distillation of the oil mixture. To obtain bio-oil, the vaporized oils are subsequently collected and either cooled or condensed (Awogbemi and Kallon 2021). The distillation method's high energy requirements are one of its disadvantages. The process of heating a liquid, turning it into vapor, and then condensing the vapor back into liquid at the liquid condenser requires a lot of energy (Thatoi et al. 2016).

1.3.2 Biological Technologies

1.3.2.1 Anaerobic Digestion

The anaerobic action of certain bacteria produces biogas under controlled conditions, with a calorific value ranging from 20 to 23 MJ/m^3. The biological degradation of organic materials under anaerobic conditions releases around 800 × 105 tons of methane into the surrounding environment. Biogas's restricted use can be attributed to its complicated composition (Kapoor et al. 2020). Methane, carbon dioxide, and other gases that make up 1–5% of hydrogen make up the majority of biogas, which is a mixture of several gases. While CO_2 reduces the calorific value and restricts the ability to transport biogas due to its abnormal combustion, methane makes it possible for biogas to burn easily. Other components of biogas, like siloxanes, hydrogen sulfide, and water vapor corrode the mechanical components, lowering the heating value. Due to the calorific biogas value, which can rise to 35.7 MJ/m^3, and other corrosive components, it is crucial to guarantee the separation of CO_2 in order to expand and boost the use of biogas (Sahota et al. 2018). Anaerobic digestion, aerobic composting, and incineration (thermochemical process) are the three primary biogas production systems utilized globally (Tshikovhi and Motaung 2023). Given the intricacy of the factors influencing the different (dis)advantages, choosing the best technology can be difficult.

1.3.2.2 Composting in an Aerobic Environment

Microorganisms break down biomass in aerobic composting/degradation by using oxygen from the air to produce heat, CO_2, and a solid product (Tekin et al. 2014). It is frequently employed in the treatment of biomass waste and uses microorganisms in particular external conditions to transform organic waste biomass into stable humus (Jia et al. 2021; Wu et al. 2022). During the composting process, the bacteria readily transform carbon and nitrogen compounds into energy and protein sources, generating heat, CO_2, H_2O, NH_3, organic acids, and finished compost products in the end (Vakili et al. 2015). Among the many advantages of aerobic composting are the reduction of animal manure quantity, pathogen destruction, and the conversion of manure waste into nutrient-dense, stable organic matter (such as compost) that may improve soil fertility. However, composting processes emit too many greenhouse gases (GHGs), including CO_2, CH_4, and N_2O, which lowers the agronomic value and adds to environmental pollution (Lu et al. 2019).

1.3.2.3 The Process of Fermentation

In the presence of microorganisms (i.e., bacteria, fungi, and yeast), the fermentation process converts biomass into desired bio-products like bioethanol, bio-hydrogen, and biogas (Hossain 2019). The fermentation of biomass is fraught with difficulties. This is because a simultaneous saccharification co-fermentation (SSCF) phase is required to convert pentose and a large number of hexose sugars to ethanol. Arabinose sugars, glucose, and xylose can be fermented by natively occurring Pichia and Saccharomyces yeast species, Klebsiella, Zymomonas, and *Escherichia coli* bacteria (Mielenz 2001). The following advantages of the fermentation process are well known:

- The entire biomass, including lignin, may be utilized regardless of its quality.
- Expensive enzyme removal and difficult pretreatment procedures.
- Greater selectivity of the biocatalyst.
- The H2: CO ratio for the independence of bioconversion.
- Because syngas is produced at high temperatures, the fermentation process is aseptic.
- Bioreactor functioning in typical circumstances.
- Metal poisoning has serious problems (Munasinghe and Khanal 2010).

1.3.3 Thermo-Chemical Technologies

Thermochemical conversion—which encompasses combustion, torrefaction, pyrolysis, hydrothermal liquefaction, and gasification—is the most practical technique for converting biomass into biofuel (Asadullah et al. 2014). The main goal of thermochemical conversion is to eliminate undesirable byproducts by modifying process variables (Hoang et al. 2021). The development and application of bio-based feedstocks has become more important in recent years as a potential substitute for energy sources derived from fossil fuels. The energy systems powered by biomass reduced dependency on fossil fuels and could potentially help eradicate the causes of global warming (Ahmadi et al. 2020). The simplest way to produce CO_2, tar, water vapor, heat, alkaline ash particles, and smoke as a byproduct of the biomass reaction is through combustion (Sadaka and Johnson 2000). Remarkably, about 90% of the biodegradable materials used as energy derivatives are processed by combustion. Direct combustion may have been used to generate the energy required for heating, steaming, or drying agricultural items throughout the prehistoric era (Michaelides 2012). In 1989, the USA saw the introduction of the first biomass-based power plant with a 6 GW power supply (Van den Broek et al. 1996). It was originally believed that organic waste could be removed by direct combustion using biomass as fuel. The low-efficiency boiler caused by the use of these inexpensive fuels has opened up new technological options for energy production. Additionally, regional concerns like taxes, electricity prices, and pollution control regulations (environmental laws) have affected the efficiency vs. investment trade-off.

Fuel, energy, and environmental considerations interact to form the three fundamental combustion applications. When biomass fuel is burned in a boiler, flammable vapors are released, which volatilize and burn like flames. Three components make up this unpredictable deteriorating combustion product: a condensable fraction of H_2O, aldehydes, and ketones; a part made up of alcohols and tar; and a vaporous component of CO, CO_2, and H_2. This causes the residue from the carbon char to burn in more air. Depending on a few more variables, the heat produced by combustion may be used as a resource for other conversion processes that produce electrical energy. Additionally, burning biomass is not always the best option for producing bioenergy because it might result in large carbon dioxide generation in some situations. However, in practice, depending on the source of the biomass and the type of combustion reaction, other gases such as COX, SOX, NOX, and CH_4 are also emitted. Such emissions must be properly regulated since the generated gases have the potential to influence both the environment and human health.

1.3.4 Biochemical Transformation

Using particular microbial populations, biochemical transformation produces fermentable carbohydrates that can be transformed into gaseous molecules like methane or liquid fuels like ethanol and butanol (Gautam et al. 2020). One essential and economical technique for creating fuels, chemicals, and minerals from biomass is biochemical conversion (Wang et al. 2019). The hydrolysis of lignocellulosic polysaccharides, such cellulose and hemicellulose, into monosaccharides and subsequent fermentation into fuels like ethanol is the most prevalent type of biochemical conversion. Because of its compact and inflexible structure, which prevents bacterial and other microbes from breaking it down, this technique is essential for the commercial use of lignocellulose as a raw material (Rastogi and Shrivastava 2017). This natural hardness makes it difficult to convert lignocellulosic biomass into sugars and, eventually, biofuels at a reasonable cost. Anaerobic digestion produces biogas, whereas ethanol fermentation and dark fermentation/photobiology create ethanol and biohydrogen from fermentable biomass, respectively (Osman et al. 2023a).

1.3.5 Hybrid Methodology

Research is also conducted on hybrid or combined systems that use a combination of various technologies, such as biochemical and thermochemical technologies. However, each bio-refinery technique can produce a specific fuel depending on the type and availability of the feedstock. Therefore, if these techniques could be combined under the concept of an integrated waste bio-refinery, a variety of feedstocks that are heterogeneous might be processed to produce a wide range of products, including fuel, power, heat, valuable chemicals, and fodder (Nizami et al. 2017). Systems with hybrid configurations produce the least amount of hazardous emissions and the highest possible energy generation efficiency (Hameed et al. 2021) (Fig. 1.6).

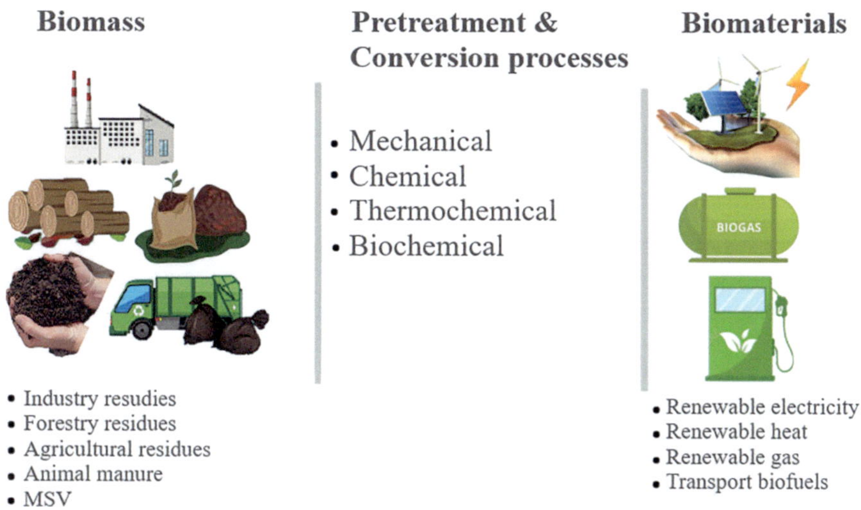

Fig. 1.6 Overview of biomass conversion into biomaterials

1.4 Applications and Benefits

The world energy distribution system is shifting from the source of pollution and depletion of fossil fuels to a system that deals with eco-friendly and renewable sources. This progression has been made possible by cutting the production cost of variable renewable energy (VRE). This pattern seems to continue, leading to more and more share of variable renewable energy in power sector (Kempener et al. 2015; Müller et al. 2014). Bioenergy manifests some distinctive properties in comparison to other types of renewable energy as it is convenient to store and transport (Szarka et al. 2013). Diverse modes of bioenergy are available and are being used in energy, transport, and heating sectors, while additional applications are being developed and scaled up stage (Bioenergy 2017).

1.4.1 Consumer Applications

The major use of bioenergy hybrid systems is the household heating area, which serves as a good example of small-scale manifestation of technology. The investments for this have major motivation like enhanced self-sufficiency, environment friendly, easy to use, and reliability. Typically centralized air or water heating systems or pellet boilers are used other than base load heating system in household. In the bioenergy sector, biomass is used in base load system in hybrid systems. This system has various benefits like it is source of additional heat to manage the high demand periods and serves as source of comfort. The limitation is choice of technologies and scaling of resources (Hakkarainen et al. 2019).

1.4.2 Industrial Applications

The adoption of renewable energy in industries is because of reduced production cost, lower CO_2 emission, and enhanced security of supply (Scheubel et al. 2017). With respect to lower scale installations, industries demand wider options of hybrid setups. In Finland, wood-derived fuels are consumed 37% in industrial sectors. Energy use in manufacturing by source is discussed by Statistics Finland (2015) [Online]. IEA Bioenergy Task 37 highlights biogas from slaughterhouse waste as a step towards energy self-sufficiency. According to data from Task 37 (2009), biogas and landfill gas produce approximately 4.7 MWh of energy and 7 MWh of heat per day, while also reducing disposal costs (Hakkarainen et al. 2019). The key players in industrial roundwood production are United States, Russia, and European Union. In EU, the major markets of industrial roundwood are situated in Finland, Sweden, Germany, Poland, Latvia, and Estonia (Proskurina et al. 2019). Torrefaction, a moderate thermal treatment of biomass at 200–300 °C, enhances biomass properties for energy applications. The torrefied biomass being dense and capable of heating in comparison to raw wood helps transportation sector and it is economical as well (Niu et al. 2019; Proskurina et al. 2017). According to IEA Task 40, world torrefied biomass need in industrial applications could theoretically exceed more than seven Mt. in 2025. Different situations indicate that total torrefied biomass need may exceed 10–18 Mt by 2030 (Proskurina et al. 2017). In Sub-Saharan Africa, bioenergy has the capability to support developing commercial energy requirements in major industries like cement, livestock, agricultural processing, and horticulture (Röder et al. 2022). Bioenergy, sourced from biomass, offers a sustainable alternative to fossil fuels in various industrial applications. It can be used for power generation, transportation fuels, and heating (Morales Pedraza 2015).

1.4.2.1 Paper and Pulp Industry

Biomass-based fibers are extensively used in the paper processing industry as they are versatile, strong, and sustainable (Pydimalla et al. 2023). These cellulose fibers that are obtained from biomass comprising of wood pulp, lignocellulosic waste, and special fibers like hemp and kenaf give the foundation for paper production. These fibers possess built-in qualities that increase the paper quality to be used in various applications. The fibers are known for their strength in paper and pulp industry because it enables them to make a durable and long lasting paper products (Karthika et al. 2022; Long et al. 2017).

Utilizing wood waste, agri-waste, and specialty fibers in its production processes helps the paper industry save natural resources and lessen its impact on the environment. Utilizing fibers derived from biomass also promotes the circular economy by reducing the material consumption loop and recycling organic waste outside of landfills (Kacprzak et al. 2022). The adoption of bioenergy on industrial scale minimizes dependence on fossil fuels and it could majorly affect the energy consumption trends (Clerman et al. 1981).

1.4.2.2 Textile Industry

Biomass-derived fibers have a prominent impact on upgrading the clothing sector by providing a sustainable eco-friendly alternative for traditional resources (Arif et al. 2023; Nguyen and Le 2023). These fibers that are acquired from biomass are being utilized in textile industry to make a variety of clothing items as per consumer's demand for their aesthetic satisfaction and environmentally derived. Due to their remarkable characteristics consisting of breathability, sustainability, smoothness, and natural fibers such as hemp, cotton, and bamboo are in demand in textile industry for a long time (Hoang et al. 2020; Tran et al. 2024). However, the consumers are more inclined towards the cloths derived from biomass-based fibers due to their biodegradability and sustainability. This has minimized the impacts on environment as compared to conventional textiles.

The new groundbreaking strategies like lyocell and production of viscose are being utilized to get cellulose materials from lignocellulosic biomass and convert them into high-quality fabrics that provide extraordinary drape, durability, and moisture-absorbent properties. The properties of natural fabric are present in these resynthesized cellulose fibers along with extra benefits of sustainability and enhanced productivity (Ramamoorthy et al. 2015).

1.4.3 Biodiesel

Biodiesel can be obtained from three different sources (Gilman 2007). The first source is residual fat left over after cooking food. One well-known example of residual fat from deep-frying is the car that Willie Nelson drives, which runs on grease-trap biodiesel. Lipids released from oil rich plants, such as sunflower, Jatropha, and soybeans, are the other source of biodiesel. Fats taken from photosynthetic microorganisms, such as algae and cyanobacteria, constitute the third source (Rittmann 2008).

1.4.3.1 By Used Cooking Oil

Another evolving product traded for energy is waste or used cooking oil (UCO) that is distinctively used for biodiesel production as an alternative to fossil fuels. This is used in a way of used cooking oil methylester and HVO, the hydro-treated vegetable oil. It is also applied for the production of bio-kerosene. This biodiesel is synthesized from same transesterification technology (Tsoutsos et al. 2016). According to “Expert opinion from MVO – The Netherlands Oils and Fats Industry (2017),” used cooking oil is gathered from across the world. During the past few years, exchange volumes of UCO have been increased. Analyzing the UCO trade levels in the EU-28 is difficult due to twice counting under the Renewable Energy Directive and diverse global imports. Yearly, around one million tones UCO have been utilized in the EU for biodiesel, out of which majorly collected in countries like Germany, the Netherlands, and Italy. In the Netherlands, nearly all restaurant UCO is collected, totaling about 60 k tons, with additional imports from both EU and non-EU countries (Proskurina et al. 2019).

The biodiesel manufactured from used cooking oil could cut the cost of fuel. If the shortage of UCO is there in EU, then imports will take place and it will increase production cost. Normally biodiesel has economic benefits as well as environmental benefits (Tsoutsos et al. 2016).

1.4.3.2 By Using Jatropha

Jatropha curcas L. is a promising non-edible bioenergy crop used for biodiesel production, offering a renewable alternative to fossil fuels (Nahar and Ozores-Hampton 2011). It is presently encouraged because of its adaptability to grow in extreme environments and high content (Sparsh et al. 2022). Biodiesel from Jatropha serves as second generation biofuel raw material that is remarkable alternative to edible oil crops which are currently used in biodiesel production (Neupane et al. 2021). Researches are ongoing in developing biotechnological approaches like tissue culture and genetic transformation to enhance the biomass content and improve biodiesel productivity (Al-Khayri et al. 2022).

However, challenges are still there in achieving sustainable commercial-scale biodiesel production from Jatropha due to various socio-economic, ecological, and technical restrains. Future research should focus on germplasm screening, adaptability studies, and optimizing technologies for seed processing and biodiesel production (Neupane et al. 2021).

1.4.3.3 Biodiesel from Microorganisms

Biodiesel derived from microorganisms is starting to grab the attention of investors and researchers now a day. However, there is a point needs to be addressed by researchers and investors, which microbe should be used. For this purpose, algae like Chlorella and Spirulina are prominent microbes. This focus could be because of photosynthesizing capabilities of alga. On the other hand, photosynthetic bacteria could be used like Cyanobacteria and Synechocystis. Both algae and bacteria could accumulate high contents of fats like 25% in wild strains (Ahlgren et al. 1992).

The mode of action of algae to produce and accumulate lipid content is during slow growth phase and under stress conditions. On the other hand, bacteria has opposite behavior I, it accumulates fats during rapid growth stage. The bacteria has natural advantage, but it is a prokaryote and needs to be modified genetically to produce lipids rapidly as compared to algae. For example, according to W. Vermaas and colleagues, a single gene mutant of Synechocystis was able to accumulate up to 50% of lipids to its dry weight (Vermaas 1998).

Biodiesel offers various benefits like it can be used in a blend to other fossil fuels or could be used 100% pure, by no or little alteration to engine, and it does not need any new refueling infrastructure. It gives the same efficiency like conventional diesel and also maintains the similar payload capability. Researchers and investors are exploring the use of biodiesel as aviation biofuels by substituting the jet-kerosene; it will open a new large-scale market. It will rapidly increase the utilization of bioenergy in transportation sector. It could further open a new discussion on food security and usage of land (Rosillo-Calle 2016).

1.4.4 Biogas and Biomethane

The blend of methane and CO_2, biogas, is widely used all over the world as a biomass-based biofuel. It is used in three sectors: at the household level where it is produced and used for cooking, in small-scale cottage industries, and in large-scale industrial uses. It is a significant transition from biogas alone to sustainable technology in last 20 years. This sustainability makes the combination of easy waste disposal and production of fertilizer in both developing and developed countries. Thus, the primary impetus behind biogas production is not energy generation, but rather the need to address environmental and hygiene problems, making it a sustainable and environmentally friendly alternative (Rosillo-Calle et al. 2015).

1.4.5 Aviation Biofuels

Civil aviation is a global trade sector facing a lot of problems and challenges that need solutions at world level. In 2014, there are 3.3 billion passengers travel in 100,000 flights daily. This number is expected to rise double by 2034 (Rosillo-Calle 2016). Bioenergy applications in civil aviation are grabbing attention as a sustainable substitute to fossil fuels. The aviation industry utilizes approximately 177 billion liters of kerosene yearly and shares 2% of carbon dioxide emissions (Baumi et al. 2020). Biofuels generated from algae, cyanobacteria, and other halophytes give a significant decrease in fuel replacements without having competition with food or water resources. However, the cost is still a prominent challenge, helping further research and development (Hendricks n.d.). The application of bioenergy technologies is important for the aviation sector to utilize substitute energy sources and minimize dependence on non-renewable fossil fuels (Поштаренко and Решетняк n.d.). Effective management and realistic life cycle cost assessments are essential for biofuels to become a viable long-term energy solution (Bomani and Hendricks 2016).

1.4.6 Bioelectricity

Scientists from all over the world have recorded the production of bioelectricity from products of fermentation and waste of lignocellulose (Rittmann et al. 2008). Therefore, the idea of MFC is validated and bioelectricity generation from waste organics is easy. The only limitation is that it is not fast enough to meet the capital expense of MFC by value of energy efficiency. Currently the MFCs are not that fast to be utilized commercially. Researchers have explored the area of power losses and worked for increasing the current and efficiency (Cheng and Logan 2007). According to Fan et al., highest current density till today is $12A/m^2$ which is ten times more than typical rates (Lee et al. 2008).

1.5 Limitations and Challenges

Biomass-based bioenergy has significant advantages like reduction in greenhouse gas emission and minimized dependency on conventional fuels (Bajpai 2019). However, there are few shortcomings that restrict their commercial scale utilization. These limitations comprised of high cost, reduced efficiency, and production of harmful chemicals (Danso-Boateng and Achaw 2022). Additionally, biomass production can result in deforestation, and this fuel is less efficient than conventional fuels (Bajpai 2019).

Life Cycle Assessment (LCA) data on bioenergy techniques show that there are inconsistencies in system boundaries, lack of reliability due to different functional unit definitions and limited scope incomparability of results due to varying functional unit definitions, and a lack of comprehensive impact categories (Wang et al. 2025). Furthermore, LCA-extracted data can be influenced by different suppositions, myths, inventory data, and impact assessment methodologies, prominently resulting in inaccurate decisions and policies (Hosseinzadeh-Bandbafha et al. 2021). To address these challenges, researchers suggest integrating various thermochemical techniques and improving LCA methodologies for more reliable assessments of bioenergy systems (Wang et al. 2025).

1.5.1 Emissions Reduction

Energy security, the switch to renewable energy, climate mitigation, and the socio-economic development of rural and local communities are all seen to be significantly impacted by bioenergy. But given the current situation, using renewable bioenergy resources might need to be more sustainable (Singh and Clough 2024). Using biomass for bioenergy can affect emissions and mitigation potential and provide some benefits. However, it also comes with risks like land degradation, food security issues, biodiversity loss, water scarcity, and challenges in meeting sustainable development goals (Calvin et al. 2021). These problems in GHG reporting and resulting from bioenergy can affect the reliability of bioenergy in mitigation scenarios (Norton et al. 2019; Searchinger et al. 2018). This, in turn, affects the potential of bioenergy markets. When faced with significant uncertainty, decision-makers in climate and environmental policy rely more on regulatory tools than economic ones (Bellamy 2018; Torvanger 2019).

Bioenergy can reduce greenhouse gas emissions from fossil fuels, but its potential is limited to 9–68% due to constraints on total available primary bioenergy (Staples et al. 2017).

1.5.2 Economic Challenges

The economic problems of bioenergy utilization are diverse. On the other hand bioenergy projects must be economically feasible for all entities in the value chain, and they are often uncertain and have low margins, deterring investors (Lunnan et al. 2008). The complicated production of biomass, harvesting, and transit to centralized biorefineries further adds to challenges (Balan 2014). However, bioenergy also has multiple socioeconomic benefits, such as economic development, employment, and reliability, which may validate government support (Lunnan et al. 2008). The transition from fossil fuels to bioenergy is crucial for a sustainable economic model, with biomass offering a promising solution by transforming waste into valuable resources (Mignogna et al. 2024).

Despite its capability to replace a prominent portion of fossil fuel consumption, the lack of efficient conversion technologies remains a limitation (Chung 2013). Navigating these problems requires technological modernization and supportive policies to create demand for biofuels (Balan 2014).

1.6 Future Prospects

Bioenergy has great potential as a sustainable alternative to fossil fuels. It can enhance energy security while supporting environmental sustainability (Rather et al. 2022). Current advancement in bioenergy production consists of thermochemical and biochemical processes, with thermochemical methods currently offering higher efficiency (Liu et al. 2022). Lignocellulosic biomass pathways, specially bioethanol for gasoline alternate and biorefineries for enhanced biomass utilization, show substantial promise (Dunnett and Shah 2007). However, problems such as land-use conflicts, biomass availability, and economic reliability need to be addressed.

Future prospects comprised of developing second- and third-generation biofuels, improving biotechnology for improved yield, and complementing thermochemical and biochemical processes to resolve individual problems (Liu et al. 2022). The success of bioenergy initiatives will depend on scientific innovation, diversified feedstocks, and sustainable production principles (Louime and Uckelmann 2008). Continuous technological advancement and ensuring sustainability will be necessary for the commercial viability of bioenergy in the future (Liu et al. 2022).

The International Renewable Energy Agency (IRENA) projects that by 2050, renewable energy sources could account for 60% or more of total final energy consumption in many countries. Biomass is expected to play the significant role, potentially contributing nearly half of global renewable energy consumption by 2030 and around 40% by 2050. According to the International Energy Agency (IEA), in 2030 energy use will increasingly rely on modern bioenergy technologies, with modern biomass consumption projected to rise to approximately 100 exajoules (EJ) by 2050 (Geletukha and Zheliezna 2021).

By 2050, bioenergy is projected to contribute between 64 and 313 exajoules (EJ), representing approximately 7.5–37% of the global energy mix. This marks a substantial shift in the framework of bioenergy supply, moving from the current reliance on traditional firewood to a greater emphasis on energy crops and biowaste. The technical potential for biofuel and electricity generation largely aligns with the projected trends in primary bioenergy supply. Additionally, significant reductions in greenhouse gas emissions are anticipated. The heuristic approach and transparency of the new modeling framework will support ongoing adjustments and the exploration of alternative scenarios over time (Errera et al. 2023).

1.7 Conclusion

Bioenergy derived from biomass contributes to a significant proportion of global renewable energy sources. It is capable of minimizing dependence on fossil fuels and contributes to reducing greenhouse gas emissions. The USA and Brazil are among the top countries that produce biofuels. According to statistical data, from

2007 to 2017, world biofuel production increased at a rate of 11.4% per year. The United States is at the top for producing 1917 petajoules of bioenergy, while Brazil is second, producing 1143 petajoules. Biofuels have important applications in various sectors like transportation fuels, electricity generation, waste management, and heating. Bioenergy plays a pivotal role in providing energy security and environmental sustainability. On the other hand, it has a few limitations, such as high costs, environmental impacts like deforestation, and reduced efficiency. Economic challenges also exist, and researchers are working to make it economically feasible. Although there are challenges that biofuels face, they have the potential for climate change mitigation and shifting towards the utilization of renewable energy. Future prospects include improving biotechnology, increasing biomass availability, and developing advanced biofuels with economic reliability to maximize the utilization of bioenergy as a sustainable and viable source of energy.

References

Abanades S, Abbaspour H, Ahmadi A, Das B, Ehyaei M, Esmaeilion F, El Haj Assad M, Hajilounezhad T, Jamali D, Hmida A (2022) A critical review of biogas production and usage with legislations framework across the globe. Int J Environ Sci Technol 19(4):3377–3400

Agdew SA, Cherie ME, Alemu WG, Gesese TN, Getahun AA (2025) Perspective chapter: biogas production–current status, challenges, and future perspectives. In: Biogas in the 21st century - developments and perspectives. Intechopen

Ahlgren G, Gustafsson IB, Boberg M (1992) Fatty acid content and chemical composition of freshwater microalgae 1. J Phycol 28(1):37–50

Ahmadi L, Kannangara M, Bensebaa F (2020) Cost-effectiveness of small scale biomass supply chain and bioenergy production systems in carbon credit markets: a life cycle perspective. Sustain Energy Technol Assess 37:100627

Aidonojie PA, Ukhurebor KE, Oaihimire IE, Ngonso BF, Egielewa PE, Akinsehinde BO, Kusuma HS, Darmokoesoemo H (2023) Bioenergy revamping and complimenting the global environmental legal framework on the reduction of waste materials: a facile review. Heliyon 9(1): e12860

Al-Khayri J, Sudheer W, Preetha T, Nagella P, Rezk A, Shehata W (2022) Biotechnological research progress in Jatropha, a biodiesel-yielding plant. Plants 11:1292. In: s note: MDPI stays neutral with regard to jurisdictional claims in published …

Ambaye TG, Vaccari M, Bonilla-Petriciolet A, Prasad S, van Hullebusch ED, Rtimi S (2021) Emerging technologies for biofuel production: a critical review on recent progress, challenges and perspectives. J Environ Manag 290:112627

Arif Z, Ali N, Mulyati S (2023) Investigation of mechanical properties and thermal analysis of bagasse fiber reinforced composite polymer foam. Int J Adv Sci Eng Inf Technol 13(4):1210–1216

Asadullah M, Adi AM, Suhada N, Malek NH, Saringat MI, Azdarpour A (2014) Optimization of palm kernel shell torrefaction to produce energy densified bio-coal. Energy Convers Manag 88: 1086–1093

Awogbemi O, Kallon D (2021) Impact of fourth industrial revolution on waste biomass conversion techniques. In: Proceedings of the SAIIE32 steps, Muldersdrift, South Africa, pp 4–6

Bajpai P (2019) Advantages and disadvantages of biomass utilization. In: Biomass to energy conversion technologies. Elsevier, Amsterdam

Balan V (2014) Current challenges in commercially producing biofuels from lignocellulosic biomass. Int Sch Res Notices 2014(1):463074
Baumi J, Bertosse CM, Guedes CLB (2020) Aviation fuels and biofuels. In: Renewable energy-resources, challenges and applications. IntechOpen
Bellamy R (2018) Incentivize negative emissions responsibly. Nat Energy 3(7):532–534
Bioenergy I (2017) Technology roadmap delivering sustainable bioenergy. Paris, IEA, p 89
Bomani BM, Hendricks RC (2016) Biofuels for green aviation. In: Encyclopedia of aircraft engineering, green aviation, John Wiley & Sons, pp 179–191
Calvin K, Cowie A, Berndes G, Arneth A, Cherubini F, Portugal-Pereira J, Grassi G, House J, Johnson FX, Popp A (2021) Bioenergy for climate change mitigation: scale and sustainability. GCB Bioenergy 13(9):1346–1371
Cheng S, Logan BE (2007) Sustainable and efficient biohydrogen production via electrohydrogenesis. Proc Natl Acad Sci 104(47):18871–18873
Chung JN (2013) Grand challenges in bioenergy and biofuel research: engineering and technology development, environmental impact, and sustainability. Front Energy Res 1:4. Frontiers Media SA
Clerman R, Joglekar R, Ouellette R, Cheremisinoff P (1981) Electrotechnology. Volume 8. Biotechnology and energy use. U.S. Department of Energy / Office of Scientific and Technical Information. OSTI ID: 5697434
Cortez LA, Baldassin R Jr (2016) Policies towards bioethanol and their implications: case Brazil. In: Global bioethanol. Elsevier, pp 142–162
Danso-Boateng E, Achaw O-W (2022) Bioenergy and biofuel production from biomass using thermochemical conversions technologies—a review. Aims Energy 10(4):585–647
Duarah P, Haldar D, Patel AK, Dong C-D, Singhania RR, Purkait MK (2022) A review on global perspectives of sustainable development in bioenergy generation. Bioresour Technol 348: 126791
Dunnett AJ, Shah N (2007) Prospects for bioenergy. J Biobaased Mater Bioenergy 1(1):1–18
Ebadian M, van Dyk S, McMillan JD, Saddler J (2020) Biofuels policies that have encouraged their production and use: an international perspective. Energy Policy 147:111906
Errera MR, Dias T d C, Maya DMY, Lora EES (2023) Global bioenergy potentials projections for 2050. Biomass Bioenerg 170:106721
Ersoy AE, Ugurlu A (2024) Bioenergy's role in achieving a low-carbon electricity future: a case of Türkiye. Appl Energy 372:123799
Evaristo AB, Grossi JAS, Carneiro A d CO, Pimentel LD, Motoike SY, Kuki KN (2016) Actual and putative potentials of macauba palm as feedstock for solid biofuel production from residues. Biomass Bioenergy 85:18–24
Ferrari G, Pezzuolo A, Nizami A-S, Marinello F (2020) Bibliometric analysis of trends in biomass for bioenergy research. Energies 13(14):3714
Fivga A, Speranza LG, Branco CM, Ouadi M, Hornung A (2019) A review on the current state of the art for the production of advanced liquid biofuels. Aims Energy 7(1):46–76
Gautam P, Upadhyay SN, Dubey S (2020) Bio-methanol as a renewable fuel from waste biomass: current trends and future perspective. Fuel 273:117783
Geletukha G, Zheliezna T (2021) Global prospects for bioenergy sector. Therm Power Eng 43(4): 75–82
Gilman D (2007) Fueling organ with sustainable biofuels. Oregon Environmental Council, Portland
Guo M, Song W, Buhain J (2015) Bioenergy and biofuels: history, status, and perspective. Renew Sust Energ Rev 42:712–725
Hakkarainen E, Hannula I, Vakkilainen E (2019) Bioenergy RES hybrids– assessment of status in Finland, Austria, Germany, and Denmark. Biofuels Bioprod Biorefin 13(6):1402–1416
Hameed Z, Aslam M, Khan Z, Maqsood K, Atabani A, Ghauri M, Khurram MS, Rehan M, Nizami A-S (2021) Gasification of municipal solid waste blends with biomass for energy production and resources recovery: current status, hybrid technologies and innovative prospects. Renew Sustain Energy Rev 136:110375

Hendricks R (n.d.) Synthetic and biomass alternate fueling in aviation RC Hendricks Robert. C. Hendricks@ grc nasa gov DM Bushnell
Hoang AT, Chau MQ, Le QB (2020) Parameters affecting fiber quality and productivity of coir spinning machines. J Mech Eng Res Dev 43(5):122–145
Hoang AT, Ong HC, Fattah IR, Chong CT, Cheng CK, Sakthivel R, Ok YS (2021) Progress on the lignocellulosic biomass pyrolysis for biofuel production toward environmental sustainability. Fuel Process Technol 223:106997
Hossain SZ (2019) Biochemical conversion of microalgae biomass into biofuel. Chem Eng Technol 42(12):2594–2607
Hosseini SE (2022) Fossil fuel crisis and global warming. In: Fundamentals of low emission flameless combustion and its applications. Elsevier, pp 1–11
Hosseinzadeh-Bandbafha H, Aghbashlo M, Tabatabaei M (2021) Life cycle assessment of bioenergy product systems: a critical review. e-Prime - Adv Electr Eng Electron Energy 1: 100015
Jamil F, Inayat A, Hussain M, Akhter P, Abideen Z, Ghenai C, Shanableh A, Abdellatief TM (2024) Valorization of waste biomass to biofuels for power production and transportation in optimized way: a comprehensive review. Adv Energy Sustain Res 5(10):2400104
Jeswani HK, Chilvers A, Azapagic A (2020) Environmental sustainability of biofuels: a review. Proc R Soc A 476(2243):20200351
Jia X, Qin X, Tian X, Zhao Y, Yang T, Huang J (2021) Inoculating with the microbial agents to start up the aerobic composting of mushroom residue and wood chips at low temperature. J Environ Chem Eng 9(4):105294
Kacprzak M, Attard E, Lyng K-A, Raclavska H, Singh B, Tesfamariam E, Vandenbulcke F (2022) Biodegradable waste management in the circular economy. https://doi.org/10.1002/9781119679523
Kang S, Selosse S, Maïzi N (2015) Strategy of bioenergy development in the largest energy consumers of Asia (China, India, Japan and South Korea). Energ Strat Rev 8:56–65
Kapoor R, Ghosh P, Tyagi B, Vijay VK, Vijay V, Thakur IS, Kamyab H, Nguyen DD, Kumar A (2020) Advances in biogas valorization and utilization systems: a comprehensive review. J Clean Prod 273:123052
Karthika R, Jayanthi B, Aruna A, Selvankumar T (2022) Nanocellulosic materials for papermaking and paper coating industry. In: Handbook of nanocelluloses: classification, properties, fabrication, and emerging applications. Springer, pp 1–34
Kazmi A, Sultana T, Ali A, Nijabat A, Li G, Hou H (2025) Innovations in bioethanol production: a comprehensive review of feedstock generations and technology advances. Energ Strat Rev 57: 101634
Kempener R, Assoumou E, Chiodi A, Ciorba U, Gaeta M, Gielen D, Hamasaki H, Kanudia A, Kober T, Labriet M (2015) A global renewable energy roadmap: comparing energy systems models with irena's remap 2030 project. In: Informing energy and climate policies using energy systems models: insights from scenario analysis increasing the evidence base. Springer, pp 43–67
Khan S, Nisar A, Wu B, Zhu Q-L, Wang Y-W, Hu G-Q, He M-x (2022) Bioenergy production in Pakistan: potential, progress, and prospect. Sci Total Environ 814:152872
Kumar A, Kumar N, Baredar P, Shukla A (2015) A review on biomass energy resources, potential, conversion and policy in India. Renew Sust Energ Rev 45:530–539
Kumar A, Kakran S, Sidhu A, Singh UP (2023) India's renewable energy preparedness: assessing the viability of bioethanol as a key element in the green energy transition. Manag Account J 58(6):52–55
Kumar G, Padhy R, Das D, Narayana SA (2024) Ethanol fuel blending program in India: analysis of environmental, economic, and policy aspects using system dynamics approach. J Adv Transp 2024(1):2002187

Lee H-S, Parameswaran P, Kato-Marcus A, Torres CI, Rittmann BE (2008) Evaluation of energy-conversion efficiencies in microbial fuel cells (MFCs) utilizing fermentable and non-fermentable substrates. Water Res 42(6–7):1501–1510

Letti LAJ, Sydney EB, de Carvalho JC, de Souza Vandenberghe LP, Karp SG, Woiciechowski AL, Soccol VT, Novak AC, Junior AIM, Burgos WJM (2022) Roles and impacts of bioethanol and biodiesel on climate change mitigation. In: Biomass, biofuels, biochemicals. Elsevier, pp 373–400

Liu T, Miao P, Shi Y, Tang KHD, Yap P-S (2022) Recent advances, current issues and future prospects of bioenergy production: a review. Sci Total Environ 810:152181

Long W, Fang B, Ignaszak A, Wu Z, Wang Y-J, Wilkinson D (2017) Biomass-derived nanostructured carbons and their composites as anode materials for lithium ion batteries. Chem Soc Rev 46(23):7176–7190

Louime C, Uckelmann H (2008) Cellulosic ethanol: securing the planet future energy needs, vol 9. Molecular Diversity Preservation International (MDPI), pp 838–841

Lu W, Alam MA, Luo W, Asmatulu E (2019) Integrating Spirulina platensis cultivation and aerobic composting exhaust for carbon mitigation and biomass production. Bioresour Technol 271:59–65

Lunnan A, Vilkriste L, Wilhelmsen G, Mizaraite D, Asikainen A, Röser D (2008) Policy and economic aspects of forest energy utilisation. In: Sustainable use of forest-based biomass for energy: a synthesis with focus on the Baltic and Nordic region. Springer, Dordrecht, pp 197–234

Makareviciene V, Sendzikiene E (2022) Application of microalgae biomass for biodiesel fuel production. Energies 15(11):4178

Malico I, Pereira RN, Gonçalves AC, Sousa AM (2019) Current status and future perspectives for energy production from solid biomass in the European industry. Renew Sust Energ Rev 112: 960–977

Manikandan S, Subbaiya R, Biruntha M, Krishnan RY, Muthusamy G, Karmegam N (2022) Recent development patterns, utilization and prospective of biofuel production: emerging nanotechnological intervention for environmental sustainability–a review. Fuel 314:122757

Michaelides EE (2012) Wind power. In: Alternative energy sources. Springer, pp 231–256

Mielenz JR (2001) Ethanol production from biomass: technology and commercialization status. Curr Opin Microbiol 4(3):324–329

Mignogna D, Szabó M, Ceci P, Avino P (2024) Biomass energy and biofuels: perspective, potentials, and challenges in the energy transition. Sustainability 16(16):7036

Moore K, Lamkey K, Karlen D (2014) Future prospects for corn as a biofuel crop. CRC Press

Morales Pedraza J (2015) The current situation and perspectives on the use of biomass in the generation of electricity. In: Electrical energy generation in Europe: the current situation and perspectives in the use of renewable energy sources and nuclear power for regional electricity generation. Springer, pp 371–419

Müller S, de Sisternes F, Patriarca E, Portellano A, Goeritz A, Moller J, Peter J, Holttinen H (2014) The power of transformation. In: Wind, sun and the economics of flexible power systems. International Energy Agency (IEA), Paris

Munasinghe PC, Khanal SK (2010) Biomass-derived syngas fermentation into biofuels: opportunities and challenges. Bioresour Technol 101(13):5013–5022

Nahar K, Ozores-Hampton M (2011) Jatropha: an alternative substitute to fossil fuel: HS1193/HS1193, 12/2011. EDIS 2011(12)

Neupane D, Bhattarai D, Ahmed Z, Das B, Pandey S, Solomon JK, Qin R, Adhikari P (2021) Growing Jatropha (Jatropha curcas L.) as a potential second-generation biodiesel feedstock. Inventions 6(4):60

Nguyen DT, Le HC (2023) Potential of jute fiber-reinforced composites in the manufacture of components and equipment used on ships and hulls. Int J Emerg Sci Eng 1(1):14–21

Niu Y, Lv Y, Lei Y, Liu S, Liang Y, Wang D, Hui SE (2019) Biomass torrefaction: properties, applications, challenges, and economy. Renew Sust Energ Rev 115:109395

Nizami A-S, Rehan M, Waqas M, Naqvi M, Ouda OK, Shahzad K, Miandad R, Khan MZ, Syamsiro M, Ismail IM (2017) Waste biorefineries: enabling circular economies in developing countries. Bioresour Technol 241:1101–1117

Norton M, Baldi A, Buda V, Carli B, Cudlin P, Jones MB, Korhola A, Michalski R, Novo F, Oszlányi J (2019) Serious mismatches continue between science and policy in forest bioenergy. GCB Bioenergy 11(11):1256–1263

Orisaleye JI, Jekayinfa SO, Dittrich C, Obi OF, Pecenka R (2023) Effects of feeding speed and temperature on properties of briquettes from poplar wood using a hydraulic briquetting press. Resources 12(1):12

Osman AI, Deka TJ, Baruah DC, Rooney DW (2023a) Critical challenges in biohydrogen production processes from the organic feedstocks. Biomass Convers Biorefinery 13(10):8383–8401

Osman AI, Lai ZY, Farghali M, Yiin CL, Elgarahy AM, Hammad A, Ihara I, Al-Fatesh AS, Rooney DW, Yap P-S (2023b) Optimizing biomass pathways to bioenergy and biochar application in electricity generation, biodiesel production, and biohydrogen production. Environ Chem Lett 21(5):2639–2705

Page-Dumroese DS, Franco CR, Archuleta JG, Taylor ME, Kidwell K, High JC, Adam K (2022) Forest biomass policies and regulations in The United States of America. Forests 13(9):1415

Pereira RG, Lameira VDJ, Vergura S (2013) Biofuels production and consumption: a decade of comparison, from 2000 to 2010. RE&PQJ 11(1). https://doi.org/10.24084/repqj11.205

Pirzadah TB, Malik B, Kumar M, Rehman RU (2014) Lignocellulosic biomass: as future alternative for bioethanol production. In: Biomass and bioenergy: applications. Springer, pp 145–163

Potrč S, Nemet A, Čuček L, Varbanov PS, Kravanja Z (2022) Synthesis of a regenerative energy system–beyond carbon emissions neutrality. Renew Sust Energ Rev 169:112924

Proskurina S, Heinimö J, Schipfer F, Vakkilainen E (2017) Biomass for industrial applications: the role of torrefaction. Renew Energy 111:265–274

Proskurina S, Junginger M, Heinimö J, Tekinel B, Vakkilainen E (2019) Global biomass trade for energy—part 2: production and trade streams of wood pellets, liquid biofuels, charcoal, industrial roundwood and emerging energy biomass. Biofuels Bioprod Biorefin 13(2):371–387

Pydimalla M, Chirravuri HV, Uttaravalli AN (2023) An overview on non-wood fiber characteristics for paper production: sustainable management approach. Mater Today Proc. https://doi.org/10.1016/j.matpr.2023.08.278

Qamar SH, Ali MM, ur Rehman H (2024) A comprehensive review of bioenergy sustainability: balancing economic, environ-mental, and social impacts. J Lat Am Sci Cult 6(9):47–65

Qaseem MF, Wu A-M (2021) Marginal lands for bioenergy in China; an outlook in status, potential and management. GCB Bioenergy 13(1):21–44

Rajvanshi M, Podili R, Dwivedi V, Sanyal D, Dasgupta S (2022) Waste biomass to bioenergy: a compressive review. In: Biotechnology for waste biomass utilization. Apple Academic Press, pp 1–28

Ramamoorthy SK, Skrifvars M, Persson A (2015) A review of natural fibers used in biocomposites: plant, animal and regenerated cellulose fibers. Polym Rev 55(1):107–162

Rastogi M, Shrivastava S (2017) Recent advances in second generation bioethanol production: an insight to pretreatment, saccharification and fermentation processes. Renew Sust Energ Rev 80: 330–340

Rather RA, Wani AW, Mumtaz S, Padder SA, Khan AH, Almohana AI, Almojil SF, Alam SS, Baba TR (2022) Bioenergy: a foundation to environmental sustainability in a changing global climate scenario. J King Saud Univ Sci 34(1):101734

Rittmann BE (2008) Opportunities for renewable bioenergy using microorganisms. Biotechnol Bioeng 100(2):203–212

Rittmann BE, Torres CI, Marcus AK (2008) Understanding the distinguishing features of a microbial fuel cell as a biomass-based renewable energy technology. In: Emerging environmental technologies. Springer, pp 1–28

Röder M, Welfle AJ (2019) Bioenergy. In: Managing global warming. Academic Press

Röder M, Chong K, Thornley P (2022) The future of residue-based bioenergy for industrial use in Sub-Saharan Africa. Biomass Bioenergy 159:106385
Rosillo-Calle F (2016) A review of biomass energy–shortcomings and concerns. J Chem Technol Biotechnol 91(7):1933–1945
Rosillo-Calle F, De Groot P, Hemstock SL, Woods J (2015) The biomass assessment handbook: energy for a sustainable environment. Routledge
Sadaka S, Johnson D (2000) Agriculture and natural resources: biomass combustion
Sahota S, Shah G, Ghosh P, Kapoor R, Sengupta S, Singh P, Vijay V, Sahay A, Vijay VK, Thakur IS (2018) Review of trends in biogas upgradation technologies and future perspectives. Bioresour Technol Rep 1:79–88
Saravanan A, Ragini Y, Karishma S, Hemavathy R, Jyotsna M (2025) A review on advancing sustainable energy: the role of biomass and bioenergy in a circular economy. Sustain Futures 10: 100835
Sarris D, Papanikolaou S (2016) Biotechnological production of ethanol: biochemistry, processes and technologies. Eng Life Sci 16(4):307–329
Scheubel C, Zipperle T, Tzscheutschler P (2017) Modeling of industrial-scale hybrid renewable energy systems (HRES)–the profitability of decentralized supply for industry. Renew Energy 108:52–63
Searchinger TD, Beringer T, Holtsmark B, Kammen DM, Lambin EF, Lucht W, Raven P, van Ypersele J-P (2018) Europe's renewable energy directive poised to harm global forests. Nat Commun 9(1):3741
Singh J, Clough Y (2024) Challenges and opportunities to a sustainable bioenergy utilization in climate mitigation: a global perspective. Front Sustain Energy Policy 3:1460370
Sorita GD, Favaro SP, Ambrosi A, Di Luccio M (2023) Aqueous extraction processing: an innovative and sustainable approach for recovery of unconventional oils. Trends Food Sci Technol 133:99–113
Sparsh S, Ramakrishna CS, Sahu R, Narain RS, Sohal JK, Sahu M, Sharma N, Kumar V, Sharma A (2022) Future prospects of biodiesel production from jatropha in India. Mater Today Proc 63: A22–A26
Staples MD, Malina R, Barrett SR (2017) The limits of bioenergy for mitigating global life-cycle greenhouse gas emissions from fossil fuels. Nat Energy 2(2):1–8
Subramaniam Y, Masron TA (2021) The impact of economic globalization on biofuel in developing countries. Energy Convers Manag X 10:100064
Suhartini S, Sunyoto NMS, Setyawan HY, Rohma NA, Elviliana, Pratiwi DA, Fitria, Mardawati E, Kasbawati, Melville L (2024) Biomass valorization for bioenergy production. In: Biomass conversion and sustainable biorefinery: towards circular bioeconomy. Springer, pp 67–104
Szarka N, Scholwin F, Trommler M, Jacobi HF, Eichhorn M, Ortwein A, Thrän D (2013) A novel role for bioenergy: a flexible, demand-oriented power supply. Energy 61:18–26
Tekin K, Karagöz S, Bektaş S (2014) A review of hydrothermal biomass processing. Renew Sust Energ Rev 40:673–687
Thatoi H, Dash PK, Mohapatra S, Swain MR (2016) Bioethanol production from tuber crops using fermentation technology: a review. Int J Sustain Energy 35(5):443–468
Thrän D, Deprie K, Dotzauer M, Kornatz P, Nelles M, Radtke KS, Schindler H (2023) The potential contribution of biogas to the security of gas supply in Germany. Energy Sustain Soc 13(1):12
Torvanger A (2019) Governance of bioenergy with carbon capture and storage (BECCS): accounting, rewarding, and the Paris agreement. Clim Pol 19(3):329–341
Tran MH, Paramasivam P, Le HC, Nguyen DT (2024) Biomass: a versatile resource for biofuel, industrial, and environmental solution. Int J Adv Sci Eng Inf Technol 14(1):268
Tshikovhi A, Motaung TE (2023) Technologies and innovations for biomass energy production. Sustainability 15(16):12121
Tsoutsos T, Tournaki S, Paraíba O, Kaminaris S (2016) The used cooking oil-to-biodiesel chain in Europe assessment of best practices and environmental performance. Renew Sust Energ Rev 54: 74–83

Vakili M, Rafatullah M, Ibrahim MH, Salamatinia B, Gholami Z, Zwain HM (2015) A review on composting of oil palm biomass. Environ Dev Sustain 17(4):691–709

Van den Broek R, Faaij A, van Wijk A (1996) Biomass combustion for power generation. Biomass Bioenergy 11(4):271–281

Vermaas WF (1998) [20] Gene modifications and mutation mapping to study the function of photosystem II. In: Methods in enzymology, vol 297. Elsevier, pp 293–310

Wang M, Dewil R, Maniatis K, Wheeldon J, Tan T, Baeyens J, Fang Y (2019) Biomass-derived aviation fuels: challenges and perspective. Prog Energy Combust Sci 74:31–49

Wang K, Tong R, Zhai Q, Lyu G, Li Y (2025) A critical review of life cycle assessments on bioenergy technologies: methodological choices, limitations, and suggestions for future studies. Sustainability 17(8):3415

Welfle A (2017) Balancing growing global bioenergy resource demands-Brazil's biomass potential and the availability of resource for trade. Biomass Bioenergy 105:83–95

Wu D, Wei Z, Mohamed TA, Zheng G, Qu F, Wang F, Zhao Y, Song C (2022) Lignocellulose biomass bioconversion during composting: mechanism of action of lignocellulase, pretreatment methods and future perspectives. Chemosphere 286:131635

Поштаренко А, Решетняк Л (n.d.) ОСОБЛИВОСТІ ВИКОРИСТАННЯ БІОЕНЕРГЕТИКИ В АВІАЦІЙНІЙ ГАЛУЗІ

Chapter 2
Socioeconomic Balance in Bioenergy and Nanotechnology: Present Time Scenario

Subhadra Rajpoot, Sheetal Thakur, and Rajni Garg

Abstract The world energy scene is in the process of profound changes within societies in the quest of a new balance between energy security, environmental sustainability, and social development. This evolution has made bioenergy a key constituent in the renewable energy arsenal and provides avenues to address several development challenges concomitantly. By contrast, if bioenergy systems are going to be implemented, to exploit the synergies between food, energy, and industry, there is a need to consider the social and economic aspects of the implementation in order to ensure that benefits are shared and that technological advances are put at the service of inclusive development. Bioenergy consists of different types of energy produced from organic materials, such as agricultural residuals, forest biomass, energy crops, and organic waste streams. The importance of bioenergy is not only limited to the supply of energy; but also it is considered that these systems could provide solutions to complex development issues such as rural development, energy access, waste management, and climate change. Given the complex nature of bioenergy systems, holistic assessment frameworks are required which cover more than purely technical performance or environmental impact and also address social equity and economic sustainability.

The incorporation of nanotechnology into bioenergy applications is an exciting frontier that has the potential to greatly increase the efficiency, economic viability, and environmental performance of renewable energy technologies. In this area of science, nanotechnology refers to the creation and use of materials on this scale, as the properties of materials can truly differ at this scale from that on a larger scale. Applications include catalysts with a higher surface-area-to-volume ration, batteries with faster charging and increased capacity, as well as thin films with improved catalytic properties. Nanomaterials have unique attributes such as high surface-to-

S. Rajpoot (✉) · S. Thakur
University Centre for Research & Development, Department of Biotechnology, Chandigarh University, Mohali, Punjab, India
e-mail: srajpoot@gn.amity.edu

R. Garg
Department of Applied Sciences And Humanities, Galgotias College of Engineering and Technology, Greater Noida, UP, India

N. Srivastava (ed.), *Sustainable Nanostrategies for Biorefineries*, Clean Energy Production Technologies, https://doi.org/10.1007/978-981-95-5875-9_2

volume ratio, high reactivity, and selective functionality, all of which can be exploited to overcome previous constraints of bioenergy technology and to increase the socioeconomic benefit. Yet, at the crossover of bioenergy and nanotechnology, the game is complex for securing socioeconomic equilibrium. Technology advances certainly have the power to impart efficiency and cost-savings, but there are still open questions about access to technology, whose benefits and the possibility that technological advances can exacerbate rather than alleviate pre-existing inequalities. The idea of a socioeconomic equilibrium in this sense concerns a fair distribution of benefits, costs, and risks of technological development in which change will contribute to achieving sustainable development goals rather than to the emergence of fresh sources of exclusion or dependence.

Keywords Bioenergy · Nanotechnology · Sustainability · Eco-friendly · Waste management · Biomass

2.1 Introduction

Recent reviews of bioenergy sustainability have emphasized the limited depth of our knowledge on the social and economic dimensions of impacts in contrast to our understanding of and measurement of environmental and technical aspects (Intergovernmental Panel on Climate Change 2007). A number of evaluation tools emphasize technological criteria, such as efficiency and certain environmental criteria such as greenhouse gas emissions, while only marginally addressing social aspects like local owner participation, local employment creation, and local culture. This discrepancy in evaluation methods reflects a wider problem within sustainable development—that technology fixes are frequently developed with too little regard to their social and economic consequences.

Nanotechnology for mitigation of climate change and pollution, a novel way towards renewable bioenergy production Nanotechnology as a new front of addressing environmental challenges -climate change and pollution, has created a new dimension to improve sustainability of the bioenergy (Fischer and Knutti 2015). Nanotechnology can help tackle more than one environmental issue simultaneously and at the same time lead to the development of more efficient and economical bioenergy systems. Yet there is insufficient understanding of the social and economic implications of the incorporation of nanotechnologies, including who might gain or lose from their application and how they might support or hinder development challenges. The purpose of the current work was to fill these voids by investigating the current socioeconomic equilibrium status in bioenergy networks and exploring how applications of nanotechnology could lead to more equitable and sustainable systems. Such an inquiry is of importance in the light of the greater international focus on sustainable development and the understanding that technical solutions are necessary to be supported by adequate social and economic contexts if there are positive effects sustained.

The research goals include multiple interrelated levels of analysis. First, we will try to analyze the structure of current systems for evaluating socioeconomic sustainability in bioenergy, focusing on relevant indicators and methods that can provide solid grounds for a more complete assessment of development impacts. In a second part, we review the current applications of nanotechnology related to bioenergy and water and waste management and discuss how it may help to solve the emerging social challenges. Third we evaluate bioenergy development from a current sociology of economics standpoint and highlight critical obstacles and opportunities to advance more equitable outcomes predicated on technological innovation and balanced institutional support. The analysis is focusing on the various scales of bioenergy deployments—from household applications known as improved cooking systems (ICSs) to full industrial-like biofuel production facilities. This multiscale strategy acknowledges that socioeconomic effects may differ greatly at different scales and settings and that other strategies might be necessary to achieve socioeconomic equilibrium at another spatial level. The review discusses several bioenergy routes like biogas, bioethanol, biodiesel, and biohydrogen routes and explains how nanotechnology can contribute to sustainability and equity in these various bioenergy routes.

The findings from this study have implications not only for academic research but also for policy, design of technology, and implementation. With growing policy attention by governments and international institutions to sustainable pathways to achieve such transitions in energy systems, knowing how to achieve the appropriate socio-economic balance in bioenergy systems becomes essential for effective policy development and resource investing. The knowledge from this work can also help in setting technology development priorities, financing mechanisms, and regulatory polices favoring the equitable development of bioenergy, wherever feasible, ensuring that the maximum societal benefits are achieved. The paper argues that attaining socioeconomic equilibrium in bioenergy systems calls for an integrated approach that pairs technological innovation with enabling institutional structures, mechanisms for community participation, and benefit-sharing. Here we contend that despite an immense potential for nanotechnology to enhance both economic and technical efficiencies in bioenergy system development, realizing this potential equitably demands critical attention to how technology is designed, strategies for deployment, and institutions for governance that will equitably yield broad benefits while initially mitigating negative effects. The analysis also shows that the successful socioeconomic integration within bioenergy systems can be achieved only if long-term technological capabilities are aligned with local needs, institutional capacities, and development priorities. This alignment depends on participatory approaches to the development and adoption of technologies, which involve affected communities in decision-making and ensure that technological "solutions" are context-specific and respond to locally determined development goals.

2.2 Theoretical Framework and Sustainable Development Context

The theoretical basis for the explanation of the societal balance of bioenergy systems is build up on several interwoven frameworks originating from sustainable development theory, technology assessment methodologies, and environmental management approaches. The most extensive framework for this kind of analysis is offered by UN's Sustainable Development Goals and the holistic way that it considers the dimensions of the impacts of the technological interventions on society, economy, and environment. The perspective presented by SDG framework is insightful for bioenergy systems, as it acknowledges that sustainable development cannot be achieved by the advancement of a single dimension, but rather through concerted efforts to advance all three dimensions together (viz., social, economic, and environmental). This framework has become especially appealing for bioenergy impact assessment, as these systems have been recognized as possible contributors to several of the development goals at the same time, including poverty alleviation, energy access, climate mitigation, and economic development. The holistic nature of the SDG framework is consistent with the complexity of bioenergy systems and the requirement of better methodologies for developing comprehensive assessments of their multidimensional impacts.

The theoretical underpinning of social sustainability assessment for bioenergy systems has developed substantially in the last 10 years from primarily impact assessment to more comprehensive assessments of community well-being, social equity, and participatory development. Current scales of the concept understand that social sustainability permeates through matters dealing with food security, health and well-being, energy availability, land hosting systems, community involvement, and the cultural maintenance. These dimensions illustrate an evolving recognition that social sustainability goes beyond immediate project recipients, to scale and long-term social relations.

It is also one of the most complicated and controversial issues in the context of bioenergy social sustainability, especially when debates of possible energy-food competition is carried out. The link between bioenergy production and food systems is very context-specific, and the outcomes are impacted by feedstock selection, the scale of production, and the location where the bioenergy crop is produced and integrated with current agricultural systems (Naithani 2016). Studies indicate that the effects of bioenergy on food security can be positive or negative depending on the technology and local conditions. Energy crops cultivated on poor lands within regions with strong food supply systems are usually less deleterious to food security than large-scale developments that directly displace food production or are situated in food insecure areas (Vel'asquez et al. 2018). Nevertheless, even planting of biofuel crops on marginal lands could have implications for food security via impacts on land markets, labor allocation, and rural economies. The complexity of such relationships indicates the relevance of comprehensive assessment pathways that take into account direct as well as indirect effects on the food system.

Health and well-being implications include several mechanisms through bioenergy systems that can bring improvements to community health through cleaner cooking fuels and hence air quality, lower exposure to indoor air pollution, and broad improvements in health through enhanced energy access. The potential health benefits of better bioenergy technology are especially important in developing country contexts, where traditional biomass combustion for cooking and warming exposes users to high health burdens, especially among women and children, who are disproportionately affected by indoor air pollution (Lowry et al. 2019). Studies universally show that the shift from traditional biomass combustion to improved bioenergy devices could lead to major health gains by reducing exposure to dangerous smoke. But the size of these benefits greatly depends on the technology selected, how it is implemented, and the specific conditions of a place. The translation of health benefits will depend, however, not only on better technologies but also on the ways in which they are disseminated, the training and support provided to users, and the support of the systems needed to ensure their adoption on a sustained basis.

Given that energy access is a core social sustainability concern, bioenergy systems could bring energy improvements for marginalized communities and wider infrastructure expansion. Although LCEEs "can sharpen the focus on the objective of scaling up renewable energy" (pp. 4–5), the connection between bioenergy production and local energy access is not linear as energy products may be germplasm sources and local energy access 311 exported rather than locally consumed, and even locally provided improved energy services may be economically unavailable to local residents.

Economic sustainability models have adopted a similar expanded view to include wider economic development impacts as well as direct economic output effects. Contemporary approaches acknowledge that sustainability is as much about the financial success of a project as it is about how a given project contributes to local and regional economic development: income generation, job creation, and skills development being only a few examples. The economic influences of bioenergy systems are often evaluated along a number of axes, such as productivity, employment, financial sustainability, and community economic development.

Job creation is an important economic sustainability measure that indicates the potential of bioenergy systems to generate livelihoods and local economic development. Employment bioenergy systems can provide jobs at several stages of the value chain, including feedstock production and harvest, conversion, supply, distribution, and end-use options, but the superior quality and stability of jobs vary greatly by system design and implementation (Singh et al. 2017). The employment opportunities in the two types of bioenergy systems have a wide spectrum, and they can be from agricultural work and manual labor, up to technical-operator and technical-management, with various qualification levels and nominal salaries. Studies have shown that rural areas with less access to other economic alternatives tend to benefit the most from the job opportunities created by bioenergy systems (see Gloy and Ray, 2012), making these educational gains to the local workforce development especially valuable for regional development. Nevertheless, the benefits of labor market participation need to be taken into consideration within the context of skilled labor development and training programs and nurturing institutional environment that can help maintain employment in the long run. Another important economic sustainability aspect is the financial viability, involving the project level financial soundness and the wider investment attractiveness. Economic stability of the bioenergy systems is influenced by many factors such as feedstock price, process yield, product price, market penetration, and policy mechanisms. Economic issues such as competition with conventional energy sources, capital costs, and market instability have posed a major obstacle to many bioenergy systems (Cerqueira et al. 2018).

The number of environmental indicators has traditionally been higher than social and economic indicators, and the former remains a moving target, with an increased focus on the interplay between bioenergy systems and environmental systems. Existing frameworks include measures of energy efficiency, greenhouse gas (GHG) emissions, changes to water quality, biodiversity drivers, and effects on soil health. The purpose of this integration is social- and economic-specific measures of environmental quality in combination of these indicators, reflecting the increasing awareness that environmental quality is not achieved behind the social and economic aspects. Nanotechnology for environmental applications has developed as an important research field with a potential connection to pathways of biofuel systems. Nanotechnology provides special tools for addressing environmental challenges such as removal of pollutants, waste recycling, and resource utilization in environmentally friendly ways/environmental issues such as pollution prevention, waste minimization, IFI capture, and eco-resources utilization. These novel properties of nanomaterials such as high surface area, improved reactivity, and selective functionalities could be exploited to address a range of bioenergy systems to potentially enhance their environmental and economic sustainability (Chausali et al. 2022).

Nevertheless, the development of nanotechnology in bioenergy also brings with its important issues of technology access, safety concern, and benefit-risk distribution. The vast majority of nanotechnology investment and research has been in developed countries and by large corporations and it is an open question as to whether such technologies would be available to smaller-scale and developing country contexts that are the site of much bioenergy development. The confluence of bioenergy and nanotechnology is posing novel challenges to the socioeconomic assessment community because of the complexity of the phenomena involved as well as their interactions. Conventional technology assessment approaches may not be able to adequately measure the nuanced effects of such systems, especially in terms of long-term social and economic consequences. This has raised the need for new tools for assessment capable of including more effectively the technical, social, economic, and environmental issues in holistic frameworks.

Recent studies have pointed out some methodological issues related to the evaluation of socio-economic balance of bioenergy systems: problems of defining suitable system boundaries for evaluation, limitations in finding comparable measures of socio-economic balance between different areas/technologies, and limitations in the existing data collecting and monitoring systems. These methodological issues are even more complex when one considers applications in nanotechnology, since these applications can introduce multiple layered and long-term effects, which are difficult to predict and measure using common assessment mechanisms.

2.3 Social Sustainability Indicators in Bioenergy Systems

Comprehensive frameworks are needed to assess and evaluate social sustainability in bioenergy systems, which consider complex ways in which energy technology interacts with social systems and impact on community well-being. Recent studies in social sustainability assessment have advanced from mere impact measurements toward complex frameworks capturing characteristics of social development and equity. Both frameworks acknowledge that social sustainability is about more than just avoiding negative social impacts but also about making positive contributions to social development and community capacity building. Food security is one of the most important and thoroughly researched social sustainability indicators in bioenergy production. The relationship between bioenergy and food has been extensively studied and debated in both the policy and academic communities, and the possibility of competition between food and energy production for land, water, and other agricultural inputs has been widely considered. Given the intricate interactions between food and energy, the assessment framework for food-energy interactions needs to account for direct, as well as indirect, effects on food availability, access, and utilization.

Recent research shows that the effects of bioenergy expansion on food security vary according to context and depend on a range of factors, such as the choice of feedstock and production scale, the location of production systems, market's integration of areas, and policy settings (Chausali et al. 2021). Bioenergy applications exploiting marginal lands or agricultural residues are in general considered to entail lower risks to food security than that using prime agricultural land solely dedicated to food production. However, even non-food feedstock using projects can induce food security implication by driving land and labor markets, input prices, and local economic status. The magnitude of bioenergy development emerges as an important determinant of food security impacts, and larger scale development is more likely to lead to significant impacts on local and regional food systems. Findings indicate that small-scale bioenergy projects that are community-based are more likely to be conducive to local food security interests, especially when they are planned with explicit consideration for the food-energy nexus and engagement of communities in overall planning (Subramanian et al. 2020). Geography and market context are also key considerations for food security outcomes, and projects situated in already well-connected, agriculturally productive areas will usually have lower FS risk than those in more isolated or food insecure ones. The opportunities for alternative livelihoods and strong food distribution networks may help to overcome, at least partially, any negative impacts of bioenergy development on food security at the local level.

Health and well-being is intended here in a broad sense and includes several of the channels through which bioenergy can impact on public health—from the direct effects of better energy systems on, e.g., respiratory health to the indirect effects on health through facilitation of economic development, through re-mediation of degraded environments, etc. The health consequences of bioenergy are particularly relevant in the context of developing countries, where conventional energy use contributes to significant health burdens to a great extent. UCL puts improved bioenergy on the map. Improved bioenergy has always been an attractive proposition, whether to significantly reduce indoor air pollution in developing countries, improve the lives and livelihoods of rural farming communities, or tackle climate change on a global scale. Existing literature consistently shows that solid biomass use for the household cooking and heating is linked to respiratory conditions, cardiovascular problems, and other diseases —particularly among women and children that spend more time in cooking and have higher exposure to the toxic substance emissions produced by the burning of solid biomass (Tripathi et al. 2018). Enhanced bioenergy technologies, such as improved cookstoves, biogas technologies, and advanced fuels for clean combustion, can offer considerable health benefits through the reduction of exposure to unhealthy emissions.

The extent of health benefits of such more advanced bioenergy technologies depends upon the adoptions of specific technologies, implementation pathways, and local situations. The findings are consistent with the results of other studies that point towards the most substantial health gains by transitions to clean-burning gaseous or liquid fuels, although improved solid fuel stoves also provide more modest, yet still considerable benefits compared to traditional cooking (Shaikh et al. 2021). Yet health gains from technology enhancement are achievable only if better technologies are not only developed but also effectively used and maintained over time. Access to energy is a central social sustainability concern that captures both the ability to obtain energy services and the affordability and reliability of these services for different groups of users. Bioenergy technologies could greatly contribute to improved energy access, especially in rural and off-grid areas with no or limited access to the grid. Nevertheless, the linkage between bioenergy production and local energy access is not mechanical and needs to be purposefully addressed in project planning and delivery. Much of bioenergy attention labels bioenergy projects as serving export markets, for urban or energy needs, rather than serving local energy access, thus potentially constraining impact on localist development goals. Studies indicate that the ability to achieve the energy access benefits of bioenergy development depends on factors such as the degree to which attention is paid to the specific design and energy needs at the local level, appropriate technology selection, and institutional arrangements to ensure that local communities are able to access and afford the services (Yousefi 2022). The configuration of bioenergy systems can greatly affect their role with respect to energy access, and decentralized designs on a modest scale are often most likely to meet local energy demands as opposed to large-scale centralized plants. Community ownership and engagement in bioenergy developments can similarly help to secure that the gains resulting from energy access investment will be reaped into the future.

Water access and quality criteria embody the potential benefits and concerns resulting from bioenergy systems in terms of water resources and community access to potable water. The impacts of bioenergy systems on water resources may occur through multiple mechanisms that alter the use, quality, and/or quantity of water used in feedstock production and conversion and in associated systems used for effluent management such as runoff from agriculture or process waste, as well as the potential to affect water quality (either exacerbate or abate) by reducing the potential for pollution or better watershed management. Studies indicate that bioenergy systems can be implemented in a way that offers a net benefit for water quality, especially in the context of sustainable agriculture and sound environmental management. Nonetheless, ensuring realization of these benefits needs deliberate system planning, environmental management, and monitoring that minimize potential negative impacts and maximize the positive impacts (Smith 2011).

Land tenure and access concerns, on the other hand, involve the intricate dynamics between bioenergy development, existing land-use systems, property rights, and community land management regimes. Land tenure regimes have important implications for the distribution of benefits and risks related to bioenergy development, where more secure land tenure is typically linked with more favorable social outcomes and greater community involvement in project benefits. Payment for bioenergy can act in several ways on land tenure, such as increasing the value of land, changing land use patterns, and opening up new economic activities that can interfere with traditional land management. Research indicates that bioenergy projects that operate within current land tenure systems and offer returns to current users of land are more likely to achieve positive social outcomes than those requiring major adjustments to land tenure (Rezaei 2018). Community involvement and governance are important aspects of social sustainability, including the degree and quality of community participation in bioenergy projects' planning, implementation, and benefit distribution. Authentic community participation does not consist of just the consultation of affected communities, but rather of real participation in project governance, decision-making, and uptake.

Evidence repeatedly shows that projects with robust community buy-in and local ownership are more apt to bring about positive social change and long-term sustainability than initiatives designed and managed by outsiders with little to no community representation. Reliable community participation in wind power projects requires proper institutional settings, capacity building assistance, and equitable benefit-sharing mechanisms, which enable communities to participate in governing projects and benefit from the outcomes of projects (Khan et al. 2019). Social sustainability is receiving increasing attention as to the preservation of cultural and social cohesion, recognizing the need for cultural activities, social relations, and community identity in the face of technological and economic changes. Bioenergy production can have significant impacts on customary practices, social relations, and cultural values, especially in indigenous and traditional societies where energy use is often embedded in social systems. Gender equity concerns are prevalent across various aspects of social sustainability, which are related to the unequal effects of bioenergy development on men and women in affected communities. Bioenergy systems can carry gender-equity advantages in the form of time savings from reduced fuel collection, income-generating opportunities, and increased access to energy for productive end-uses. But to obtain these gains, gender-related issues must be explicitly addressed in project design, implementation, and sharing of benefits (Dhankher and Foyer 2018).

Similarly, women carry a greater share of the energy burden under traditional energy systems—in hours collecting fuel and exposure to indoor air pollution from cooking—in energy-related decision-making processes. Properly designed bioenergy projects can also help address some of these gender disparities as well as bring significant benefits to community development and women's empowerment. The inclusion of such broad ranging indicators of social sustainability in comprehensive assessment methodologies continues to be challenging for methodological and data availability reasons and because it may lead to juxtaposed or incommensurable social dimensions. Current research reveals several priorities in the development of social sustainability assessment, such as the development of more effective methods to measure community participation and empowerment, improved methodologies for assessing cultural and social cohesion impacts, and the increased incorporation of gender considerations across all aspects of social sustainability assessment (Table 2.1).

2.4 Economic Sustainability and Community Development

Economic sustainability of bioenergy systems includes direct economic costs and broader economic implications on economic growth, employment, and community economic health. Economic sustainability assessment requires broad frameworks that can integrate the multiple dimensions of how bioenergy systems link to local, regional, and national economies and that account for both short-term financial performance and long-term economic development impacts.

Job generation and income opportunity are basic economic sustainability indicators, which show the capacity of bioenergy systems to create jobs and contribute to local economy. Bioenergy life cycles can generate employment at various stages of the value chain, from feedstock cultivation and collection, processing, distribution to maintenance and operation of energy systems. The potential for bioenergy systems for employment is particularly high in rural areas, where industrial alternatives may not exist and in which a substantial portion of bioenergy feedstock may be produced. Studies show that bioenergy systems can be good for jobs, with employment generation varying widely according to system design, scale, and technology pathway. Labor-intensive bioenergy pathways, such as feedstock collection and processing, can yield relatively large potential for jobs for less-skilled workers, whereas more technologically intensive systems may generate lower levels of work but higher-skilled jobs (Schewe et al. 2019).

The level of job quality and long-term job sustainability for the jobs that bioenergy systems generate will depend very much on wage levels, the skill

Table 2.1 Social sustainability indicators matrix for bioenergy systems

Table 1: Social Sustainability Indicators Matrix for Bioenergy Systems

Indicator Category	Related SDGs	Measurement Focus	Literature Coverage	Priority Level
Food Security & Nutrition SI-01 Trends in food output and prices vs. previous trends	SDG 2	Food price indices, production volumes, food-vs-fuel debate impacts, marginal land utilization effects	Well Covered	High*
Health & Well-being SI-02 Indoor air quality and particulate emissions	SDG 3	Mortality/disease from indoor smoke, improved cookstove adoption, respiratory health impacts	Well Covered	High*
Water Access & Quality SI-03 Access to clean water/municipal supply	SDG 6	Water quality improvements, pollution reduction, watershed protection benefits	Moderate	Medium
Energy Access & Infrastructure SI-04 Energy diversity and clean energy access	SDG 7	Local energy uptake, infrastructure development, grid integration benefits	Poorly Tracked	High*
Risk Management SI-05 Risk of catastrophic disruptions	SDG 11	Diversification benefits, wildfire risk reduction, pest outbreak management	Limited Data	High*
Land Tenure & Governance SI-06 Land allocation, tenure, and compliance	SDG 12	Land use changes, property rights, regulatory compliance, circular economy integration	Moderate	High*
Community Participation SI-07 Public opinion and information sharing	SDG 17	Stakeholder engagement, public acceptance, information access, decision-making participation	Moderate	Medium
Gender Equity SI-08 Time spent on biomass collection	SDG 3 SDG 2	Unpaid domestic work reduction, women's time savings, school attendance improvements	Well Covered	High
Cultural Preservation SI-09 Traditional practices and social cohesion	SDG 11	Cultural value protection, traditional knowledge integration, community cohesion maintenance	Emerging	Medium
Local Capacity Building SI-10 Training and skill development	SDG 7 SDG 17	Technical training provision, local expertise development, technology transfer effectiveness	Limited	Medium

Note: *Priority indicators identified by workshop participants (Mabee, 2024). Coverage assessment based on literature review of 41 studies. SDG correlation follows Blair et al. (2021) classification: High (H), Moderate (M), Low (L) correlation with bioenergy supply chains.

intensity of those jobs, working conditions, and the sustainability of that employment. Studies indicate that the sustainable improvement of employment entails the provision of training and development, adequate income and the offering of stable, long-term employment rather than short-term or seasonal work (Nutan et al. 2020). Gender mainstreaming, in particular, is important when considering women's participation in bioenergy operations, when compared to conventional energy systems where women carry a heavier burden in fuel collection and processing. Sensibly designed bioenergy systems can address these burdens while offering new economic prospects to women, via the possibility of employment in bioenergy value chains, or by saving time for women to participate in other productive activities. Financial soundness and economic performance are important sustainability aspects to include project-level financial returns as well as overall economic efficiency. Economic sustainability of bioenergy systems is based on various factors regarding capital investment, operating costs, feedstock costs, product selling price, availability of a market, and policies to support. Most of the bioenergy technologies are economically uncompetitive because of competition with commercial energy services, initial capital costs, and financial and market risks.

Studies show that the economics of bioenergy systems differ widely among technologies, scales, and markets. Low capital costs and simplified operation enable small-scale, community-based systems to reach financial viability, whereas higher capital investments and more complicated market dynamics are associated with the large-scale commercial systems (Zhao et al. 2015). The financial viability and the economic sustainability of bioenergy systems can be greatly influenced by the incorporation of existing economic activities. Systems that incrementally enhance existing agricultural or forestry practices may incur lower incremental costs and be more likely to succeed in a financial sense than do those depending upon new economic activities or infrastructure construction. Market and value chain development are key considerations for economic sustainability, which includes the development of secure markets for bioenergy products and the efficient organization of bioenergy value chains. To maintain a sustainable market, it is important to consider supply-side factors (the availability of feedstocks, the capacity to produce) and demand-side factors (quality of product, product price, and customer acceptance). Literature has shown that effective market development often takes holistic approaches to the key elements of bioenergy supply chains to include feedstock production, conversion technology, product distribution, and utilization of end use applications. The development of local and regional markets might especially be important to let the benefits of the economy reach the local level rather than be exported to distant markets (Lemma et al. 2016).

Efficient and well-organized value chains have great potential effects on the economic sustainability of bioenergy systems and in the sharing of economic benefits between various actors. Empirical evidence shows that not just general economic exigency of the value chain but also the share of value addition accruing to feedstock producers, processors, and other members of the value chain is affected by how this value chain is organized. Community economic development takes into consideration larger economic implications in addition to the project's direct effects such as multiplier effects, local and regional business development, and regional economic advancement. Bioenergy development contributes to economic development within the community. Bioenergy development can contribute to economic development within communities through activities such as supply chain and regional production, increased local economic linkages, and value addition.

The level of community economic development benefits is determined by the local content and participation in bioenergy value chains, level of local ownership and control, and integration of bioenergy development with other forms of community development. Studies have found that community-owned and community-controlled bioenergy projects tend to produce more local economic development benefits than projects with external ownership (Schauberger et al. 2017). Localities and biomass feedstock supplies play a critical role in shaping the community economic development impact of bioenergy projects. This plastic waste can further be used as a local resource, for waste collection itself, decreasing transportation costs and bringing benefits not only to the entire local community but also to the environment. This aspect refers to how well an organization can continue improving its technologies, innovating its processes, and adapting to changing market needs in order to remain economically sustainable. The continuous effort to develop new technology, transfer technology, and create an innovative environment is essential to form the foundation for long-term sustainable bioenergy development and to remain competitive and adaptable in reaction to changing markets and policies. The local capability to innovate and adapt to technology can play a key role in the long-term economic viability of bioenergy systems. Domestic and regional communities possessing superior technical competencies and innovation systems may find it easier to fine-tune locally adapted bioenergy technologies and develop innovations, which will increase its overall economic performance over time (Adger et al. 2007). Financial and investment access is key economic sustainable consideration that relates to bioenergy finance availability and terms of finance provision. The availability of adequate financing (even for such "low risk" fuel) is often a significant hurdle for bioenergy projects, particularly for small operators and community projects that may lack access to traditional financing sources.

Studies indicate that investment constraints affect bioenergy implementation and can affect the size of projects, type of ownership of projects, and the economic viability of projects. Inventive finance mechanisms, such as community investment, cooperative ownership, and public private partnerships, might help to mediate cash flow limitations while equitably redistributing economic moral hazards among diverse stakeholders (Nurkolis et al. 2022). The cost of capital, its terms and structure, contractual rates of return, repayment schedules and collateral can have a major impact on the net economic viability of bioenergy projects and on the distribution of economic risk and benefits among the parties involved. Appropriate risk-sharing financing mechanisms are more likely to contribute to sustainable economic development when those mechanisms are structured to be consistent with the cash flow merits of bioenergy projects. Economic resilience and risk management: Bioenergy systems can have economic performance under varying conditions and provide reliable tools for effective economic risk management. Economic stability is even more critical for bioenergy due to potential high feedstock price volatilities, policy environment changes, and market uncertainty.

Diversification can strengthen the economy and make it more resilient by reducing dependence on a single market, feedstock, or revenue source. It has been shown that bioenergy systems based on integration with other economic activities or producing multiple products show more economic stability than having single products or markets associated (Jansson and Hofmockel 2020). Insurance, contracts, and cooperative agreements are tools to manage economic risks and promote sustainability. Successful risk management needs to address markets risk and operational risks and risk sharing. And there need ways to distribute risks across a range of stakeholders according to their ability to absorb and manage different types of risks (Table 2.2).

2.5 Nanotechnology Applications and Environmental Solutions

The infusion of nanotechnology into bioenergy systems is an emerging and transformative strategy that exploits special nanomaterial properties to address contemporary environmental issues and to improve the sustainability and efficiency of renewable energy devices. The range of nanotechnology applications in this field is broad, including catalysis, conversion, environmental remediation, and pollution prevention, and presents new possibilities for enhancing both the technical performance and environmental benefits of bioenergy systems. The singular characteristics of nanomaterials such as high surface-to-volume ratio, improved reactivity, controllable surface chemistry, and quantum feature permit applications that can overcome classical constraints of conventional bioenergy technologies. These characteristics enable nanomaterials with potent capabilities to act as highly efficient catalysts, selective adsorbents, and multifunctional materials to co-actively tackle numerous engineering and environmental challenges (Fulekar et al. 2014).

Environmental applications of nanotechnology provide exciting opportunities to remediate pollutant contaminated or degraded ecosystems in the process of bioenergy sustainability. Application areas also include water treatment, air purification, soil remediation, and waste management, with clear implications for increasing overall bioenergy system sustainability via improved feedstock quality (which might reduce demand for fertilizer and pesticide in future bioenergy crops), reduced environmental impacts, and greater system sustainability in general. Nanotechnology-based water-treatment systems have been known to possess high efficiencies for the removal of various pollutants such as heavy metals, organic compounds, and biological contaminants. Carbon nanotube materials have exhibited great potential in water/wastewater treatment, and functionalized carbon nanotubes have achieved greater than 95% of removal efficiencies for various organics and heavy metals and also remain reusable due to some regeneration methods (Qu et al. 2013).

Graphene-based nanomaterials also have unique properties to be used in water treatment and environmental remediation such as adsorption of heavy metals and photocatalytic degradation of organic pollutants. It has been reported that graphene oxide and modified graphene could have high adsorption capacity for different kind of contaminants and, importantly, remain stable and reusable, making them especially interesting for bioenergy applications because these processes must be cost-effective and sustainable (Sadegh et al. 2016). Superparamagnetic nanoparticles offer a number of advantages for environmental remediation due to the potential for

Table 2.2 Economic sustainability metrics for bioenergy systems

Table 2: Economic Sustainability Metrics for Bioenergy Systems

Economic Indicator	SDG Link	Key Metrics	Assessment Criteria	Literature Status	Priority
Economic Productivity EC-01 Gross value added, profit per unit area	8.1.1 8.2.1	• GDP per capita growth • Economic efficiency ratios • Regional value creation • Productivity benchmarks	Positive contribution to regional economy, competitive returns vs alternatives	Well Tracked	High
Employment Creation EC-02 Direct/indirect jobs, income trends	8.5.1 8.5.2	• Jobs per GWh generated • Income per capita trends • Employment duration • Skill level requirements	1 job per 3,600-13,300 tonnes biomass/year typical range (Mabee, 2024)	Well Tracked	High
Financial Viability EC-03 Project-level economic performance	9.3.1 9.b.1	• ROI and payback periods • NPV calculations • Sensitivity analysis • Market competitiveness	Positive NPV, competitive with fossil alternatives, acceptable risk profile	Well Tracked	High*
CO_2 Abatement Cost EC-04 Environmental cost-effectiveness	9.4.1	• \$/tonne CO_2 equivalent • Lifecycle cost analysis • Carbon credit potential • Mitigation efficiency	Cost range US\$89-300/tonne CO_2e reported (Mabee, 2024)	Well Tracked	High*
Community Economic Benefit EC-05 Local economic development impact	10.4.2	• Local procurement rates • Multiplier effects • Business development • Economic diversification	Positive local economic multipliers, reduced inequality, community ownership	Poorly Tracked	High*
Energy Cost & Accessibility EC-06 Consumer affordability metrics	7.1.2	• Energy price per unit • Price volatility measures • Affordability indices • Access rates	Competitive with conventional fuels, stable pricing, improved access	Poorly Tracked	High*
R&D Investment EC-07 Innovation and development spending	9.5.1	• R&D expenditure/GDP • Patent applications • Technology transfer • Capacity building	Sustained investment levels, increasing innovation output, knowledge transfer	Limited Data	Medium
Land Price Trends EC-08 Land market impacts	1.4.2	• Land value changes • Tenure security • Access equality • Market stability	Stable land markets, secure tenure, equitable access for small holders	Poorly Tracked	Medium
Gender Economic Equity EC-09 Women's economic participation	5.4.1	• Time savings monetization • Women's employment rates • Income by gender • Economic decision-making	Reduced unpaid work burden, improved income opportunities, equal participation	Well Tracked	High
Market Development EC-10 Value chain efficiency and stability	8.2.1	• Market penetration rates • Value chain integration • Price stability • Supply security	Growing market share, stable supply chains, predictable pricing	Moderate	Medium
Workforce Development EC-11 Training and skill enhancement	4.1.2	• Training completion rates • Skill level improvements • Educational investment • Certification programs	Improved skill levels, sustainable employment, career advancement	Limited Data	Medium
Regional Economic Integration EC-12 Supply chain and market linkages	9.1.1	• Regional sourcing rates • Cross-sector linkages • Economic clustering • Infrastructure development	Strong regional integration, diversified economic base, infrastructure improvements	Moderate	Medium
Financial Risk Management EC-13 Economic resilience and stability	1.5.1	• Risk assessment scores • Insurance coverage • Diversification metrics • Emergency reserves	Low financial risk, adequate insurance, diversified revenue streams	Moderate	Medium
Circular Economy Integration EC-14 Resource efficiency and waste reduction	12.5.1	• Waste valorization rates • Resource efficiency ratios • Circular material flows • By-product utilization	High resource efficiency, minimal waste, integrated circular flows	Emerging	Medium

Key: *Workshop priority indicators (Mabee, 2024). SDG indicators follow UN Statistics Division framework. Assessment based on literature coverage of 14 economic indicators across bioenergy systems. Economic productivity and employment data show regional variations from €159-319 (Senegal) to C\$330-550M NPV (Canada).

removal and recovery from treatment systems using an external magnetic field. The materials are easy to functionalize for selective pollutant removal and can be recycled multiple times, thereby providing an economically viable technology, particularly for bioenergy applications where economic viability is of high importance (Guan et al. 2017).

Another important environmental application of nanotechnology to climate change mitigation and sustainability is the management of greenhouse gases. Nanomaterials can be used for improving the capture and storage of greenhouse gas through mechanisms like the increase of adsorption capacity, selective gas separation, and the improved chemical reactivity for conversion processes. Metal-organic frameworks (MOFs) are one of the more promising class of photonic-nanostructures for mitigating greenhouse gases, thanks to their high surface areas, adjustable pore architectures, as well as selective adsorption characteristics. Studies have shown that different MOF compositions additionally offer a possibility for high capacity and selectivity for carbon dioxide sequestration—some showing promising performance exceeding existing capture technologies (Bankole et al. 2017). Carbon nanotube materials have also garnered considerable interest, for the capture of greenhouse gases, especially as both single- and multi-walled carbon nanotubes exhibit high adsorption capacities for CO2 (Tripathi et al. 2018) and also for methane. Functionalized carbon nanotubes show promise as materials that can achieve higher selectivity and capacity with the associated stability under operation conditions in a diluted gas stream representative of a bioenergy (Ahamad et al. 2019).

Nano-zeolite materials can present alternative options as well in the field of research directed at greenhouse gas management, with potential applications in bioenergy systems. These materials could have extremely high adsorption capacities for a number of greenhouse gases, along with possessing thermal stability and renderability, which are important for practical applications (Luan et al. 2019). Applications of nanotechnology to air purification will further enhance the environmental biocompatibility of bioenergies and will support public health, which is important in society-based sustainability. Photocatalytic nanoparticles, such as TiO2, can destroy airborne pollutants and ameliorate air quality in the vicinity of bioenergy plants, yielding direct health advantages to nearby populations (AlOthman et al. 2019). Therefore, biofuel production enhancement with nanotechnology has the potential to increase conversion efficiency, lower processing costs, improve product quality, and possibly reduce ecological consequences. These applications exploit the catalytic properties of nanomaterials for improving different aspects of biofuel production including feedstock pretreatment, conversion, and product separation.

Carbon nanotubes in biofuel production have proven to provide significant increase in enzyme immobilization and conversion efficiency. Enzyme immobilized on functionalized carbon nanotubes exhibits high reaction activity by recycling to lactic without the need to cultivate expensive enzyme at a higher cost for process economy (Kariim et al. 2020). Biofuels are promising alternative energy sources, and utilizing magnetic nanoparticles has advantages for biofuel production, including improved enzyme immobilization and separation. The magnetic characteristics of these materials are beneficial to recycling expensive enzymes and reduce the processing cost and the secondary pollution (Zhao et al. 2021). Metal oxide nanoparticles have been proposed as catalysts for a multitude of biofuels processes, such as transesterification for biodiesel synthesis and for fermentation enhancement in bioethanol production. Such materials can have higher catalytic activity and may provide certain advantages with respect to catalyst recovery and re-use (Yang et al. 2021).

The employment of nanotechnology in bioenergy systems further facilitates new methodologies for process monitoring and control, increasing the efficiency and environmental performance of processes. One important application of nanosensors is the in situ control of processing conditions, finished product quality, and environmental conditions, which helps in performance optimization and quick troubleshooting of the system (Bera et al. 2016). The use of nanotechnology in bioenergy systems, however, brings up substantial concerns related to bioenergy nanomaterial safety, environmental fate, and risk issues. The state-of-the-art current research suggests the necessity to use life cycle assessment frameworks that highlight broader environmental and health impacts, from the production to disposal, of nanomaterials (Wan et al. 2016).

The environmental destiny and the environmental transport of nanomaterials are important issues in relation to the need for the net environmental benefits stemming from nanotechnology. Evidence indicates that some nanomaterials may stay in environmental systems or have intricate environmental interactions that need to be cautiously evaluated before their application to avoid unwanted exposures (Ashour et al. 2017). The above-described synthesis and processing of nanomaterials may also have environmental consequences, which need to be considered when evaluating the sustainability of a process. The energy consumption and adverse environmental effects of nanomaterial synthesis can overlook some of the gains made from their application; thus, it is important to synthesize nanomaterials using environmentally friendly routes and to optimize them. Cost-effectiveness of nanotechnology applications is also an important factor for effective utilization in bioenergy processes. While nanomaterials typically show better performance properties than traditional materials, their increased cost could limit adoption unless some part of the cost, in terms of improved efficiency, decreased operating expense, or added product value (Diraki et al. 2018).

The literature has shown that improvements in cost-effectiveness of nanomaterials can be achieved with strategies such as material recycling, integration with current available systems and processes, optimization of nanomaterials synthesis, and application process. Cheap synthesis methods and exploiting abundant resources can also contribute to a more economical nanotechnology application. The scale-up of nanotechnology applications is a key consideration to enable the use of these materials in bioenergy technologies. There are many nanotechnology applications that have been shown at a laboratory, or prototype, scale but may not yet be feasible at a commercial scale due to technical or economic limitations (Rajpoot S 2019).

To scale applications of nanotechnology successfully, the manufacturing processes, quality, supply chain, and economics will have to be addressed. Standardization of fabrication and quality specification may be useful to support the transition between research and commercial use. The interface of commercial bioenergy systems/infrastructure with the applications of nanotechnologies is also critical, powerful, for practical applications. Connecting nanotechnology systems to existing systems might be less costly than implementing entirely new system; however, these two technologies may need to be carefully integrated and checked for compatibility.

Successful integration will require an understanding of the technological needs of nanotechnology requirements as well as the dynamic performance of existing bioenergy systems. It may mean the development of interfaces, changes to the way operating takes place, and training for those who have to work in enhanced nanotechnology systems.

2.6 Present Scenario Analysis: Challenges and Opportunities

Socioeconomic balance in the current bioenergy system with added value by nanotechnology is a composite picture of technological opportunities, implementational problems, and a set of emerging prospects. The contemporary setting is marked by unprecedented advances in biorefinery and nanotechnology, as well as lingering challenges in realizing equitable benefits, sharing and steering technological advancements towards inclusive development goals. Technical access is one of the most challenging issues in establishing social balance in bioenergy systems enabled by nanotechnology. On the one hand, such applications of nanotechnology present great promises of efficiency improvement and in comparison, of mitigating the impact of industrial activity on the environment; on the others, the development and commercialization of these technologies has so far been very much centered in industrialized nations and large companies. This brisk pace raises questions about whether nanotechnology gains will be available to smaller scale operators and developing country contexts (where most bioenergy development is and where the socio-economic gains are very much sought).

Nanotechnology costs may be high and can prevent adoption, and particularly community bioenergy initiatives and small operators may not have sufficient budget to invest in new technologies. Research indicates that TSC for the nanotech-enriched bioenergy systems could be substantially higher than TSC for the traditional systems, which could make these energy technologies out-of-reach for some of the very communities that stand to gain the most from better energy technologies (Abou Chacra et al. 2018). Another pervasive issue to overcome for a fair distribution of the benefits of nanotechnology is the lack of institutional capacity. Effective deployment of nwBA-based bioenergy systems is difficult in developing country settings due to lack of technical know-how, institutional support mechanisms, and regulatory systems. The passage explains that introducing new or advanced bioenergy (NWBA-based) technologies in developing countries is challenging because the fundamental support systems needed for successful adoption are often weak or missing. (Rajpoot S 2023).

Regulatory and safety issues are other challenging factors for commercializing nanotechnology for bioenergy systems. Environmental and health implications of nanomaterials are yet to be fully understood and in many cases law and regulations for the use of nanotechnology lag. This lack of regulation leads to market barriers to adoption and concerns of risks to human health and environmental systems (Cao et al. 2019). The scale of nanotechnology applications can also induce access to reliance of foreign technical support and special knowledge that hinder building of local capacity and transfer of technology. Unlike most of the traditional bioenergy technologies that can be adapted and operated by local capabilities and resources, the applications of nanotechnology can necessitate special training and technical support which are not frequently available in many settings. Nevertheless, the current context also offers valuable opportunities for reaching a better socioeconomic equilibrium through novel methodologies in technology creation, dissemination, and regulation. New directions in nanotechnology R&D recognize greater emphasis on sustainability concerns and social considerations, thereby exposing opportunities in support of more inclusive technology development methodologies.

Open innovation and collaborative research paradigms are being initiated, which might contribute to countering the accessibility issues of technology, by allowing developing country researchers and institutions to participate in the process of technology development in the area of nanotechnology. Globalization and technology transfer programs are allowing for opportunities to share knowledge and promote capacity building which if properly implemented could result into democratization of access to the benefits of nanotechnology (Sundaran et al. 2019). The possibility of fabricating organic nanomaterials via low-cost methods using indigenous resources is also an area with potential for enhancing technology availability. Investigation into bio-based synthesis routes and the utilization of agricultural waste for nanomaterials manufacture may minimize cost and simultaneously generate extra rural economic opportunities (Riaz et al. 2019).

Innovations adopted by the community approaches present additional opportunities for the management to ensure that nanotechnology for applications is tailored to meet the local conditions and needs. Participatory technology design and testing, conducted in fellow community member spaces, can help guarantee that technological solutions engage with the priorities of local communities while also enhancing local capacity to engage with technology. On the other hand, the exploitation of nanotechnology applications in combination with conventional bioenergy value chains and systems offers the potential for evolutionary rather than revolutionary growth solutions. This approach can lower both the implementation costs and have access to previous knowledge and infrastructure, contributing to facilitating the access to new technology advances to smaller operators.

These policy innovations such as technology transfer agreements, public private initiative, cooperative arrangements for international technology transfer are generating fresh momentum to deal with the issue of technology access. These mechanisms can help mitigate the cost and risk of nanotechnology uptake and distribution of these benefits more fairly among a range of stakeholders (Nguyen and Juang 2019). The increasing focus on sustainable and socially responsible technology development is sourcing markets for the development of nanotechnology applications that provide solutions to social and environmental issues. This trend might contribute to moving innovation toward more accessible and suitable solutions for developing country settings and at the same time scale commercially.

Innovative financing models such as blended finance, impact investment, and community ownership are expected to pave way for the promotion of nanotechnology-enabled energy development in neglected markets. Such financing methods may also relieve capital constraints and provide community input into project design and execution. The current situation also identifies the possibilities of using nanotechnology to combat some longstanding bioenergy development issues. The performance enhancements and cost reductions that are possible through nanotechnology implementation may help to make bioenergy systems more competitive with traditional energy systems, while simultaneously decreasing environmental footprints and enhancing societal benefits.

2.7 Case Studies and Implementation Examples

We can see what "balanced" bioenergy technology for socio-economic development, which benefits from nanotechnologies, would look like in practice when we look at "works in progress" cases or practical implementation stories which mirror both the potential and barriers in the synthesis of these technologies. There is still a lack of detailed long-term study of the (potential) implementation of nanotechnology-assisted bioenergy systems, partly as a result of the relative novelty of these technologies, though some illustrative case examples offer significant insights into issues of how such systems might best be realized and the related outcomes. Biogas community plants incorporated with MNPs for increased methane production are therefore one of the most interesting applications to reach the balance between socioeconomic and ecosystem values. Studies have shown that iron nanoparticles can increase the biogas production by 120% higher than conventional systems and improve the system stability and reduce starting-up time by (Chen et al. 2020)). The economic feasibility of small-scale biogas systems could be greatly improved by these advances, contributing to the energy security of rural populations.

Implementation of nanoparticle enriched biogas systems in developing country context has shown both some good success stories and challenges. Although these technical advances are impressive, there are barriers to its adoption, including the expense of nanoparticle incorporation and technical know-how for the optimization of systems. They have also been successful when attention is given to purification processes and capacity building, enabling local operators to manage advanced systems effectively. Nanomaterial-mediated filtration devices for water purification, when coupled with bioenergy plants, are another possibility through which several sustainability goals could be met simultaneously. Supplementary file 3 includes case studies of bioenergy facilities that have integrated nanomaterial-based water treatment and have realized decreased effluent quality with decreased treatment costs and water reuse in production processes (Bu et al. 2020).

Combining water treatment with bioenergy production may also offer a source of relief for the economics of renewable fuel and power generation, because of savings on water bills and the revenue possible from water treatment services. Yet, to get there, you have to design the system correctly and train the operators and ensure the system can be maintained over the life of it.

Nano catalyst applications in biodiesel synthesis have shown good prospects for enhanced conversion efficiency and lower cost of processes. Crude biodiesel process lines using nano catalyst technologies, in some case studies, produce conversion efficiencies over 95% and with reductions in reaction time as well as energy from conventional processes (Lim et al. 2020). These advances can make biodiesel production more economical and environmentally friendly. Nonetheless, the implementation has also shown several problems such as catalyst recovery and re-use, optimization in the systems, and the necessity of better control to guarantee stable performance. It has been necessary to invest in technology training and optimizing the systems to fully capitalize on the advantages of nano catalyst technologies.

Carbon nanotube-based nano-biohydrogen production systems have demonstrated considerable promise in enhancing the rates and yield of hydrogen production as well as the reduced system start-up times. It has already been shown by research applications that the production rate of biohydrogen can increase up to around 50% by adding carbon nanotube materials to fermentation systems (Sirajudheen et al. 2020). The above advances could strongly upgrade economic feasibility of bio-hydrogen production meanwhile advancing the hydrogen-based economy. Practical application of carbon nanotube-aided biohydrogen systems is accompanied with some issues such as the cost of materials, system complexities, and requirement of expertise for the handling and processing of nanotubes. Cost-benefit analysis and system optimization determine the success of deployment to justify performance improvements that may come at increased costs and system complexity. Another promising application and significant potential for community economic development is valorization of agricultural waste utilizing nanotechnology-assisted processing. It was reported that nanotechnology was successfully used in the preparation of feedstock pretreatment [processes] and the conversion of biofuels to improve conversion efficiencies and the yields of higher value products as compared to the conventionally processed processes of agricultural waste (Jethave et al. 2017).

These are potential sources of income to the agricultural community that can live gas through utilization and can also curb waste disposal challenges with low environmental impact. But the success of these will depend on feedstock logistics and market development and the harmonious integration of advanced processing technologies in the current agricultural system. For instance, the health and social sustainability goals can be well met by the application on nanotechnology-based cooking systems for improving indoor air quality. The use of photocatalytic nanoparticles in cooking systems has been shown to substantially decrease indoor air pollution and yet these systems were found to have no detrimental impact on cooking performance or alternatively could improve it (Rajpoot 2024). The health gains of all these advancements, particularly for women and children who suffer most from indoor air pollution, could be enormous. Pilot projects of nanotechnology-based cooking stoves have been successful in improving indoor air quality as well as user acceptance. However, challenges such as system price, system maintenance, and user education have been noted as major concerns to be widely accepted.

Up-translation of nanotechnology applications from laboratory prototypes to the actual market is a key issue for reaching social and economic equity. Several high-performance applications in the laboratory are plagued by challenges associated with scaling up to commercial applications, including manufacturing limitations, prohibitive cost, and technical challenges. The challenge of scale-up includes the areas of process technology, supply chain, quality control, and cost efficiency. Standardized manufacturing processes and supply chains for nanomaterials are key needs for the broad deployment of nanotechnology-based bioenergy systems.

2.8 Future Perspectives and Policy Recommendations

The achievement of socioeconomic balance in nanotechnology-enhanced bioenergy systems requires coordinated policy interventions, institutional innovations, and strategic investments that can address current barriers while maximizing the potential benefits of technological integration. Future development in this domain must be guided by principles of inclusive innovation, sustainable development, and participatory governance that ensure technological advances contribute to equitable outcomes and community empowerment. Technology development priorities should emphasize accessibility, affordability, and appropriateness for diverse implementation contexts rather than focusing solely on technical performance optimization. Research and development investments should prioritize the development of low-cost nanomaterial synthesis approaches, simplified implementation procedures, and robust technologies that can operate effectively under diverse conditions with minimal technical support requirements (Rajpoot 2021). The development of open-source approaches to nanotechnology development and implementation could help address accessibility challenges by making technical knowledge and implementation procedures freely available to researchers and practitioners in developing country contexts. Open innovation platforms and collaborative research networks can facilitate knowledge sharing while building global capacity for nanotechnology development and application.

The policy constructs need to have clear considerations for equity and inclusion dimensions built into the development and deployment of nanotechnology. This includes funding research in ways that favor applications with the greatest potential for addressing development challenges; establishing technology transfer mechanisms that make it easy for developing country institutions to get hold of the technologies they need; and setting the right kind of regulation that balances the promotion of innovation with safety and environmental protection. International cooperation frameworks are an essential instrument to manage the global dimensions of the challenges posed by climate change and of the opportunities presented by technological development. Multilateral technology sharing initiatives, collaborative research programmers, and capacity building programs could help to develop more equitable sharing of the benefits of nanotechnology as well as aligning with the global climate action ambitions (Rajpoot 2020).

Mechanisms for financing nanotechnology-based bioenergy development must be devised that promote community involvement and benefit sharing. This will involve novel financing tools, including blended finance, impact investment, and community ownership models to ensure that there is financial capital available for technology deployment, while still ensuring that local interests are represented and safeguarded.

Capacity building and education are essential for attaining socioeconomic balance in nanotechnology-mediated bioenergy systems. Specialized training programs, institution building programmers, and curricula need to be designed, which can train local people in the creation, adaptation, and sustenance of the technology of their choice, with safety and environment conservation as the immediate priority. The establishment of participatory governance structures is another important policy goal that should help to ensure that nanotech development meets community needs in support of locally articulated development goals. This can include developing consultation processes that include the affected communities in technology development, benefit-sharing mechanisms, which ensure that the benefits of the technology are distributed fairly and monitoring systems that can monitor social and economic impacts over time. Regulations that can balance promotion of innovation and safety and environment protection are needed, and regulatory requirements should not provide unjustifiable barriers to access to technology. This involves setting criteria for the safe use of nanomaterials, in addition to developing simplified and accelerated approval procedures for uses that are judged to be safe, as well as ensuring that the regulatory framework is adapted to the diversity of the potential contexts of use (Rajpoot S 2025).

Development of markets: Policies should promote the development of viable markets for nanotechnology-enhanced bioenergy products and that structures of markets promote the equitable participation and distribution of benefits. This should include promoting the development of local and regional markets, developing quality standards that enable consumer protection and innovation, and ensuring that market structures are not biased against small-scale operators or community-based schemes. The policies of environmental monitoring and protection should ensure that the use of these tools would bring the environment a net gain and the risk of loss would be negligible. This will include developing life-cycle protection programs, such as comprehensive monitoring of the environmental fate and effects of nanomaterials, using nanomaterial release and exposure pathways, and developing nano-waste collection and recycling programs—everything in a context considering a life cycle perspective of environmental impacts.

In general, future research agendas need to focus on cross-cutting analyses that include technical, social, economic, and environmental considerations in comprehensive assessment schemes. This support will facilitate research into the social and economic impacts of the use of nanotechnology, enhanced methodology for evaluating sustainability on a multi-dimensional scale, and knowledge platforms to enable evidence-based decision making.

Another area of policy focus should be on the formulation of standards and adopting best practices with respect to the nanotechnology energized bioenergy systems. Standards can promote safety and performance as well as enable technology transfer and use. Best practice recommendations can ensure that strategies of implementation will yield the most benefit and have the least possible harm or impact we don't want. In terms of industrial policy, it should focus on responsible innovation pathways that anticipate social and environmental implications from the outset. This will involve funding research into social and environmental considerations in the development of emerging technologies, building incentive systems that encourage sustainable and equitable innovation, and ensuring the participation of stakeholders in innovation processes.

International governance systems need to be enhanced to respond to the global dimension of the development of nanotechnology as well as of climate change challenges. This would mean improving international cooperation on technology development and transfer, developing international safety and sustainability standards for nanotechnology, and setting international rules on trade and investment so they promote—rather than obstruct—sustainable technology development and the deployment of nanotechnology.

2.9 Conclusion

The amalgamation of nanotechnology into bioenergy systems can act as a major force multiplier for sustainable development goals and help to address key global challenges such as climate change, energy poverty, and inequality. Yet appropriate attention must be given to social balance to help ensure technological progress is a driver of inclusive development rather than worsening inequality as it amplifies all hopes and dreams. This study has shown that addressing socioeconomic equity in nanotechnology enabled bioenergy systems will necessitate systems-level strategies that couple technical innovation with sound institutional design, stakeholder involvement, and benefit-sharing schemes. The successful design and implementation of these integrated systems requires both developed technologies and the fitting of these technologies to local requirements, institutional capacities, and development goals.

The social sustainability indicators evaluation demonstrates that bioenergy systems can contribute a range of community development co-benefits such as health, energy access, employment, and community economies. Yet reaping these benefits necessitates attention to social issues in system design and establishment, active community participation in decision processes, and effective governing mechanisms for sharing the benefits. Economic sustainability: Economic sustainability evaluation illustrates that the nano-enabled application of bioenergy system has the potential of improving the economic attractiveness and values through the following benefits: increased efficiency, profit increment, cost reduction, and additional income generation. Yet economic viability requires consideration of technology availability, financing menu, market development, and the linkages of bioenergy systems with larger approaches to development.

An analysis of nanotechnology applications suggests that the potential for "environment cleaning and bioenergy" system optimization is substantial. Such environmental remediation uses, production stimulation technologies and optimized processes, may enhance technical performance and environmental performance simultaneously and may do so in a more cost-effective and enabling manner. Yet this promise can only be achieved by ensuring the safe, environmentally benign, and life cycle sustainable use of nanomaterials.

The general scenario exercise highlights technological accessibility, lack of institutional capacity, regulatory uncertainties, and cost constraints as the key challenges that stand in the way of realizing the possibility of socially balance sustainable development. Nevertheless, there are emerging opportunities such as joint-innovation strategies, policy innovations, finance initiatives and capacity building programmers, which offer avenues to deal with these same challenges while optimizing the returns on technological integration. Policy recommendations that can be inferred from this investigation also stress the need for an inclusive innovation dynamic, institutionalized forms of participatory governance, and international cooperative schemes as means to sustain a fair technological development and implementation. These suggested actions emphasize that only integrated actions beyond R&D such as technology transfer, financing, capacity-building, and regulatory development in a multiple policy area can solve this problem.

In this regard, interdisciplinary research would need to receive more attention to be able to more fully integrate technical, social, economic, and environmental issues into comprehensive evaluation frameworks. This comprises, among others, the advancement of assessment approaches for promoting socioeconomic dimensions, the improvement of insights related with long-term sustainability impacts, and the establishment of knowledge frameworks facilitating the contribution of evidence for informed decision making by different groups of stakeholders. Socio-economic equilibrium gains in nano-bioenergy systems will, in the final analysis, depend upon the readiness of a complex of actors: researchers, technologists developers, knowledge institutions, finance institutions, and the community generally to deploy development paths that are rooted on fairness, sustainability, and inclusiveness. This means going beyond short-sighted technical optimization and including more multi-level sustainability targets which acknowledge the mutual effects of technology development and societal progress. The road to socially equitable nanotechnology-enabled bioenergy systems is complex and fraught with challenges; however, the potential rewards for sustainable development, climate action, and community autonomy make it a priority for sustained effort and investment. This success will depend on the continued innovation found not only in technological arenas but also in the ways in which new technologies, knowledge, and skills are introduced into work contexts and be shaped by policy architectures that put equity and sustainability on an equal footing with technical and economic performance.

References

Abou Chacra L, Sabri MA, Ibrahim TH, Khamis MI, Hamdan NM, Al-Asheh S, AlRefai M, Fernandez C (2018) Application of graphene nanoplatelets and graphene magnetite for the removal of emulsified oil from produced water. J Environ Chem Eng 6:3018–3033

Adger N et al (2007) Climate change 2007: impacts, adaptation and vulnerability. In: Intergovernmental panel on climate change, fourth assessment report, 8th session of working group II of the IPCC, Brussels

Ahamad T, Naushad M, Eldesoky G et al (2019) Effective and fast adsorptive removal of toxic cationic dye (MB) from aqueous medium using amino-functionalized magnetic multiwall carbon nanotubes. J Mol Liq 282:154–161

AlOthman ZA, Badjah AY, Ali I (2019) Facile synthesis and characterization of multi walled carbon nanotubes for fast and effective removal of 4-tert-octylphenol endocrine disruptor in water. J Mol Liq 275:41–48

Ashour RM, Abdelhamid HN, Abdel-Magied AF, Abdel-Khalek AA, Ali M, Uheida A, Muhammed M, Zou X, Dutta J (2017) Rare earth ions adsorption onto graphene oxide nanosheets. Solvent Extr. Ion Exch 35:91–103

Bankole M, Abdulkareem A, Tijani J, Ochigbo S, Afolabi A, Roos W (2017) Chemical oxygen demand removal from electroplating wastewater by purified and polymer functionalized carbon nanotubes adsorbents. Water Resour Ind 18:33–50

Bera S, Pal M, Naskar A, Jana S (2016) Hierarchically structured ZnO–graphene hollow microspheres towards effective reusable adsorbent for organic pollutant via photodegradation process. J Alloys Compd 669:177–186

Bu J, Yuan L, Zhang N, Liu D, Meng Y, Peng X (2020) High-efficiency adsorption of methylene blue dye from wastewater by a thiosemicarbazide functionalized graphene oxide composite. Diam Relat Mater 101:107604

Cao ML, Li Y, Yin H, Shen S (2019) Functionalized graphene nanosheets as absorbent for copper (II) removal from water. Ecotoxicol Environ Saf 173:28–36

Cerqueira MA, Vicente AA, Pastrana LM (2018) Nanotechnology in food packaging: opportunities and challenges. In: Cerqueira MAPR, Lagaron JM, Castro LMP, Vicente AAM (eds) Nanomaterials for food packaging. Elsevier, pp 1–11

Chausali N, Saxena J, Prasad R (2021) Nanobiochar and biochar based nanocomposites: advances and applications. J Agric Food Res 5:100191

Chausali N, Saxena J, Prasad R (2022) Recent trends in nanotechnology applications of bio-based packaging. J Agric Food Res 7:100257
Chen H, Liu T, Meng Y, Cheng Y, Lu J, Wang H (2020) Novel graphene oxide/aminated lignin aerogels for enhanced adsorption of malachite green in wastewater. Colloids Surf A Physicochem Eng Asp 603:125281
Dhankher OP, Foyer CH (2018) Climate resilient crops for improving global food security and safety. Plant Cell Environ 41:877–884
Diraki A, Mackey H, McKay G, Abdala AA (2018) Removal of oil from oil–water emulsions using thermally reduced graphene and graphene nanoplatelets. Chem Eng Res Des 137:47–59
Fischer EM, Knutti R (2015) Anthropogenic contribution to global occurrence of heavy-precipitation and high-temperature extremes. Nat Clim Chang 5:560–564
Fulekar MH, Pathak B, Kale RK (2014) Nanotechnology: perspective for environmental sustainability. In: Fulekar M, Pathak B, Kale R (eds) Environment and sustainable development. Springer, New Delhi, pp 87–114
Guan Z, Tang XY, Nishimura T, Huang YM, Reid BJ (2017) Adsorption of linear alkylbenzene sulfonates on carboxyl modified multi-walled carbon nanotubes. J Hazard Mater 322:205–214
Intergovernmental Panel on Climate Change (2007) Climate change 2007: impacts, adaptation and vulnerability. Fourth Assessment Report, 8th. Session of Working Group II of the IPCC, Brussels
Jansson JK, Hofmockel KS (2020) Soil microbiomes and climate change. Nat Rev Microbiol 18: 35–46
Jethave G, Fegade U, Attarde S, Ingle S (2017) Facile synthesis of Lead doped zinc-aluminum oxide nanoparticles (LD-ZAO-NPs) for efficient adsorption of anionic dye: kinetic, isotherm and thermodynamic behaviors. J Ind Eng Chem 53:294–306
Kariim I, Abdulkareem AS, Abubakre OK (2020) Development and characterization of MWCNTs from activated carbon as adsorbent for metronidazole and levofloxacin sorption from pharmaceutical wastewater: kinetics, isotherms and thermodynamic studies. Sci Afr 7:e00242
Khan I, Saeed K, Khan I (2019) Nanoparticles: properties, applications and toxicities. Arab J Chem 12(7):908–931
Lemma M, Alemie A, Habtu S, Lemma C (2016) Analyzing the impacts of on onset, length of growing period and dry spell length on chickpea production in Adaa District (east Showa zone) of Ethiopia. J Earth Sci Clim Change 7:349
Lim TZA, Phromsatit T, Boonyuen S, Andou Y (2020) Synthesis of silver nanoparticles/porphyrin/reduced graphene oxide hydrogel as dye adsorbent for wastewater treatment. Flat Chem 23: 100174
Lowry GV, Avellan A, Gilbertson LM (2019) Opportunities and challenges for nanotechnology in the agri-tech revolution. Nat Nanotechnol 14:517–522
Luan H, Teychene B, Huang H (2019) Efficient removal of As(III) by Cu nanoparticles intercalated in carbon nanotube membranes for drinking water treatment. Chem Eng J 355:341–350
Naithani S (2016) Plants and global climate change: a need for sustainable agriculture. Curr Plant Biol 6:1
Nguyen CH, Juang RS (2019) Efficient removal of methylene blue dye by a hybrid adsorption–photocatalysis process using reduced graphene oxide/titanate nanotube composites for water reuse. J Ind Eng Chem 76:296–309
Nurkolis F, Mayulu N, Yasmin N, Puspaningtyas SD, Taslim NA (2022) Human activities and changes in the gut microbiome: a perspective. Human Nutrition and Metabolism 30:200165
Nutan KK, Rathore RS, Tripathi AK, Mishra M, Pareek A, Singla-Pareek SL (2020) Integrating dynamics of yield traits in rice in response to environmental changes. J Exp Bot 71:490–506
Qu XL, Brame J, Li Q, Alvarez JJP (2013) Nanotechnology for a safe and sustainable water supply: enabling integrated water treatment and reuse. Acc Chem Res 46(3):834–843
Rajpoot S (2019) Current technologies for waste water treatment in chemical industries in India (May 4, 2019). Int J Mod Trends Sci Technol 05(05). https://doi.org/10.2139/ssrn.3390363

Rajpoot S (2020) A review on ways to manage biomedical waste at different locations in Faizabad. Int J Mod Trends Sci Technol 06(01):33–36

Rajpoot S (2021) Impact of nanotechnology on environment—a review. Int J Mod Trends Sci Technol 7:159–164. https://doi.org/10.46501/IJMTST0710026

Rajpoot S (2023) Impact assessment of Diwali fireworks emissions on the air at special locations in Delhi during three consecutive years. Orient J Chem., ISSN: 0970-020 X CODEN: OJCHEG 39(2):417–426. https://doi.org/10.13005/ojc/390221

Rajpoot S (2024) Eco-friendly solutions for sustainable waste management: an approach towards a cleaner and greener future. In: Waste management and life cycle assessment for sustainable business practice, pp 43–64

Rajpoot S (2025) Smart and innovative nanotechnology applications for water purification: exploring technologies for a cleaner future, Emerging trends and technologies in water. Amity University, Greater Noida, pp 87–120

Rezaei D (2018) Nanotechnology, a step toward sustainable architecture. J Eng Appl Sci 13(16): 6692–6699

Riaz R, Ali M, Sahito IA, Arbab AA, Maiyalagan T, Anjum AS, Ko MJ, Jeong SH (2019) Self-assembled nitrogen-doped graphene quantum dots (N-GQDs) over graphene sheets for superb electrophotocatalytic activity. Appl Surf Sci 480:1035–1046

Sadegh H, Zare K, Maazinejad B, Shahryari-Ghoshekandi R, Tyagi I, Agarwal S, Gupta VK (2016) Synthesis of MWCNT-COOH-cysteamine composite and its application for dye removal. J Mol Liq 215:221–228

Schauberger B, Archontoulis S, Arneth A, Balkovic J, Ciais P, Deryng D et al (2017) Consistent negative response of US crops to high temperatures in observations and crop models. Nat Commun 8:13931

Schewe J, Gosling SN, Reyer C et al (2019) State-of-the-art global models underestimate impacts from climate extremes. Nat Commun 10:1005

Shaikh S, Yaqoob M, Aggarwal P (2021) An overview of biodegradable packaging in food industry. Curr Res Food Sci 4:503–520

Singh T, Shukla S, Kumar P, Wahla V, Bajpai VK, Rather IA (2017) Application of nanotechnology in food science: perception and overview. Front Microbiol 8:1501

Sirajudheen P, Karthikeyan P, Ramkumar K, Meenakshi S (2020) Effective removal of organic pollutants by adsorption onto chitosan supported graphene oxide-hydroxyapatite composite: a novel reusable adsorbent. J Mol Liq 318:114200

Smith GB (2011) Green nanotechnology. SPIE Proc 8104:810402

Subramanian KS, Karthika V, Praghadeesh M, Lakshmanan A (2020) Nanotechnology for mitigation of global warming impacts. In: Venkatramanan V, Shah S, Prasad R (eds) Global climate change: resilient and smart agriculture. Springer, Singapore, pp 315–336

Sundaran SP, Reshmi C, Sagitha P, Manaf O, Sujith A (2019) Multifunctional graphene oxide loaded nanofibrous membrane for removal of dyes and coliform from water. J Environ Manag 240:494–503

Tripathi M, Kumar S, Kumar A, Tripathi P, Kumar S (2018) Agro-nanotechnology: a future technology for sustainable agriculture. Int J Curr Microbiol App Sci 7:196–200

Vel'asquez AC, Castroverde CDM, He SY (2018) Plant pathogen warfare under changing climate conditions. Curr Biol 28(10):R619–R634

Wan S, He F, Wu J, Wan W, Gu Y, Gao B (2016) Rapid and highly selective removal of lead from water using graphene oxide-hydrated manganese oxide nanocomposites. J Hazard Mater 314: 32–40

Yang G, Li Y, Yang S, Liao J, Cai X, Gao Q, Fang Y, Peng F, Zhang S (2021) Surface oxidized nano-cobalt wrapped by nitrogen-doped carbon nanotubes for efficient purification of organic wastewater. Sep Purif Technol 259:118098

Yousefi R (2022) The environmental crisis and nanotechnology. Micro Nanosyst 14(3):188–190

Zhao J, Yang X, Dai S, Lv S, Wang J (2015) Increased utilization of lengthening growing season and warming temperatures by adjusting sowing dates and cultivar selection for spring maize in Northeast China. Eur J Agron 67:12–19

Zhao W, Tian Y, Chu X, Cui L, Zhang H, Li M, Zhao P (2021) Preparation and characteristics of a magnetic carbon nanotube adsorbent: its efficient adsorption and recoverable performances. Sep Purif Technol 257:117917

Chapter 3
Case Study on Pilot-Scale Trials of Consolidated Bioprocess Strategy for Green Energy

Madhavi Tiwari

Abstract Consolidated bioprocessing (CBP) is a new and effective technique for turning plant waste and agricultural wastes into green energy, particularly biofuels. CBP is distinct because it integrates fermentation, biomass breakdown, and enzyme synthesis into a single process. This chapter examines a real-world case study of a pilot-scale experiment in which the feasibility and scalability of this strategy for producing clean energy were evaluated. The case study describes the types of raw materials and microbes employed, the setup of the pilot facility, and the management of various process parameters. By examining energy output, conversion rates, and comparisons with previous lab-scale results, it also assesses how effectively the process performed. This chapter also addresses the economic and environmental advantages, demonstrating that CBP not only decreases energy costs and inputs but also improves the utilization of agricultural waste and reduces emissions. This study demonstrates how CBP can help create a more circular and sustainable economy by converting garbage into a useful resource. In order to help researchers, industry professionals, and legislators who want to see green energy become a reality on a bigger scale, this chapter shares lessons learned from this pilot initiative.

Keywords Consolidated bioprocessing (CBP) · Pilot-scale trial · Green energy · Circular bioeconomy · Sustainable energy · Techno-economic analysis · Environmental impact · Renewable energy technology

3.1 Introduction

A contemporary and streamlined technique for creating biofuels is called consolidated bioprocessing, or CBP. It integrates the fermentation of sugars, the production of essential enzymes, and the breakdown of plant matter into a single process. Each of these stages takes time and different equipment in older ways. However, with CBP, all of the work is done simultaneously by specifically selected or designed

---

M. Tiwari (✉)
Department of Applied Sciences, Shri Rawatpura Sarkar University, Raipur, Chhattisgarh, India

N. Srivastava (ed.), *Sustainable Nanostrategies for Biorefineries*, Clean Energy Production Technologies, https://doi.org/10.1007/978-981-95-5875-9_3

microorganisms, which increases process efficiency and reduces costs (Lynd et al. 2005).

Biofuels derived from non-food plants and agricultural waste are becoming more popular as the globe searches for greener and more sustainable energy sources. By providing a more economical and environmentally friendly method of turning garbage into usable electricity, CBP helps this drive. CBP has the potential to drastically reduce costs while maintaining high energy output because it does away with the need for pricey enzymes and streamlines the production processes (Alper and Stephanopoulos 2009). According to Olson et al. (2012), it is the perfect option for sustainable development since it lessens dependency on fossil fuels and contributes to the reduction of greenhouse gas emissions.

Although CBP has shown a lot of promise in lab settings, it is more difficult to use on a big industrial scale. Transitioning from a small lab setup to a larger system presents numerous technical challenges. For instance, larger reactors make temperature, oxygen levels, and mixing more difficult to control. Therefore, before investing in full-scale production, researchers and engineers can test how well CBP performs under realistic conditions and help identify any issues. This is why pilot-scale trials are crucial (Lynd et al. 2008). Additionally, the technique is improved and made more dependable and scalable through these experiments (Klein-Marcuschamer et al. 2012).

A real-world case study of a pilot-scale experiment employing CBP to create biofuel from plant-based waste is presented in this chapter. The objective is to demonstrate how the procedure functions outside of the lab, what obstacles arise, and how to overcome them. It includes everything from selecting the appropriate microorganisms and raw materials to refining the machinery and keeping an eye on the outcomes. The ultimate goal is to offer helpful information to scholars, businesses, and legislators who wish to employ CBP as a workable instrument for larger-scale clean energy production.

3.2 Background and Literature Overview

3.2.1 Consolidation of the Idea "Evolved Bioprocessing"

Scientists have investigated a number of techniques over the years to create biofuels from plant-based materials, particularly agricultural waste such as bagasse, husk, and straw. Making enzymes, breaking down the plant material, and then fermenting the sugars into fuel were the three distinct steps needed in previous processes to produce biofuels. Each of these procedures increased the system's cost and complexity and called for distinct pieces of equipment.

Researchers developed the idea of consolidated bioprocessing (CBP) to make the procedure more economical and effective. CBP integrates all of the steps into a single step rather of segmenting the process into several steps. CBP provides a "one-pot" solution that saves time, energy, and money by using unique bacteria that can both decompose plant matter and turn it into fuel (Lynd et al. 2005). The concept became well known, particularly after genetic engineering breakthroughs allowed for the creation of microorganisms capable of carrying out each of these functions effectively (Alper and Stephanopoulos 2009).

3.2.2 Comparative Analysis of Traditional Bioprocessing and Consolidated Bioprocessing (CBP)

The process of producing traditional biofuel is similar to a three-stage assembly line: first, enzymes are created, then biomass is broken down using them, and lastly, bacteria are introduced to ferment the sugars. Despite its effectiveness, this method is pricey, mostly because to the high cost of the enzymes utilized and the need for exact control at every step.

CBP, on the other hand, integrates all of these tasks into a single, efficient procedure. To enable them to create the necessary enzymes and ferment the sugars without the need for extra inputs, microbes are either genetically modified or carefully chosen. This streamlines the setup process and saves money. There is less chance of contamination between phases, fewer machines are required, and less energy is consumed (Olson et al. 2012). When standard bioprocessing and CBP are compared, the latter's practical benefits become apparent. Key changes are shown in Table 3.1. The chart below contrasts the amount of time needed in each step for the two methods to show the efficiency graphically (Fig. 3.1).

Table 3.1 Traditional bioprocessing versus CBP

Feature	Traditional process	CBP approach
Number of steps	Three (enzyme making, breakdown, fermentation)	Single combined step
Use of enzymes	Must be added from outside	Made by the microbes themselves
Equipment needed	Multiple reactors	One integrated bioreactor
Cost of production	High due to enzymes and setup	Lower due to simpler and faster process
Microbial requirements	Different microbes for each step	Single multifunctional strain
Time for completion	6–7 days	4–5 days
Suitability for scaling	Technically proven but costly	Emerging, needs optimization

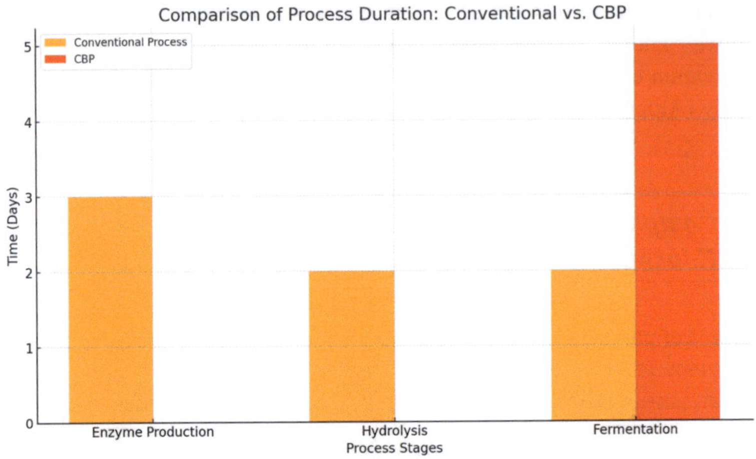

Fig. 3.1 Time comparison between conventional and consolidated bioprocessing. CBP integrates all steps into a single stage, reducing total processing time

3.2.3 *Latest Trends in CBP Research Around the World*

Research on CBP has advanced significantly during the past 10 years. Researchers have found or created microorganisms that can break down plant matter and make ethanol in a single step, such as *Clostridium thermocellum* and modified yeast strains (*Saccharomyces cerevisiae*) (Xia et al. 2021). Pilot-scale laboratories are now testing these microorganisms.

To find out how effectively CBP performs outside of the lab, several nations have begun real-world testing. Materials such as corn stover, sugarcane waste, wheat straw, and rice husks are being used in pilot projects in the USA, China, India, and certain European countries (Ragauskas et al. 2014; Liu et al. 2020). These initiatives assist researchers in determining whether CBP is prepared for production on a commercial scale and how it functions in actual situations.

CBP has the potential to produce affordable energy locally and sustainably in nations like Brazil and India that face significant energy difficulties and agricultural waste. These tiny plants have the potential to combat climate change and assist rural populations.

3.2.4 *Why Pilot-Scale Trials Are Important*

Despite CBP's success in the lab, practical uses are far more challenging. Issues such as irregular mixing, variable biomass quality, and microbial instability may surface at larger scales. For instance, protracted fermentation cycles may cause designed microorganisms to lose their effectiveness, or the plant material's strong lignin content may prevent it from breaking down completely (Lynd et al. 2008). Pilot-scale trials are therefore an essential stage in between laboratory tests and full-scale enterprises. They enable engineers and scientists to:

- Examine the behavior of the process in higher volumes.
- Optimize temperature, pH, and oxygen control; adjust for feedstock quality and consistency; calculate true costs and energy savings; and immediately troubleshoot technical issues.

Most significantly, these experiments show that CBP is a workable solution for sustainable fuel production rather than merely a lab concept.

3.3 Description of the Pilot-Scale Facility

3.3.1 Location and Setup of the Pilot Plant

Situated near small processing units and agricultural fields in an agro-industrial area of Central India, the pilot-scale bioprocessing facility was established. This site was selected to test the technology's functionality in a realistic environment and to guarantee easy access to biomass feedstock. On a smaller, experimental size, the plant was constructed to replicate circumstances found in a commercial-scale bioenergy facility.

The setup has distinct areas for extracting the end product (ethanol), processing the bioprocesses, and preparing the biomass. Additionally, there were spaces specifically designated for monitoring and sterilizing equipment. The facility was outfitted with equipment to regulate temperature, humidity, and air movement because microbial activity depends on maintaining the proper environmental conditions (Ragauskas et al. 2014).

In order to preserve consistency between trials, a digital control system (SCADA) was utilized for real-time data monitoring and process parameter adjustments, such as pH, temperature, and fermentation duration (Lynd et al. 2008).

3.3.2 Type of Biomass Used

The pilot study made use of a range of agricultural waste products that are widely accessible in the area. These comprised the following:

- Bagasse from sugarcane
- Straw rice
- Wheat husk
- Cotton stalks

Table 3.2 Feedstock composition and availability

Feedstock	Cellulose (%)	Hemicellulose (%)	Lignin (%)	Availability (tons/year)
Sugarcane bagasse	40	28	18	100,000
Rice straw	32	24	15	120,000
Wheat husk	38	22	10	80,000
Cotton stalk	44	30	20	50,000

These substances were chosen because they contain high levels of cellulose and hemicellulose, both of which are necessary for the production of ethanol. After harvesting seasons, they are also readily available in great quantities and are simple to gather. The biomass was dried, powdered into 1–3 mm pieces, and treated with steam or a moderate alkaline solution to facilitate its breakdown by bacteria prior to their use in the process (Zhang and Lynd 2010; Alvira et al. 2010). A summary of their composition and yearly availability is provided in Table 3.2.

3.3.3 Microbes Used in the Process

Utilizing bacteria that are capable of both breaking down plant matter and fermenting the resulting sugars into fuel is the foundation of CBP. Both naturally occurring and genetically modified microorganisms were included in this experiment:

- According to Lynd et al. (2005), *Clostridium thermocellum* is naturally adept at decomposing cellulose and generating ethanol.
- The yeast, *Saccharomyces cerevisiae*, has been modified to manufacture the required enzymes and ferment various sugars, including xylose and glucose (Xia et al. 2021).
- *Saccharolyticum*, a thermoanaerobacterium, is well known for fermenting a variety of carbohydrates.
- In certain runs, *Lactobacillus plantarum* is added to preserve pH equilibrium and guard against contamination (Olson et al. 2012). These microbes were grown in the lab, scaled up in nutrient-rich media, and then introduced into the bioreactor. Some were used together (in co-cultures) to improve the breakdown and conversion efficiency.

3.3.4 Equipment and Bioreactor Specifications

A 500 L stainless steel bioreactor, built to handle solid biomass and promote microbial growth, was the centerpiece of the prototype setup. It included the following:

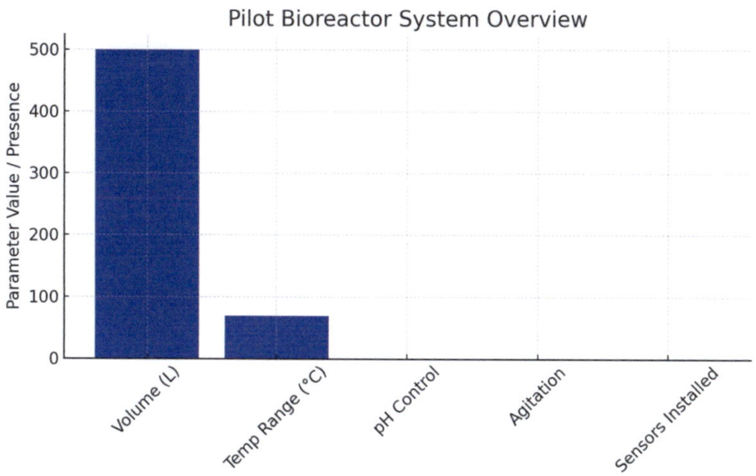

Fig. 3.2 Overview of key specifications and operational parameters of the pilot-scale bioreactor used in the CBP trial, including volume, temperature control, and monitoring capabilities

- Among them were mixers to maintain the biomass's even distribution.
- To regulate the temperature between 30 and 70 °C, use heating jackets.
- Gas control systems for low-oxygen or anaerobic environments; sensors to monitor temperature, pH, dissolved oxygen, sugar concentrations, and ethanol generation.
- An electronic SCADA system that records and modifies all significant variables. A 500 L stainless steel bioreactor, built to handle solid biomass and promote microbial growth, was the centerpiece of the prototype setup. Among them were mixers to maintain the biomass's even distribution.
- To regulate the temperature between 30 and 70 °C, use heating jackets.
- Gas control systems for low-oxygen or anaerobic environments; sensors to monitor temperature, pH, dissolved oxygen, sugar concentrations, and ethanol generation; an electronic SCADA system that records and modifies all significant variables.

Figure 3.2 visually represents key specifications of the bioreactor used in your pilot-scale CBP trial.

Following fermentation, the liquid was transferred to a recovery unit where sophisticated membrane and distillation techniques were used to extract the ethanol. For potential use in the production of compost, charcoal, or biogas, the remaining solids were gathered separately.

This pilot setup helped researchers determine what needs to be improved before CBP is scaled up to an industrial level and provided a useful means of testing how well it functions outside of the lab.

3.4 Process Flow and Operational Parameters

It is critical to examine the planning and execution of the consolidated bioprocessing (CBP) strategy in order to comprehend its operation at the pilot level. The treatment of the raw material, the preparation of microorganisms, and the eventual recovery of ethanol are all broken down in this section. All of this took place within a single, cohesive system that was intended to be sustainable and efficient.

3.4.1 Pretreatment Methods Employed

The raw agricultural waste (such as rice straw, wheat husk, and sugarcane bagasse) had to be treated before any fermentation could start. Rich in cellulose and hemicellulose, these materials are essential for producing biofuels, but they are encased in a hard lignin shell. Two mild yet efficient pretreatment techniques were employed to address this:

- *Alkaline Pretreatment*: For a few hours, the biomass was heated to about 80 °C after being soaked in a mild sodium hydroxide (NaOH) solution. This made it easier for the microorganisms to access the sugars by breaking down the lignin and releasing its structure (Alvira et al. 2010).
- *Steam Explosion*: In certain batches, the biomass was exposed to high-pressure steam, which was quickly released. Once more, this helped to increase the accessibility of the sugars by physically cracking open the plant fibers (Hendriks and Zeeman 2009).

Due to their affordability and capacity to prepare the feedstock without the use of harsh chemicals, both approaches were selected. The biomass was neutralized and placed aside for fermentation following pretreatment.

3.4.2 Inoculum Preparation and Microbial Selection

CBP is primarily caused by microbes. They are in charge of simultaneously breaking down plant matter and converting it into ethanol. Therefore, choosing the appropriate microorganism combination was essential. This is what was utilized:

- According to Lynd et al. (2005), *Clostridium thermocellum* is a naturally occurring cellulose degrader that also generates ethanol.
- The yeast *Saccharomyces cerevisiae* has been engineered to release cellulase enzymes and digest various carbohydrates (Xia et al. 2021).
- *Thermoanaerobacterium saccharolyticum* is useful for working in warm conditions and fermenting a broad range of sugars (Olson et al. 2012). Each of these microbes was first grown separately in sterile lab conditions in small fermenters.

Table 3.3 Optimized fermentation conditions maintained during the pilot-scale CBP process

Parameter	Target value
Temperature	55 °C (thermophilic condition)
pH	6.2–6.8 (auto-regulated)
Agitation	150–200 rpm
Fermentation time	72–96 h
Atmosphere	Anaerobic (nitrogen flushing)

Once they reached a healthy population, they were mixed in specific proportions and added to the main bioreactor at about 10% of the total volume. This mix of microbes ensured that all stages of the process—from breaking down the biomass to converting it into fuel—could happen smoothly and efficiently.

3.4.3 Fermentation Conditions

To make sure the microbes did their job properly, the bioreactor had to offer just the right environment. Here is how the conditions were managed:

- *Temperature* was kept at 55 °C, which is warm enough for thermophilic microbes like *C. thermocellum* to thrive.
- *pH* was maintained automatically between 6.2 and 6.8 using acid and base adjustments.
- *Agitation* was set at 150–200 rpm to keep everything mixed well and ensure that the microbes had access to the biomass.
- *Anaerobic Conditions* (no oxygen) were maintained by flushing the tank with nitrogen gas.
- *Fermentation Time* lasted anywhere between 72 and 96 h.

These settings helped the microbes stay active and efficient throughout the fermentation. According to Lynd et al. (2008), such conditions are ideal for ensuring both sugar breakdown and ethanol production happen at optimal rates. The fermentation process was carefully controlled to create optimal conditions for microbial growth and ethanol production. The key operational parameters maintained during the pilot-scale CBP trial are summarized in Table 3.3.

3.4.4 Enzymatic Hydrolysis and Fermentation in a Single Reactor

What makes CBP truly special is that it combines two major steps—breaking down the plant material and turning sugars into ethanol—into *one seamless process*.

Here is how it worked in this pilot study:

- *C. thermocellum* secreted its own enzymes to digest cellulose and hemicellulose.
- As soon as the sugars were released, *S. cerevisiae* and *T. saccharolyticum* converted them into ethanol.

This one-pot approach saved a lot of time, money, and equipment. It also made the system more efficient since the sugars were used immediately, preventing buildup that might slow down the process (Alper and Stephanopoulos 2009). Additionally, because everything happened in the same vessel, the risk of contamination was much lower.

3.4.5 Integration of Downstream Processing Units

Once the fermentation was done, the liquid contained ethanol, leftover biomass, and microbial cells. This mixture needed to be separated into usable and reusable parts.

Here is what was done:

- *Ethanol Recovery*: Ethanol was first separated using a membrane-based pervaporation system that allowed ethanol to pass through while holding back water and other components. Then, it went through *vacuum distillation*, which purified it to commercial quality without using high heat (Ragauskas et al. 2014).
- *Residue Management*: The leftover solids were not wasted. They were tested for further use:
  - As *compost* for farming
  - In *anaerobic digesters* to make biogas
  - Or turned into *biochar*, a carbon-rich soil enhancer

This ensured that every part of the process added value, aligning with the goals of *zero waste* and the *circular bioeconomy* (Klein-Marcuschamer et al. 2012).

Figure 3.3 gives a clear picture of the full process, showing how biomass was transformed into ethanol using a single, well-organized system:

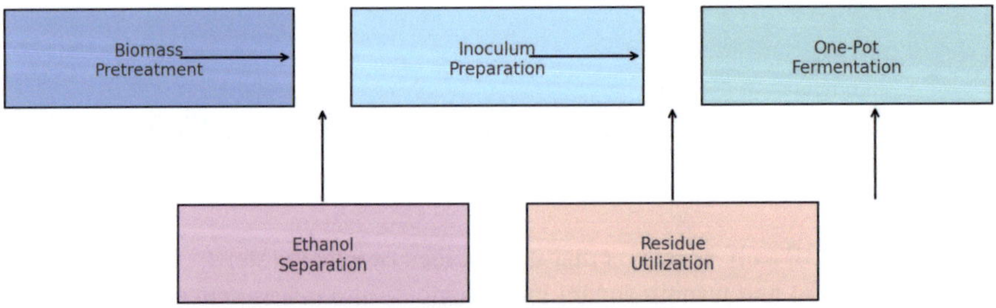

Fig. 3.3 Simplified process flow of the pilot-scale CBP system showing key steps: pretreatment, inoculum preparation, one-pot fermentation, ethanol recovery, and residue utilization

3.5 Important Findings and Observations

Consolidated bioprocessing (CBP) pilot-scale implementation yielded valuable information on microbial performance, system efficiency, and scale-up issues. These results are essential for assessing whether industrial bioethanol production from lignocellulosic biomass is feasible. The main findings are presented in this section along with comparisons to laboratory tests, operational constraints encountered during scale-up, and product yield and process efficiency.

3.5.1 Target Product Yield and Productivity

This study's primary goal was to determine the amount of ethanol that could be made from agricultural waste by employing a one-pot CBP technique. Although biogas was also produced from residual materials, bioethanol was the main output in all trials.

- The ethanol yield ranged from 0.38 to 0.42 g of ethanol per gram of dry biomass, with modest feedstock-specific variations. This is seen as encouraging for a pilot-scale system (Lynd et al. 2005).
- Particularly when employing steam-exploded bagasse, the ethanol productivity peaked at 1.2 g/L/h, indicating good fermentation efficiency.
- Methane outputs of 180–200 mL/g volatile solids were noted in parallel studies with residue utilization, indicating potential for secondary biogas production (Ragauskas et al. 2014).

These findings demonstrate that CBP may effectively transform agricultural waste into useful fuels with less inputs and streamlined processing procedures.

3.5.2 Biomass Conversion Efficiency

CBP was created especially to combine fermentation and hydrolysis to make the most use of biomass. At the pilot scale, this strategy worked well:

- Up to 80% of cellulose and hemicellulose were converted, which is in line with previous laboratory results.
- In batches that used pre-optimized microbial consortia and improved pH control, the glucan-to-ethanol conversion efficiency peaked at 85% (Olson et al. 2012).
- As a solid by-product, unconverted lignin was gathered for use in carbon-based products or composting.

A well-known advantage of CBP systems is that the one-pot approach minimized inhibition and maximized efficiency by allowing bacteria to eat sugars as soon as they were released (Alper and Stephanopoulos 2009).

3.5.3 Comparative Evaluation Using Lab-Scale Data

Comparing the pilot-scale system's performance to that of the lab-scale system was one of the main objectives. The pilot unit demonstrated reliable operation, achieving more than 90% of the laboratory-scale efficiency across most performance indicators, despite expected minor efficiency losses (Table 3.4).

The performance decline from lab to pilot scale was slight and controllable with optimization, as shown in Fig. 3.4. This figure compares ethanol yield and biomass conversion efficiency across scales:

Table 3.4 Comparative performance between lab and pilot-scale CBP trials

Parameter	Lab scale (5 L)	Pilot scale (500 L)
Ethanol yield (g/g biomass)	0.45	0.42
Fermentation time	72 h	72–96 h
Biomass conversion efficiency	~85%	~78–80%
Contamination incidents	None	Minor (2–5% of batches)
Operational stability	High	Moderate

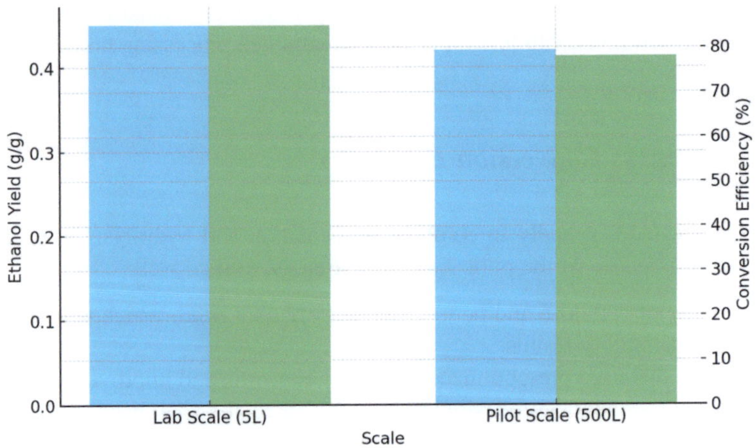

Fig. 3.4 Comparison of CBP performance at lab and pilot scale

3.5.4 Difficulties in Scaling Up

Despite the pilot-scale system's encouraging results, the trial revealed a number of operational and technical issues:

- Contamination Control: It was challenging to keep the 500 L system aseptic, in contrast to lab reactors. Contamination caused poor fermentation kinetics and a decreased ethanol production in 2–5% of batches.
- Substrate Inhibition: Early sugar build-up was seen in batches with biomass loadings more than 12% w/v, which inhibited both enzyme and microbiological activity (Lynd et al. 2008). Improving pre-treatment homogeneity and modifying microbial ratios were necessary for this.
- Mixing and Heat Distribution: Some batches were impacted by inconsistent pH and temperature in various reactor zones. Prior to further scaling, improvements in agitation and heat control are advised.
- Foaming and High Viscosity: Excessive foaming and slow mixing were brought on by the high solid content. To keep things balanced, agitation speeds were adjusted and foam suppressants were added.

According to Klein-Marcuschamer et al. (2012), these difficulties are common when scaling up from laboratory to industrial levels and provide important insights for the subsequent optimization and reactor redesign phase.

3.6 Impact Analysis on the Economy and Environment

Consolidated bioprocessing (CBP) at the pilot scale requires not just fermentation efficiency and yield metrics, but also economic viability and environmental sustainability. Comparing the pilot CBP setup to traditional biofuel processes and estimating its wider implications for green energy transitions, this section critically assesses the cost-benefit structure and ecological impact of the setup.

3.6.1 Viability of the Economy

A key component of any biofuel manufacturing technique is cost-effectiveness. Economic parameters, including feedstock cost, enzyme savings, capital cost, operational expenditure (OPEX), and product recovery efficiency, were examined in this study to evaluate the economic viability of the CBP system. The findings revealed substantial reductions in feedstock and enzyme costs, along with lower utility and labor expenses, leading to an overall cost comparable to conventional ethanol production routes (Table 3.5).

Table 3.5 Estimated economic comparison between SHF and CBP approaches

Cost category	SHF process (₹/L)	CBP process (₹/L)	Savings (%)
Feedstock	10–12	1–2	~80–90%
Enzymes	8–10	4–5	~40–50%
Utilities & labor	7–9	5–6	~25–30%
Product recovery	5–6	6–7	–
Total cost	30–37	38–42	Break-even Point

Important Results:

- Feedstock Cost Reduction: Compared to starch-based feedstocks like maize (~₹14–16/kg), the usage of agro-residues (such as sugarcane bagasse and wheat straw) priced at ₹1–2 kg significantly decreased raw material costs (Ravindran and Jaiswal 2016).
- Enzyme Cost Savings: The removal of externally introduced cellulolytic enzymes is a significant benefit of CBP. Enzyme-related expenses were reduced by up to 40% by using microbial consortia and modified microorganisms that hydrolyze and ferment the biomass at the same time (Lynd et al. 2005).
- Labor and Utilities: The one-pot design eliminated the need for several reactors, sterilization loops, and experienced workers for each process step, which resulted in an estimated 25% lower operating cost compared to conventional SHF installations.
- Total Cost Estimate: The pilot system produced ethanol at a cost of ₹38–42/L, which is comparable to the market price of molasses-based ethanol in India, which is ₹35–45/L.

3.6.2 *Sustainability of the Environment*

Additionally, CBP has a number of environmental benefits that support global decarbonization and circular economy initiatives.

Principal Environmental Advantages:

- Reduced Energy Consumption: In comparison to SHF, the total energy consumption was lowered by 30–35% with fewer reactors and fewer heating/cooling cycles (Jin et al. 2020).
- Decreased Water Use: Because there was only one significant sterilizing process and less cleaning in between unit operations, water consumption was decreased.
- Carbon Emission Offsets: Taking into account both the combustion of ethanol and the decreased methane emissions from trash burning, the system offsets

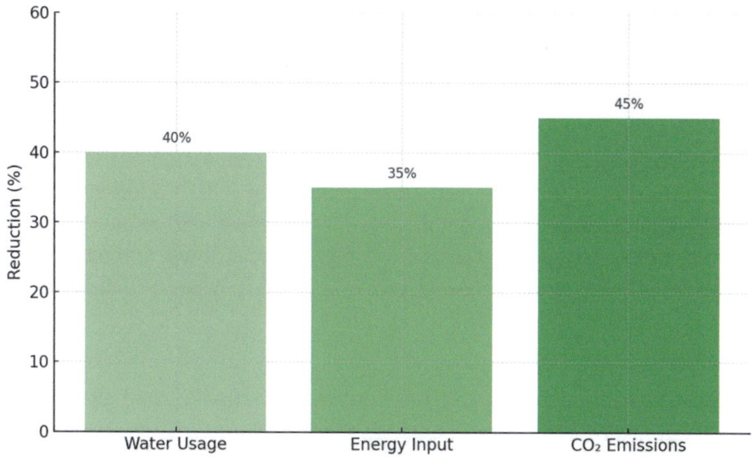

Fig. 3.5 Percentage reduction in key environmental parameters–water usage, energy input, and CO_2 emissions–when adopting consolidated bioprocessing (CBP) instead of traditional multi-step bioprocessing. The data demonstrate significant sustainability benefits, with CBP enabling up to 45% lower carbon emissions and notable decreases in resource consumption

roughly 1.2–1.5 tons of CO_2 equivalent for every tonne of dry biomass processed (IEA Bioenergy 2018).

- Waste Valorization: To close the loop and reduce landfill inputs, the solid leftovers were recycled into biogas and biochar.

As compared to conventional methods, Fig. 3.5: showcases the Environmental Impact Reduction Achieved by Pilot-Scale CBP System illustrates the percentage decrease in energy input, water consumption, and CO_2 emissions per liter of ethanol.

When compared to traditional biofuel production methods, the pilot-scale consolidated bioprocessing (CBP) system significantly reduced the environmental effect by using 40% less water, 35% less energy, and 45% less CO_2 per liter of ethanol (Fig. 3.5).

3.6.3 *The Circular Economy and Lifecycle Potential*

CBP is in line with the circular bioeconomy's tenets, which include recycling energy, repurposing residuals, and obtaining value from waste biomass. The system's regional sustainability could be improved by integrating it with food processing facilities, sugar mills, or farmer cooperatives. Biogas, CO_2, and lignin residues are examples of co-product streams that can be fed into supply chains for fuel, fertilizer, or soil amendments. CBP systems have the potential to be essential in rural bioeconomy hubs if properly scaled, enhancing energy access while assisting with livelihood creation and climate goals.

3.7 Final Thoughts and Prospects

3.7.1 Final Thoughts

The sustainable generation of bioethanol from lignocellulosic agro-waste was significantly advanced by the consolidated bioprocessing (CBP) pilot-scale trial. CBP significantly decreased process complexity, energy inputs, and reliance on commercial enzymes by combining fermentation, substrate breakdown, and enzyme synthesis into a single bioreactor system (Lynd et al. 2005; Alper and Stephanopoulos 2009).

The trial was successful in:

- Up to 0.42 g/g of biomass can be converted to ethanol, and the biomass-to-ethanol conversion efficiency is close to 80%.
- Elimination of distinct fermentation and hydrolysis processes results in significant labor and enzyme cost savings (Olson et al. 2012).
- Significant environmental advantages, such as decreased greenhouse gas emissions and energy and water use (Jin et al. 2020).

Nevertheless, the study also identified a number of scale-up issues that need to be resolved for full-scale commercial use, such as irregular mixing, contamination in longer fermentations, and substrate inhibition at high solids loadings (Lynd et al. 2008; Klein-Marcuschamer et al. 2012).

3.7.2 Prospective Paths

Research and implementation in the following areas are crucial to advancing CBP from pilot-level innovation to full industrial deployment:

1. *Enhancements to Reactor Design*
   Bioreactors with improved heat distribution, foam suppression, and mixing dynamics are needed to scale CBP. Computational fluid dynamics (CFD) is one technique that can be integrated to increase design performance and accuracy.
2. *Improved Microbial Varieties*
   Increasing ethanol tolerance, improving sugar utilization (particularly pentoses), and sustaining mixed cultures in co-fermentation conditions should be the main goals of next-generation microbial engineering (Xia et al. 2021).
3. *Integration of Biorefineries*
   Modular biorefineries that additionally generate value-added products like biogas, organic acids, xylitol, or charcoal ought to incorporate CBP. Waste reduction and economic viability will be improved by this co-production strategy (Ragauskas et al. 2014).
4. *Market Enablement and Policy*

Commercialization will be greatly aided by supportive policies, particularly in developing nations with abundant biomass, such as carbon credit systems, renewable energy subsidies, and technology incubation platforms (IEA Bioenergy 2018).

5. *A Thorough Evaluation of the Economy and Environment*

   To confirm environmental sustainability and profitability under varied feedstocks, reactor sizes, and locations, future research should carry out full-scale lifecycle assessment (LCA) and techno-economic analysis (TEA) (Ravindran and Jaiswal 2016).

   To sum up, CBP has a great chance of becoming a vital part of the bioeconomy. This technology has the potential to increase the sustainability of renewable fuels while also lowering their cost and increasing their accessibility globally, provided it is further optimized and interdisciplinary cooperation is maintained.

References

Alper H, Stephanopoulos G (2009) Engineering for biofuels: exploiting innate microbial capacity or importing biosynthetic potential? Nat Rev Microbiol 7(10):715–723. https://doi.org/10.1038/nrmicro2186

Alvira P, Tomás-Pejó E, Ballesteros M, Negro MJ (2010) Pretreatment technologies for an efficient bioethanol production process based on enzymatic hydrolysis: a review. Bioresour Technol 101(13):4851–4861. https://doi.org/10.1016/j.biortech.2009.11.093

Hendriks ATWM, Zeeman G (2009) Pretreatments to enhance the digestibility of lignocellulosic biomass. Bioresour Technol 100(1):10–18. https://doi.org/10.1016/j.biortech.2008.05.027

IEA Bioenergy (2018) Greenhouse gas emissions from the use of biofuels in transport. Task 38. International Energy Agency (IEA). Retrieved from https://www.ieabioenergy.com/publications/greenhouse-gas-emissions-from-the-use-of-biofuels-in-transport/

Jin M, Gunawan C, Balan V, Dale BE (2020) Consolidated bioprocessing (CBP) of lignocellulosic biomass for bioethanol production: current advances, challenges and perspectives. Bioresour Technol 302:122828. https://doi.org/10.1016/j.biortech.2020.122828

Klein-Marcuschamer D, Oleskowicz-Popiel P, Simmons BA, Blanch HW (2012) The challenge of enzyme cost in the production of lignocellulosic biofuels. Biotechnol Bioeng 109(4): 1083–1087. https://doi.org/10.1002/bit.24370

Liu Z-H, Ying Y, Li X, Wyman CE (2020) Coordinated development of advanced biofuels in China: opportunities and challenges. Renew Sust Energ Rev 119:109573. https://doi.org/10.1016/j.rser.2019.109573

Lynd LR, van Zyl WH, McBride JE, Laser M (2005) Consolidated bioprocessing of cellulosic biomass: an update. Curr Opin Biotechnol 16(5):577–583. https://doi.org/10.1016/j.copbio.2005.08.009

Lynd LR, Laser MS, Bransby D, Dale BE, Davison B, Hamilton R, Himmel M, Keller M, McMillan JD, Sheehan J, Wyman CE (2008) How biotech can transform biofuels. Nat Biotechnol 26(2):169–172. https://doi.org/10.1038/nbt0208-169

Olson DG, McBride JE, Shaw AJ, Lynd LR (2012) Recent progress in consolidated bioprocessing. Curr Opin Biotechnol 23(3):396–405. https://doi.org/10.1016/j.copbio.2011.11.026

Ragauskas AJ, Beckham GT, Biddy MJ, Chandra R, Chen F, Davis MF et al (2014) Lignin valorization: improving lignin processing in the biorefinery. Science 344(6185):709–720. https://doi.org/10.1126/science.1246843

Ravindran R, Jaiswal AK (2016) Exploitation of food industry waste for high-value products. Trends Biotechnol 34(1):58–69. https://doi.org/10.1016/j.tibtech.2015.10.008

Xia C, Zhang Y, Zhang R, Qian Y (2021) Engineering Saccharomyces cerevisiae and Clostridium thermocellum for consolidated bioprocessing of lignocellulosic biomass to bioethanol. Biotechnol Adv 50:107768. https://doi.org/10.1016/j.biotechadv.2021.107768

Zhang YHP, Lynd LR (2010) Toward an aggregated understanding of enzymatic hydrolysis of cellulose: noncomplexed cellulase systems. Biotechnol Bioeng 88(7):797–824. https://doi.org/10.1002/bit.20282

Chapter 4
Nanotechnology—Present Status

Jahnavi Alwala, Rani Padmini Velamakanni, Vagdevi Pally, Farheen Naaz, and Ramchander Merugu

Abstract Nanotechnology has emerged as one of the most transformative scientific domains of the twenty-first century, offering applications spanning medicine, energy, environment, agriculture, and industry. Since its conceptual introduction by Norio Taniguchi in 1974 and further advancements through discoveries such as the scanning tunnelling microscope and fullerene structures, nanotechnology has redefined material science by enabling manipulation at atomic and molecular levels. The distinctive physicochemical, biological, and catalytic properties exhibited by nanoparticles within the 1 to 100 nm range have allowed the creation of novel systems and devices with enhanced functionality and precision. This chapter presents a comprehensive analysis of nanotechnology's present status, highlighting its wide-ranging applications, from targeted drug delivery, diagnostic imaging, and theragnostics in biomedicine to catalytic water purification, pollutant remediation, advanced fuel cells, and solar energy harvesting. Further emphasis is placed on its role in food safety, active packaging, textiles, and consumer products. While nanotechnology offers immense opportunities in designing cleaner, sustainable, and cost-effective solutions, concerns related to nanoparticle toxicity, ecological disruption, and long-term health risks remain significant. Thus, fields such as nanotoxicology and nanomedicine are evolving to address the dual challenge of maximizing benefits while minimizing adverse effects. Indicators such as workforce expansion, research publications, patents, and global funding demonstrate their rapid growth, while transitions from basic discovery to complex integrated systems signify its evolution. Overall, nanotechnology represents a revolutionary interdisciplinary frontier,

J. Alwala · R. Merugu (✉)
Department of Biochemistry, Chaitanya Deemed to be University, Hyderabad, Telangana, India

R. P. Velamakanni · V. Pally
Department of Biochemistry, Mahatma Gandhi University, Nalgonda, Telangana, India

F. Naaz
Center for Gene Regulation in Health and Disease, Department of Biological, Geological, and Environmental Sciences, College of Arts and Sciences, Cleveland State University, Cleveland, OH, USA

N. Srivastava (ed.), *Sustainable Nanostrategies for Biorefineries*, Clean Energy Production Technologies, https://doi.org/10.1007/978-981-95-5875-9_4

capable of reshaping economies, industries, and environmental strategies, provided that its deployment is balanced with robust safety protocols and responsible innovation.

Keywords Nanotechnology · Nanoparticles · Inter-disciplinary applications · Environmental remediation · Advanced materials

4.1 Introduction

In 1974, Japanese scientist, Professor Norio Taniguchi, used the word "nanotechnology" for the first time during an international symposium on industrial production that took place in Tokyo. He used the term to explain the nanoscale ion beam milling and semiconductor modification approach as encompassing the dissociation mechanisms, combination, and material deformation (Tawkif 2014). The discovery of Buckminsterfullerene C60 (Buckyballs) by Robert Curl, Richard Smalley, and Harry Kroto significantly advanced this era (Kroto et al. 1985). The "era of excellence" of nanotechnology started in the year 1981 when Gerd Binning and Heinrich Rohrer discovered the Scanning Tunneling Microscope (STM), which enabled the recognition of individual atoms ("Scanning the past", 2013). One of the important contributors to the progression of nanotechnology was Eric Drexler of Massachusetts Institute of Technology (MIT). In 1986, Drexler gave nanotechnology a definition in a commercial sense in his book, "Engines of Creation". To broaden public understanding about nanotechnology ideas and implications, Drexler founded the Foresight Institute following the publication of the book. Drexler referred to the future era of nanotechnology as engineering on the billionth of a meter scale (Drexler 1986). Throughout the late 1980s and early 1990s, significant breakthroughs continued to drive the progress of nanotechnology. These improvements resulted in important developments in the study of nanotechnology, design, and publication output, molding its trajectory for future expansion and advancement.

The scientific world is fascinated by nanoscience—the majority of current definitions, attention, and discussions center on nanotechnology, which is a conceptual approach. As a result, it is a general term that exemplifies the pinnacle of man's unwavering quest for knowledge with potential applications. Any technology that operates at the nanoscale and has practical uses, such as using individual atoms and molecules to create functional structures, is referred to as nanotechnology (Feynman 1960). Nanotechnology focuses on designing, developing, and applying physical, chemical, and biological systems whose structural elements range from single atoms or molecules up to submicron scales. It also includes integrating these nanostructures into larger functional systems (Taniguchi 1974; Drexler 1986). "Technology on the nanoscale" is the most basic definition of nanotechnology. Nanoscale is a scale ranging from 1 to 100 nm and can be briefly defined as an "atomically precise technology" or "engineering with atomic precision" (Feynman 1960). Nanotechnology is related to systems and materials; the parts and structures, due to their nanoscale dimensions, exhibit new and significantly better chemical, physical, and biological qualities, processes, and phenomena. "The design, analysis, production, shape and size-controlled application of matters in the "nanoscale" is the definition of nanotechnology given in the dictionary (Royal Society and Royal Academy of Engineering 2004). An alternative definition in the dictionary is "the careful and accurate adjustment, precision placement, modeling, measurement, and manufacture of materials at the nanoscale in order to make matters, systems, and devices with basically unique features and functionalities" (Royal Society and Royal Academy of Engineering 2004). Nanotechnology is a multidisciplinary domain that falls within the broader scope of science, chemistry, physics, biology, and several other scientific fields containing the study of phenomena at the nanoscale (Taniguchi 1974). When the dimensions of a material are minimized below 100 nm, striking changes may occur in their features (Drexler 1986).

Materials can be engineered at the nanoscale to enhance specific performance traits or to introduce entirely new properties beyond the typical changes associated with size and structural differences. Such types of macromolecules, and the particles formed of a small number of molecules, at sizes ranging from 1 to 50 nm, have varied physicochemical properties (Gleiter 2000). Compared to larger materials, nanoparticles (NPs) have properties of enhanced activity when they make use of the same applications. The definition of an NP is a cluster of atoms connected via an average radius ranging from 1 to 100 nm, i.e., often comprising 10 to 10^5 atoms (Schmid 1992). An important application of NPs is the production of a novel category of catalysts, recognized as nanocatalysts (Somorjai and Park 2008). Important discoveries in the field are advancing the production and various functions of NPs as catalysts for the enhanced efficiency of chemical reactions (Astruc et al. 2005). A catalyst's capacity is largely dependent on the size and dispersion of its particles. NPs' chemical characteristics, such as their wider surface-to-volume ratio, surface morphology, and electronic properties give them advantages as catalysts (Taniguchi 1974; Bell 2003).

The development of innovative nanostructured materials and nanosystems is made possible by the invention of new materials as well as the advancement in theoretical and experimental research approaches. There are many current and expected advancements in nanoscale science and nanotechnology with respect to their advantages in agriculture, electronics, medicine, energy, and more. These improvements are being made at an increasing rate (Somorjai and Park 2008; Roco 2003b). Nanotechnology can play a vital role in the extension of evolution methods used to frame novel products, to replace present production equipment, to reformulate new materials and chemicals towards better performance, causing declined material and energy utilization, to decrease harm to the atmosphere, and even to pollution control (Karn et al. 2009). Though a reduced absorption of matter and energy supports the environment, nanotechnology offers promising opportunities to solve problems in a wide, sustainable way. Environmental benefits (Fig. 4.1) of nanotechnology include the improvement of solutions to present environmental challenges, measures to represent both the ensuing problems from interactions of material and energy with the environment, and the possible risks accompanied by nanotechnology (Karn et al. 2009).

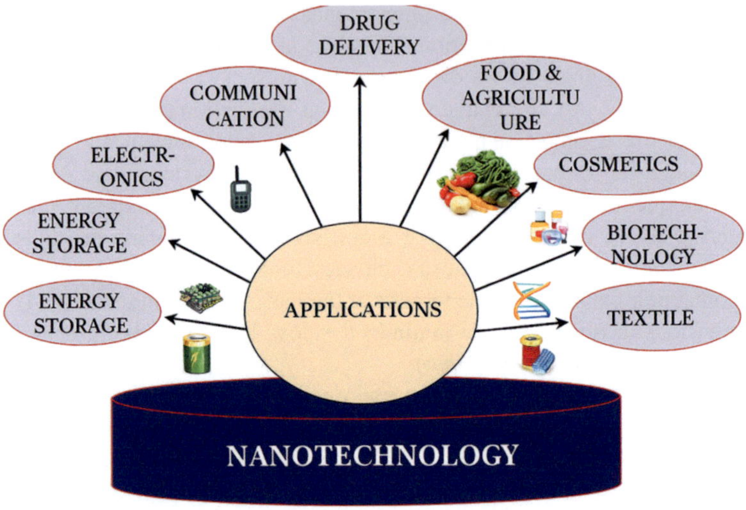

Fig. 4.1 Latest advancement in nanotechnology

4.2 Nanotechnology's Present and Future Applications

According to expert forecasts, with important contributions to financial growth and employment creation, nanotechnology is a developing field set for exponential growth. Several studies have highlighted the risks, and many promises, of nanotechnology in different technological fields. By utilizing nanoparticles, nanotechnology has changed its methods related to bioavailability, manufacture, packaging, and product shelf life in industries such as food and cosmetics. Food-borne germs can be removed with the use of nano-sensors such as zinc quantum dots, which show improvement in food safety and quality (Das and Ganguly 2023).

Moreover, by using designed nanoparticles to foster a world free of pollution, nanotechnology has made significant improvements in pollution cleanup. These tools have also made medical operations more accessible, safer, lower cost, and more comfortable (Afzal et al. 2022; Ma and Xia 2021; Ramesh et al. 2023). In the medical field and even across research areas, nanotechnology has broader applications in biotechnology. Medical treatments have been evolving with the development of nanoparticle-based medical devices such as drug delivery systems and diagnostic biosensors.

Nanotoxicology and nanomedecine have been developed to tackle these concerns. Nanotoxicology focuses on studying and reducing the adverse effects of nanoparticles on both human health and the environment. The widespread exposure of humans to nanoparticles has raised crucial concerns about health as well as environmental risks, regardless of their widespread adoption and substantial advantages. Nanomedicine examines the benefits and potential risks of applying nanomaterials in medical fields such as diagnostics, bioimaging, and tissue

Table 4.1 Key areas and applications of nanotechnology

Area	Description	Applications	References
Nanomedicine	Use of nanoparticles for diagnosis, drug delivery, and therapy at the molecular level	Targeted drug delivery, cancer therapy, imaging	Kumar and Singh (2024)
Nanoelectronics	Miniaturization of electronic components using nanomaterials	Faster processors, memory devices, sensors	Wang et al. (2023)
Nanomaterials	Development of materials with nanoscale properties like strength and conductivity	Stronger composites, coatings, catalysts	Zhao et al. (2023)
Environmental Nanotech	Use of nanotechnology for pollution control, water treatment, and environmental monitoring	Water purification, pollutant sensors	Patel and Kumar (2023)
Agricultural Nanotech	Nanoparticles to enhance fertilizers, pesticides, and crop yield	Smart delivery systems, soil health monitoring	Singh and Sharma (2023)
Nanofabrication	Techniques to create nanoscale structures and devices	Nanochips, nanopatterned surfaces	Lee and Chen (2024)
Energy Nanotech	Nanotech to improve energy storage and generation	Solar cells, batteries, supercapacitors	Smith et al. (2023)

engineering. The benefits of medical nanomaterials include enhanced drug delivery, early cancer cell detection, and decreased inflammation. (Table 4.1). Nevertheless, ongoing research is crucial to fully understand and manage the potential risks involved with nanotechnology (Tiple et al. 2020; Akçan et al. 2020; Tharkar et al. 2019; Choi et al. 2010).

4.3 Impact of Nanotechnology

Scientific developments have provided a wider scope for the research community, enabling them to explore deeper and more, and provide technologies in every field possible that are cleaner, more cost-effective and less toxic for the benefit of mankind. Nanotechnology is therefore a promising solution for the future in every field, including environment, agriculture, medicine, industry, etc. Nanoparticles are produced both synthetically and naturally. The creation and release of nanoparticles offer distinct advantages and drawbacks across industries. Researchers consider nanotechnology a breakthrough field, as physical and chemical properties depend on particle morphology. The expanding knowledge of nanotechnology and its uses is delivering benefits such as environmental protection, new medical therapies, and more accurate diagnostics.

Positive impact The developments in the field of nanotechnology are paving the way for innovative and promising technology such as designing molecules that can mimic the living system. Nanotechnology can create materials with specific functions that are also smart such as appliances and machines or gadgets that are fully automated at the molecular and atomic level (Malik et al. 2023). Nanocatalysts are used for environmental remediation and can accurately remove contaminants. Nanoproducts, such as nano-enabled filtration membranes and nanocatalysts can be especially useful in removing pollutants from contaminated water. The advantage of using nanocatalysts is that they have the ability to selectively degrade a specific pollutant. Nano-based sensors are magnificent devices that can be used for checking and scanning environmental pollution. These sensitive devices can be used to precisely measure the quality of the air, water, and soil. They are convenient and portable, measure even the slightest changes and function as real-time devices that have a huge impact in providing environmental data (Rani et al. 2024).

Negative impact Nanoparticles are released by various sources such as traffic, industry, agriculture, natural disasters such as volcanic eruptions, and so on. As they are released into the environment, they can pollute air, land, and water, posing a threat to life. They can have serious negative impacts on various living entities such as animals, humans and plants. On humans, the negative impacts could be in the form of skin issues, damage to organs such as the respiratory system, gastrointestinal system, and others as a result of long-term exposure. The nature and type of nanoparticle determine the severity of the illness. At the atomic level, they can alter proteins, lipids, and cell DNA and can introduce harmful mutations (Taghavi et al. 2013; Chang et al. 2012). Studies suggest that silver nanoparticles that are dissolved in the water can be harmful to aquatic life such as fish and algae. The effect significantly influences the shape, size, and the pH of the particles responsible for pollution. Iron and copper oxide nanoparticles can also be detrimental to soils (Frenk et al. 2013). The consequences can be growth retardation, alteration of genetic material, reduction of yield, and minimized absorption of minerals and nutrients from the soil (Xia et al. 2015). In plants grown in nanoparticle-polluted soil, cytotoxicity, genotoxicity, and oxidative stress can occur, impairing plant growth. Therefore, it is necessary to follow safety measures, such as wearing a mask when working with nanomaterials. It is essential to study and understand the nanomaterial before it is used for application in any field. Nanoparticles can be a serious threat to the environment and pose serious, complex problems, such as toxicity. It can be a serious threat to the environment. Though they are largely used in the purification and removal of pollutants from air and water, there is little impact on the ecosystem (Rani et al. 2024).

4.4 Indicators for the Development of Nanotechnology

Roco's studies suggest four key indicators that describe the development of nanotechnology. These indicators include: (1) the size of the manual work force employed in nanotechnology; (2) nano-related research, publication of papers, and filing of patents; (3) products available in the market; (4) funding provided by the government and private organizations for nano-related work. It is estimated that the government and private organizations workforce increased by approximately 25% from 2008 to 2015 in one or both areas of nanotechnology (Roco 2003a). An increase in the Science Citation Index was also observed during the same period. The average growth rate is observed as 23% worldwide (Chen and Roco 2009). Between 1991 and 2008, the number of patents filed in nanotechnology and related fields increased by 34% (Dang et al. 2010) while, within a span of 5 years, from 2019 to 2023, the increase in filing patents and publishing articles in the area of nanotechnology is observed to be up to 40% (Khatoon and Veli Dandi 2025). The use of nanotechnology-based products has increased significantly from around 25% in 2000 to 2008 (Roco and Bainbridge 2001). The availability of three-dimensional nanomaterials is expected to be higher by 2020, which is almost 72% of the various types of nanomaterials in differentindustrial divisions. Worldwide investment in nanotechnology is 35% as research and development funding from both the public and private sectors and the venture capital investment is observed as 30% from 2000 to 2008, (Fig. 4.2) while the rise is notably high until 2024, the proposed budget from 2021 to 2027 for nanotechnology and its related firms in both public and private sectors in Europe alone constitutes €5.4 billion (Khatoon and Velidandi 2025).

4.5 Transition in the Development of Nanotechnology

Several factors have contributed to the development of nanotechnology. It evolved from simple, passive structures to more complex systems. The first phase concentrated on the discovery, properties, synthesis, and future prospects. Development and

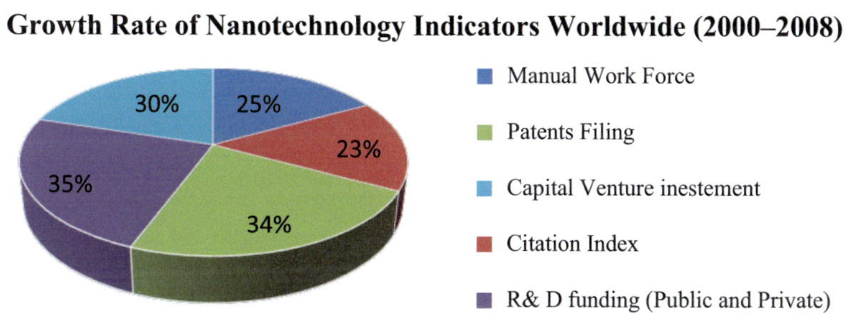

Fig. 4.2 Growth rate of nanotechnology indicators worldwide (2000–2008)

advancement were the defining criteria of this phase, which was primarily science-and knowledge-centric. The second phase focused on integrating existing knowledge of nanosystems into science and engineering to produce more specific and advanced outcomes. Research and design interests shifted toward the complex nanosystems that had improved utility. This phase was driven largely by socio-economic considerations. After 2020, nanotechnology expanded into nearly every field, showing divergent trends from atomic and molecular to macromolecular scales, incorporating assembly, automation, and evolutionary approaches (Roco 2011).

4.6 Applications of Nanotechnology

Nanotechnology has transformed the world through its wide range of applications. The pace of this nano revolution has been rapid, with nanotechnology becoming deeply integrated across scientific fields. Below are some of the most widely recognized applications of nanotechnology.

Biomedical Applications: Nanotechnology is improving the standard of human life by conquering wide-ranging applications in medicine, diagnostics, and therapeutics. Nanomedicine has enabled biologists to better understand disease mechanisms and develop advanced approaches across healthcare, nanotechnology-based genomics, gene therapy, and tissue engineering. Some of the key biomedical applications are discussed below.

Drug Delivery Systems: Nanoparticles are well-suited for delivering therapeutic agents due to their biocompatibility and ability to encapsulate drugs. They can enhance drug solubility, stability, and target delivery to specific tissues or cells, thereby reducing side effects and improving efficacy.

Targeted Drug Delivery and Therapeutics: As multifunctional drug carriers, NPs can be engineered to deliver multiple drugs simultaneously, leveraging their tunable surface properties. This capability can be used to co-deliver synergistic drug combinations for complex diseases like cancer (Afzal et al. 2022).

Imaging Agents: Nanoparticles, such as gadolinium, gold, and iron oxide, synthesized by different methods can serve as contrast agents in various imaging modalities (e.g., MRI, CT, and optical imaging). Their tunable properties enable sensitive and specific detection of diseases like cancer. These show enhanced properties and can be best worked when they are in a coordinated complex with acyclic, cyclic chelate agents, by minimizing the toxicity of the metal ion in its free state.

Contrast Agents: Nanoparticles can enhance contrast in imaging techniques such as MRI and positron emission tomography (PET), aiding in the early detection and monitoring of diseases.

Theragnostic: Green-synthesized nanoparticles can be engineered for theragnostic applications, where they combine diagnostic imaging with therapeutic capabilities. For example, nanoparticles loaded with both imaging agents and therapeutic drugs enable real-time monitoring of treatment efficacy (Pellico et al. 2019).

Environmental Sensing and Remediation

Sensitive Detection of Pollutants: Nanotechnology-based sensors can be incorporated into sensors to detect trace amounts of environmental pollutants. Their high surface area and enhanced reactivity improve the sensitivity and selectivity of these sensors.

Catalytic Water Treatment: Nanocatalysts such as titanium dioxide (TiO2) and zinc oxide (ZnO) can catalyze the breakdown of organic contaminants in water, offering a sustainable solution for wastewater treatment. For example, palladium-gold BNPs can be used to remove harmful organic compounds through catalytic oxidation.

Water Purification: Nanostructured membranes and adsorbents such as graphene and activated carbon have shown promise in removing heavy metals, organic pollutants, and pathogens from contaminated water sources. Nanoparticles like iron oxide and silver possess excellent adsorption and antimicrobial properties.

Air Filtration: Nanoparticles embedded in filter materials can effectively capture particulate matter and harmful gases from both indoor and outdoor air, contributing to improved air quality (Rani et al. 2024).

Renewable Energy Technologies

Advanced Fuel Cells: Metal nanoparticles such as Au, Pd. Fe, Ni, Si, etc. are being used to enhance the efficiency and durability of hydrogen fuel cells. These fuel cells are critical for clean energy production and reducing reliance on fossil fuels (Merugu et al. 2022).

Solar Energy Harvesting: Nanomaterials such as graphene can improve the efficacy of photovoltaic cells by enhancing light absorption and reducing electron-hole recombination. These cells offer 25% productivity, which is a good improvement in the generation of energy using solar cells (Belkhode et al. 2024).

Wearable and Flexible Electronics

Conductive Inks: Metallic NPs can be used in conductive inks for printed electronics, enabling the fabrication of flexible, lightweight, and high-performance electronic devices. These inks are crucial for developing wearable electronics, flexible displays, and electronic textiles (Fernandes et al. 2020).

Sensors and Actuators: NPs enhance the performance of sensors and actuators in flexible electronics, making them more responsive and reliable for applications such as health monitoring and soft robotics.

Food Safety and Preservation

Active Packaging: NPs made with metals such as silver with antimicrobial properties can be incorporated into food packaging materials to prevent microbial growth and extend the shelf life of perishable goods. This application is particularly important for reducing food waste and ensuring food safety. Sensors based on NPs can detect spoilage indicators or contaminants in food, providing real-time information about food quality and safety.

Food Packaging: Green-synthesized nanoparticles can extend the shelf life of perishable food items by inhibiting microbial growth and oxidative reactions. They also serve as active packaging materials to monitor food quality.

Agrochemicals: Nanoparticles loaded with fertilizers or pesticides can be targeted to specific plant tissues, reducing the number of agrochemicals required and minimizing environmental impact (Altammar 2023).

Catalysis and Energy Applications

Catalytic Converters: Green-synthesized nanoparticles are used as catalysts in environmental applications, including the conversion of harmful gases such as nitrogen oxides into harmless substances. In particular, bimetallic nanoparticles show improved catalytic efficiency as a result of the synergistic interactions between the two constituent metals.

Energy Storage and Conversion: Nanoparticles synthesized through green methods play a vital role in energy storage devices like batteries and supercapacitors. Their high surface area and conductivity improve the efficiency and stability of energy conversion processes (Altammar 2023).

Textiles and Consumer Products

Antimicrobial Textiles: Green-synthesized nanoparticles, particularly those containing silver, are incorporated into textiles to impart antimicrobial properties. This application reduces bacterial growth and odor formation in clothing and medical textiles (Chakrabarty and Jasuja 2022).

Sunscreen and Cosmetics: Nanoparticles like zinc oxide and titanium dioxide, synthesized using green approaches, are used in sunscreens and cosmetics for their UV-blocking properties. Nanoparticle-based sunscreen prevents the skin from being chalky. Liposomes and nanosomes are involved in skin care formulations; their efficiency is enhanced by transporting the active ingredients to the site of action.

Nutricosmetics: It is a combination of nutrition and cosmetics. These include the cosmetics that can be supplemented orally. Vitamins, minerals, and other needed compounds that benefit the skin, hair, and nails function by removing free radicals, protecting against the oxidative stress, and so on (Pandey et al. 2024).

4.7 Challenges and Limitations

Nanotechnology is the science that studies and utilizes materials at the nanometer scale, enabling precise manipulation of atoms and molecules. Nanotechnology or nanoscience is an emerging field, and it is expected it will play a significant role in upcoming decades, especially as a powerful tool for drug delivery. Despite challenges in advancing nanotechnology and its applications, its potential for rapid growth in the industrial sector is substantial. Nanotechnology has positive impacts on diverse fields, including physics, chemistry, biology, materials science, and engineering, with applications in medicine, chemistry, biology, materials science, food, environmental management, electronics, and energy.

Present Status of Nanotechnology
Drug delivery system as nano biosensors and liposomal drugs, therapeutic, imaging and diagnostics in healthcare and medical fields. In the electronics field, carbon nanotubes and quantum dots are revolutionizing transistors, memory chips, and display tools in today's digital world. With advances in nanotechnology, energized nanomaterials offer greater efficiency in solar power, hydrogen storage, and batteries for energy fuels. Environmentally, the use of nanofilters and catalysts is emerging in water purification and waste management services. In agriculture, nanofertilizers and nanopesticides (Fig. 4.3) help improve crop yields and reduce the use of hazardous chemical.

In nanotechnology, the present era is characterized by technological advancements that have led to the development of quantum dots and nanocrystal particles, which show excellent conductive properties and can be physically manipulated at the nanometer scale. High accuracy techniques are used for the magnification of nanoparticles at the atomic level are scanning electron microscopy, magnetic resonance microscopy, transmission electron microscopy, and electron microscopy. From a strategic perspective, the electronic industry has economically advanced upgraded lithographic methods for steady-state minimization in size and cost. Based on scale and precise measurements, materials have been observed, manipulated, and developed using advanced technologies. At the nanoscale, multiple fields of science and technology have introduced unique properties that differ significantly from bulk materials, leading to innovative growth. This review of recent advances in nanotechnology highlights its applications, opportunities, and challenges across various fields. Based on products identified across different generations in nanotechnology, new evidence has emerged, including the both passive and active nanostructures.

The passive properties of these nanostructures offer several advantages. For example, titanium-based nanolayers in sunscreens effectively absorb UV light. Carbon nanotubes, which are stronger than steel, are used in products like tennis rackets to provide stiffness without adding extra weight. In addition, the nanolayer of each material retains its unique properties and structural stability, allowing it to maintain its function when encapsulated within a product.

Active nanostructures are more advanced than passive ones due to their complex manufacturing and innovative implementation in drug delivery. The use of these active nanostructures enables more precise and predictive control in the field of drug delivery. Nanoparticles target cancer cells quickly by delivering electromechanical nanostructures that facilitate drug release. Once they've been inserted, they release agents that disrupt and destroy cancerous cells.

Nano systems are the nanotools that serve as the final goal for the main components in the network and for data exchange in the developmental process. As the

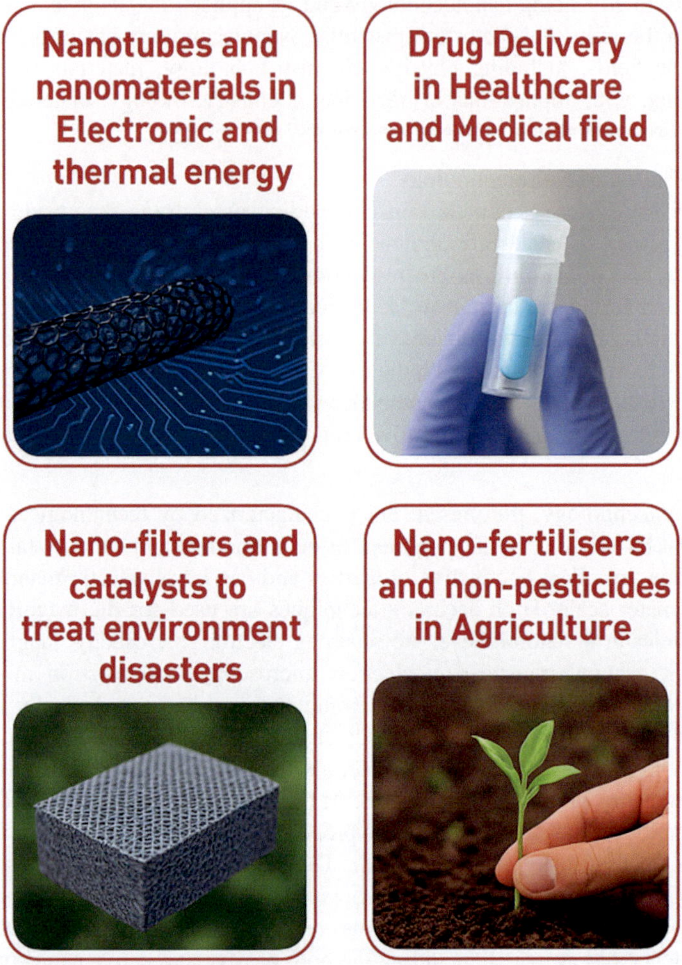

Fig. 4.3 Present status of nanotechnology in various fields

nanostructures are growing, they instantly assemble into reticulates on the bone and tissues. Electromechanical nanomaterials or devices explore cancerous cells and their development. The advancements in the fields of robotics, biotechnology, agriculture, electrical, and next-generation technology will release upgraded products as a significant aspect.

Nanostructured materials play a significant role in nanoceramics in both powdered form and liquid diffusions (silica, titanium dioxide, alumina, iron oxides, zinc oxide and zirconium oxide, nanotubes as conductors and semiconductors with good conductivity, nanocomposites as polymer-based nanosized filters.

Nanotechnology involves the manipulation of materials at atomic, molecular, and macromolecular scales to harness their unique properties. It encompasses the design, characterization, production, and application of structures, devices, and systems by precisely controlling their shape and size at the nanoscale. This exemplifies their unique physical, chemical, electrical, biological, and optical properties, highlighting their potential for future applications across various fields.

As an interdisciplinary field,nanotechnology has expanded into numerous areas, including materials, electronics, agriculture (Tawkif 2014), environment, metrology, energy, robotics, security, healthcare, and medicine. It encompasses innovations such as nanomaterials, nanoelectronics, nanostructured materials, nanosized filters, and nanobiotechnology (Kroto et al. 1985; Drexler 1986; Feynman 1960) as well as applications in IT, biometrics, pharmaceuticals, manufacturing, agriculture, construction, and food processing. Nanotechnology plays a significant role in drug delivery (Taniguchi 1974) and in treating various diseases. Its rapid growth across industries, healthcare, and medicine is expected to have a positive impact on future quality of life. Applications in various fields are shown in the Table 4.2 with examples.

Limitations of Nanotechnology and its Applications in Different Fields
Nanotechnology faces several challenges, including technical, economic, ethical, regulatory, and environmental issues. Technically, it is difficult to scale up production from the laboratory level to industrial-scale manufacturing. There is also a lack of global standards for nanomanufacturing, measurement, and testing tools, and the long-term health effects of nanoparticles remain unclear. In particular, their ability to penetrate cells and affect organs raises concerns about toxicity and the adequacy of existing safety standards. Economically, cost-effective nanotechnology production is expensive and there is a lack of infrastructure. Ethical concerns include the use of nanosensors for surveillance, (which raises privacy issues), as well as limited public awareness and acceptance. Regulatory challenges involve the lack of clear legal frameworks for use and disposal of nanomaterials. Environmental risk factors like accumulation of nanoparticles in soil and water can harm ecosystems, and there is a need for effective protocols for managing and disposing of nanowaste.

4.8 Conclusion

Nanotechnology is shaping the future of advanced development, with remarkable progress across fields ranging from medicine and energy to electronics. While it offers innovative possibilities, it also brings important considerations regarding health, safety, and ethics. Research in nanotechnology highlights both its potential and its challenges, emphasizing the need for interdisciplinary collaboration and a comprehensive approach. Despite these challenges, nanotechnology continues to make a substantial global and economic impact across various fields. In the twenty-first century, nanotechnology has become one of the most exciting areas of scientific advancement. Developments at this nano-level have led to the rapid creation and improvement of new tools and instruments. As a transformative technology,

Table 4.2 Applications of nanotechnology in different fields

S. No	Field & its applications	Examples
1.	Drug Delivery systems	Encapsulated forms like nanocapsules, nanobiomagnets, liposomes, new formulations for drug delivery and gene therapies
2.	Energy & Production	Carbon nanotubes for coating membrane filters, enhancing the batteries based on electrode materials and in the film coating for solar cells Quantum dots are used for photovoltaic and organic light-emitting devices Nanocatalysts for generation of hydrogen Hybridized protein-polymer bio membranes
3.	Health care monitoring	Nanotubes and nanoparticles for healthcare to ensure the monitoring of various health issues by using sensors, etc.
4.	Diagnostic & screening of diseases	Nanosensor arrays made from carbon nanotubes Nanoliter volumes used for bioassays as lab-on-a-chip devices, single cell analysis, droplets for PCR amplification, miniature diagnostic kits used for nano-scale liters for the detection of COVID-19, malaria, cancerous cells Quantum dots for various disease diagnoses Magnetic nanoparticles used as nanosensors Nanoscale bioconjugates used as multifunctional dendrimers for enabling applications in the medical field to diagnose of HIV-1 and cancer Nanowired & nanobelt nanoensors are miniatured sensing devices at nanoscale and are used to detect biological, chemical, and physical stimuli at high ratios Nanoparticles enhance medical imaging
5.	Agriculture Sector	Nanoporous zeolites enable molecular sieving, greenhouse gas mitigation, proper dosage of water and fertilizers to plants, nutrients and drugs for livestock Nanocapsules enable herbicide delivery Nanosensors to monitor plant growth and Soil contaminants are removed by nanomagnets Nanoliter volumes are used for pathogen detection and pesticide residue analysis Vector detection by nanosensors, and pest control by nanoparticles
6.	Water Management	Nano-based bio membranes for water purification and treatments. Nano-based zeolites, porous polymers and palygorskite for water purification Nanosensors for detecting pathogens Magnetic nanoparticles are used for wastewater treatment Catalytic degradation of titanium-based nanoparticles is used to degrade water pollutants
7.	Pollution remediation	Nano zero-valent iron is used to reduce toxic metals for degrading chlorinated compounds in ground water Titanium based nanoparticles are used for photocatalysis to degrade organics in water and air

(continued)

Table 4.2 (continued)

S. No	Field & its applications	Examples
		Carbon nanotubes for adsorption of dyes and metals for selective permeability in water and air Magnetic nanoparticles to remove the adsorbents in water and soil Nanoparticles and nanosensors of different metals are used in detecting disinfection by antibacterial activity, photocatalytic degradation, dye adsorption Nanoclays are used to adsorb heavy metals Gas separation nanodevices for remediation processes
8.	Food industries	Nanosensors to detect pathogens and color Nanomaterials to improve shelf life and food safety Enhanced sensory properties Nanocomposites for plastic film in food packaging Nanoemulsions in decontamination of food packaging as an antimicrobial Nanotechnology-based antigen detecting biosensors for identification of pathogen contamination

nanotechnology is advancing across multiple industries, offering a wide range of innovative applications. New developments in nanotechnology are transforming materials used in electronics, medical products, and other raw materials, making them more efficient and cost-effective. Its applications are expanding in diverse fields such as food, medicine, computing, energy, robotics, agriculture, and diagnostics, providing ethical and practical benefits to humanity. Current research shows that nanotechnology has immense transformative potential, although there are still challenges in safety, regulation, and affordability. Nanotechnology is a global hub and is advancing through sustainable investment, active patents, and rapid progress across multiple sectors.. Its commercialization is reshaping industries and addressing critical challenges in healthcare, agriculture, energy, environment, and computing, Realizing this potential requires inclusive international collaboration, careful regulation, and ethical oversight. As of 2025, nanotechnology stands at a pivotal point, offering a promising potential to drive global innovation and impact.

References

Afzal O, Altamimi ASA, Nadeem MS, Alzarea SI, Almalki WH, Tariq A, Mubeen B, Murtaza BN, Iftikhar S, Riaz N, Kazmi I (2022) Nanoparticles in drug delivery: from history to therapeutic applications. Nanomaterials (Basel) 12(24):4494. https://doi.org/10.3390/nano12244494. PMID: 36558344; PMCID: PMC9781272

Akçan R, Aydogan HC, Yildirim M, Taştekin B, Sağlam N (2020) Nanotoxicity: a challenge for future medicine. Turk J Med Sci 50(4):1180–1196

Altammar KA (2023) A review on nanoparticles: characteristics, synthesis, applications, and challenges. Front Microbiol 14:1155622. https://doi.org/10.3389/fmicb.2023.1155622. PMID: 37180257; PMCID: PMC10168541

Astruc D, Lu F, Aranzaes JR (2005) Nanoparticles as recyclable catalysts: the frontier between homogeneous and heterogeneous catalysis. Angew Chem Int Ed 44(48):7852–7872

Belkhode P, Giripunje M, Dhande M, Gajbhiye T, Waghmare S, Tupkar R, Gondane R (2024) Nanomaterials applications in solar energy: exploring future prospects and challenges. Mater Today Proc. https://doi.org/10.1016/j.matpr.2024.04.035

Bell AT (2003) The impact of nanoscience on heterogeneous catalysis. Science 299(5613): 1688–1691

Chakrabarty S, Jasuja K (2022) Applications of nanomaterials in the textile industry. In: Nanoscale engineering of biomaterials: properties and applications, pp 567–587. https://doi.org/10.1007/978-981-16-3667-7_20

Chang Y-N, Zhang M, Xia L, Zhang J, Xing G (2012) The toxic effects and mechanisms of CuO and ZnO nanoparticles. Nature 5:2850–2871

Chen H, Roco M (2009) Mapping nanotechnology innovations and knowledge. Global and longitudinal patent and liter ature analysis series. Springer, Berlin

Choi YE, Kwak JW, Park JW (2010) Nanotechnology for early cancer detection. Sensors (Basel) 10(1):428–455

Dang Y, Zhang Y, Fan L, Chen H, Roco MC (2010) Research, trends in worldwide nanotechnology patent applications: 1991 to 2008. J Nanopart Res 12(3):687–706

Das TK, Ganguly S (2023) Revolutionizing food safety with quantum dot–polymer nanocomposites: from monitoring to sensing applications. Foods 12(11):2195

Drexler E (1986) Engines of creation: the coming era of nanotechnology. Knopf Doubleday Publishing Group, p 298

Fernandes IJ, Aroche AF, Schuck AP, Lamberty, Peter CR, Hasenkamp W, Rocha TLAC (2020) Silver nanoparticle conductive inks: synthesis, characterization, and fabrication of inkjet-printed flexible electrodes. Sci Rep 10:8878. https://doi.org/10.1038/s41598-020-65698-3

Feynman RP (1960) There's plenty of room at the bottom. Eng Sci 23(5):22–36. SpringerLink+2rd.springer.com+2ScienceDirect+2

Frenk S, Ben-Moshe T, Dror I, Berkowitz B, Minz D (2013) Effect of metal oxide nanoparticles on microbial community structure and function in two different soil types. PLoS One 8:84441

Gleiter H (2000) Nanostructured materials: basic concepts and microstructure. Acta Mater 48(1): 1–29

Karn B, Kuiken T, Otto M (2009) Nanotechnology and in situ remediation: a review of the benefits and potential risks. Environ Health Perspect 117(12):1823–1831

Khatoon UT, Velidandi A (2025) An overview on the role of Government initiatives in nanotechnology innovation for sustainable economic development and research progress. Sustainability 17(3):1250. https://doi.org/10.3390/su17031250

Kroto HW, Heath JR, O'Brien SC, Curl RF, Smalley RE (1985) C60: Buckminsterfullerene. Nature 318:162–163

Kumar V, Singh P (2024) Nanomedicine: recent advances in nanoparticle-based drug delivery and diagnostics. Int J Nanomedicine 19:57–80. https://doi.org/10.2147/IJN.S351234

Lee H, Chen Y (2024) Nanofabrication techniques for nanoscale devices. Nano Fabric Lett 9(4): 200–215. https://doi.org/10.1016/j.nanofab.2023.101230

Ma T, Xia T (2021) Nanoparticle-based activatable probes for bioimaging. Adv Biol (Weinh) 5(1): 2000193

Malik S, Muhammad K, Waheed Y (2023) Nanotechnology: a revolution in modern industry. Molecules 28(2):661. https://doi.org/10.3390/molecules28020661. PMID: 36677717; PMCID: PMC9865684

Merugu R, Gothalwal R, Velamakanni RP, Velamakanni RS, Chitturi KL, Naz F (2022) Green synthesis of nanoparticles by plants and their renewable energy applications. In: Srivastava M, Malik MA, Mishra P (eds) Green Nano solution for bioenergy production enhancement. Clean

energy production technologies. Springer, Singapore. https://doi.org/10.1007/978-981-16-9356-4_9
Pandey AS, Bawiskar D, Wagh V (2024) Nanocosmetics and skin health: a comprehensive review of nanomaterials in cosmetic formulations. Cureus 16(1):e52754. https://doi.org/10.7759/cureus.52754. PMID: 38389646; PMCID: PMC10882253
Patel R, Kumar S (2023) Environmental applications of nanotechnology: pollution control and water treatment. Environ Nanotechnol 18(1):25–42. https://doi.org/10.1016/j.envnano.2023.100234
Pellico J, Ellis CM, Davis JJ (2019) Nanoparticle-based paramagnetic contrast agents for magnetic resonance imaging. Contrast Media Mol Imaging 2019:1845637. https://doi.org/10.1155/2019/1845637. PMID: 31191182; PMCID: PMC6525923
Ramesh M, Janani R, Deepa C, Rajeshkumar L (2023) Nanotechnology-enabled biosensors: a review of fundamentals, design principles, materials, and applications. Biosensors 13(1):40
Rani NU, Sharma P, Sharma RK, Shanmuga Selvan KCR, Pal SS (2024) The role of nanotechnology in environmental remediation opportunities and challenges. 6(10):4360–4380. https://doi.org/10.33472/AFJBS.6.10.2024.4359-4380
Roco MC (2003a) Broader societal issues of nanotechnology. J Nanopart Res 5(3–4):181–189
Roco MC (2003b) Broader societal issues of nanotechnology. J Nanopart Res 5(3-4):181–189
Roco MC (2011) The long view of nanotechnology development: the National Nanotechnology Initiative at 10 years. J Nanopart Res 13:427–445. https://doi.org/10.1007/s11051-010-0192-z
Roco MC, Bainbridge W (eds) (2001) Societal implications of nanoscience and nanotechnology. Springer (formerly Kluwer Academic Publishers), Boston
Royal Society & Royal Academy of Engineering (2004) Nanoscience and nanotechnologies: opportunities and uncertainties. The Royal Society, London
Schmid G (1992) Large clusters and colloids. Metals in the embryonic state. Chem Rev 92(8): 1709–1727
Singh A, Sharma R (2023) Applications of nanotechnology in agriculture: enhancing crop production and protection. Agric Nanotechnol Rev 12(3):123–140. https://doi.org/10.1016/j.agnanotech.2023.101234
Smith J, Johnson L, Brown M (2023) Nanotechnology in energy: advances in storage and generation. Energy Nanotechnol 22(1):80–98. https://doi.org/10.1016/j.energynano.2023.100123
Somorjai GA, Park JY (2008) Molecular surface chemistry by scanning tunneling microscopy and sum frequency generation vibrational spectroscopy. Chem Soc Rev 37(9):2155–2162
Taghavi SM, Momenpour M, Azarian M, Ahmadian M, Souri F, Taghavi SA, Sadeghain M, Karchani M (2013) Effects of nanoparticles on the environment and outdoor workplaces. Electron Physician 5(4):706–712
Taniguchi N (1974) On the basic concept of 'Nano-Technology'. Proceedings of the international conference on production engineering. Tokyo, Part II, Japan Society of Precision Engineering
Tawkif S (2014) Detection: from electrochemistry to spectroscopy with chromatographic techniques, recent trends with nanotechnology. Detection 2(4):27–32
Tharkar P, Varanasi R, Wong WSF, Jin CT, Chrzanowski W (2019) Nano-enhanced drug delivery and therapeutic ultrasound for cancer treatment and beyond. Front Bioeng Biotechnol 7:324
Tiple AD, Badwaik VJ, Padwad SV, Chaudhary RG, Singh NB (2020) A review on nanotoxicology: aquatic environment and biological system. Mater Today Proc 29:1246–1250
Wang X, Liu Z, Zhao Q (2023) Advances in nanoelectronics: materials and devices. Nano Electron Rev 11(2):100–115. https://doi.org/10.1016/j.nanorev.2023.100115
Xia B, Chen B, Sun X, Qu K, Ma F, Du M (2015) Interaction of TiO2 nanoparticles with the marine microalga Nitzschia closterium: growth inhibition, oxidative stress and internalization. Sci Total Environ 508:525–533
Zhao Y, Wang Q, Chen J (2023) Nanomaterials for advanced composites and catalysis. Mater Sci Eng Rep 75:55–72. https://doi.org/10.1016/j.mser.2023.100456

Chapter 5
Green Nanotechnology: Status, Advantages, and Existing Limitations

Maryam Kainat, Muhammad Sajjad Ahmad, Hafiz Abdullah Shakir, Muhammad Khan, Yuliya Kulikova, Olga Babich, Marcelo Franco, and Muhammad Irfan

Abstract Green nanotechnology integrates the principles of sustainability with precision of nano science. It offers environment friendly solutions across multiple sectors. These sectors include healthcare, agriculture, energy and environmental remediation. This approach prioritizes low toxicity, energy efficiency, and biodegradable nanomaterials. So, it emphasizes biological and biotechnological perspectives. For this purpose, nanomaterials are synthesized through plant extracts, microbial systems, enzymes, and renewable bioresources. These biosynthetic methods align closely with goals of circular economy and sustainable development. These enable applications ranging from nano-based wastewater treatment to biocompatible drug delivery systems. Despite its growing relevance, green nanotechnology faces various challenges. These include variability in bio-based synthesis, limited scalability, regulatory problems, and knowledge gaps across interconnected disciplines. Advanced tools from biotechnology are emerging to enhance reproducibility, pathway prediction, and material customization. Recent developments demonstrate its increasing industrial adoption and international support. But broader implementation requires coordinated research, policy innovation, and interdisciplinary collaboration. Green nanotechnology stands at a critical juncture. It can contribute to safe material design, sustainable innovation, and long-term ecological and human health outcomes.

Keywords Green technology · Nanotechnology · Current status · Advantages

M. Kainat · M. S. Ahmad · M. Irfan (✉)
Department of Biotechnology, University of Sargodha, Sargodha, Pakistan
e-mail: irfan.ashraf@uos.edu.pk

H. A. Shakir · M. Khan
Institute of Zoology, University of the Punjab New Campus, Lahore, Pakistan

Y. Kulikova · O. Babich
Institute of Living Systems, Immanuel Kant BFU, Kaliningrad, Russia

M. Franco
Department of Exact Science, State University of Santa Cruz, Ilheus, Brazil

N. Srivastava (ed.), *Sustainable Nanostrategies for Biorefineries*, Clean Energy Production Technologies, https://doi.org/10.1007/978-981-95-5875-9_5

5.1 Introduction

Nanotechnology is the science of manipulating matter at the atomic and molecular scale. It typically involves sizes below 100 nm. It has emerged as one of the most transformative technology of the twenty-first century. Nanotechnologies contribute to almost every field of science, including physics, materials science, chemistry, biology, computer science, and engineering (Bayda et al. 2019). It has revolutionized these fields by enabling the design and engineering of materials with novel properties. These properties are increased surface area, enhanced reactivity, and unique optical or electronic behavior. The fields include medicine, energy, agriculture, electronics, and environmental science (Haleem et al. 2023; Soni et al. 2024). However, concerns regarding their environmental and health impacts rise as the production and application of nanomaterials increase. This has led to the emergence of green nanotechnology. It is a sub-discipline that integrates the principles of sustainability with the innovations of nanoscience (Narayanan and Bhaskar 2024).

Green nanotechnology includes the development of nanomaterials and nanodevices through environmentally friendly methods. It involves the use of renewable resources and the minimization of energy consumption and toxic by-products. It is based on principles of green chemistry and green engineering. It also emphasizes waste reduction, safe chemical design, energy efficiency, and the use of nontoxic solvents and reagents (Alqarni et al. 2022). Conventional nanotechnology often relies on hazardous synthesis routes and persistent materials. On the other hand, green nanotechnology tries to harmonize technological advancement with ecological responsibility as shown in Fig. 5.1. Its goal is to mitigate the negative consequences of nanoscale materials and also to actively contribute to environmental

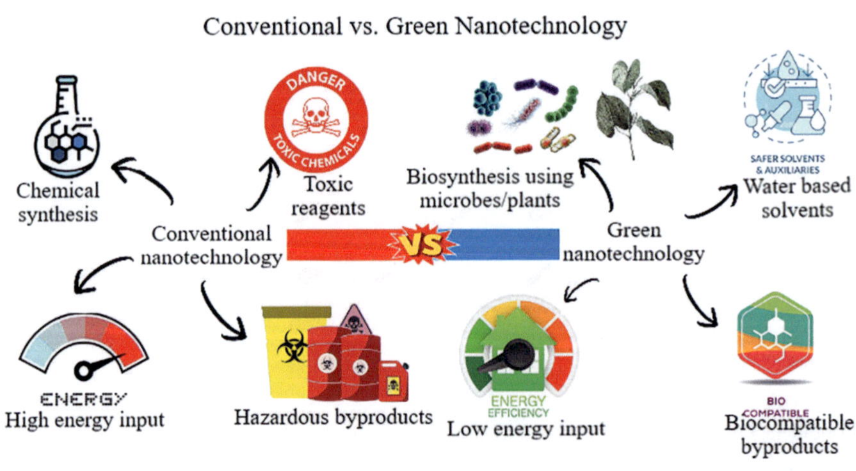

Fig. 5.1 A comparison between conventional nanotechnology and green nanotechnology showing the benefits of green approach

remediation and sustainable development (Maksimović and Omanović-Mikličanin 2017).

Sustainability has become a central concern in the design of modern technologies. The growing pressures of climate change, environmental degradation, and biodiversity loss and resource depletion require the development of cleaner, safer, and more energy-efficient systems. Emerging technologies are evaluated for their performance and also for their life cycle impacts, carbon footprint, and contribution to sustainable goals. In this context, green nanotechnology plays a pivotal role (Khan 2019). It offers a pathway to synthesize functional nanomaterials using low impact methods and to use them in ways that address key sustainability challenges, for example, purification of polluted water, enhancement of crop production, and delivering of targeted therapeutics with minimal side effects (Afonso et al. 2024).

Green nanotechnology has an interdisciplinary foundation. It covers advances in chemistry for reaction pathways and material characterization. It includes biology and biotechnology for biosynthesis using microorganisms, enzymes, and plant extracts. It also covers materials science for functionalizing nanomaterials for specific applications. Moreover, it includes environmental engineering for deploying nanomaterials in remediation and pollution control (Srivastava and Bhargava 2021). Biosynthetic approaches using bacteria, fungi, or plant metabolites have demonstrated not only lower toxicity and energy demands but also enhanced biocompatibility of the formed nanomaterials. These biological platforms show how biotechnology can be used to advance the goals of green nanotechnology without compromising efficacy (Nain et al. 2025).

Moreover, the relevance of green nanotechnology is not confined to laboratory innovation. It has implications across several sectors facing global sustainability pressures. In climate change mitigation, nanomaterials are being explored for use in carbon capture systems, efficient energy storage, and lightweight materials for renewable energy devices (Elezz et al. 2025). In healthcare, green-synthesized nanocarriers are improving the safety and precision of drug delivery and diagnostic systems (Khafaga et al. 2023). In agriculture, nanofertilizers and nanopesticides produced via green methods enable controlled nutrient release and reduced environmental contamination (Kekeli et al. 2025). And in pollution control, nanomaterials serve as effective agents for removing heavy metals, organic pollutants, and pathogens from soil, air, and water (Ali et al. 2023).

The objective of this chapter is to provide a comprehensive overview of the current status, advantages, and limitations of green nanotechnology. It particularly focuses on the sustainable synthesis methods and their practical applications; discusses guiding principles and frameworks that define green nanotechnology; and examines the ecofriendly synthesis strategies, both chemical and biological, in detail. Next it explores the broad spectrum of green nanotechnology applications across environmental, agricultural, biomedical, and energy sectors. Key advantages include reduced toxicity, enhanced biodegradability, and alignment with circular economy models. Limitations include scalability challenges, regulatory problems, and gaps in eco-toxicological data. In summary, green nanotechnology represents a promising and responsible pathway forward. High performance materials can be developed without compromising the ecological integrity of our planet. Green nanotechnology can shape a more sustainable technological future.

5.2 Principles and Frameworks of Green Nanotechnology

Green nanotechnology operates at the intersection of nanoscale innovation and sustainable practice. It is guided by established frameworks such as green chemistry, green engineering, and life cycle assessment (LCA) (Pati et al. 2014). These principles provide a scientific and ethical foundation for designing nanomaterials and nanoprocesses that are not only functionally effective but also environmentally benign and biologically compatible (Khan 2019).

5.2.1 Green Chemistry and Green Engineering Principles

Green chemistry focuses on the design of chemical products and processes that reduce or eliminate the use and generation of hazardous substances. This includes principles such as designs for energy efficiency, use of renewable feedstocks, avoidance of toxic reagents, and enhancement of atom economy. Atom economy is the concept according to which all atoms in a reaction should ideally be incorporated into the final product. In biological contexts, these principles align closely with the goals of biocompatibility, biosafety, and low impact synthesis (Ahluwalia and Kidwai 2004; Srivastava and Bhargava 2021). On the other hand, green engineering extends these ideas to process and system design. It focuses on resource efficiency, minimization of life cycle impacts, and the integration of environmental and health considerations into engineering of nanomaterials. For example, applications such as biosensors, drug delivery systems, or nano-biocomposites involve green engineering principles. These promote processes that are scalable, nontoxic, and compatible with living systems (García-Serna et al. 2007).

5.2.2 Atom Economy, Waste Minimization, and Benign Solvents

Atom economy is among the core metrics of green nanotechnology. It evaluates the efficiency of reactions. It measures that how well starting materials are converted into useful products. For biologically derived nanomaterials, this principle is particularly relevant when using enzymes, microbial pathways, or plant extracts. They often operate under mild conditions with minimal by-products (Singh et al. 2018). Similarly, waste minimization is important in both laboratory and industrial settings.

It reduces the environmental footprint of nanomaterial production. Many conventional nanoparticle synthesis methods generate hazardous waste or they require purification steps which are harsh. In contrast, biological synthesis routes often occur in aqueous media. They occur at ambient temperatures and produce fewer harmful by-products (Singh et al. 2023). The use of benign solvents is another essential component of green nanotechnology. It usually involves water or bio-based alternatives, e.g., plant oils and ethanol. These solvents support enzyme stability, cell viability, and biological compatibility. It makes them ideal for nanomaterials that are to be used for environmental or medical purposes (Shanker et al. 2016).

5.2.3 Safe by Design and Life Cycle Assessment

Safe by design approach has become a central strategy to ensure that nanomaterials do not cause unforeseen hazards on ecology or health. It involves assessing the toxicological profile, environmental impact, and bioaccumulation potential of nanomaterials during the early stages of development. This principle is important when designing materials for interaction with cells, tissues, or ecosystems (Schmutz et al. 2020). Life cycle assessment (LCA) complements safe by design principle. It offers a quantitative tool to evaluate the environmental impact of nanomaterials across their life cycle. It evaluates their impact from raw material extraction and synthesis to use, disposal, and degradation. LCA supports decision-making in research and policy. It identifies stages with the greatest environmental burden and opportunities for sustainable improvement (Ke et al. 2025).

5.3 Green Synthesis of Nanomaterials

The green synthesis of nanomaterials lies at the core of sustainable nanotechnology. It reflects a growing shift away from traditional methods that rely on toxic chemicals and high energy inputs. On the other side, the approach of green synthesis of nanomaterials aligns with the principles of green chemistry. It promotes the use of environmentally benign substances, renewable feedstocks, and minimum energy consumption. Green synthesis offers a viable path for biotechnology and life science applications. It helps toward the development of functional nanomaterials that are not only effective but also biocompatible and ecofriendly (Naikoo et al. 2021).

5.3.1 Physical and Chemical Approaches

Conventional methods of nanoparticle synthesis include sol gel processing, chemical vapor deposition, thermal decomposition, and microwave-assisted and ultrasound-mediated techniques. These methods have been extensively used for their ability to control the shape, size, and crystallinity of nanomaterials (Abady et al. 2025). But these methods frequently involve toxic solvents and hazardous reducing agents, for example, use of sodium borohydride and hydrazine. They also demand high temperature and pressure conditions. Efforts are being put to "green" these techniques, for example, by using ethanol or water as solvents or reduce reaction times through microwave heating. But still their sustainability remains limited (Vinukonda et al. 2025). Moreover, they are often cost intensive and energy demanding. Also the purification of the final product can be challenging. These limitations have prompted increased interest in biological and bioinspired methods. Biological methods offer safer, simpler, and more sustainable alternatives for nanomaterial synthesis. They are also especially useful in the context of biological applications (Abady et al. 2025).

5.3.2 Biological Approaches

Biological synthesis methods are gaining prominence. They are viable alternatives for making nanomaterials in a sustainable and environmentally responsible manner. These approaches use the metabolic and enzymatic capabilities of biological entities. These include use of plants, microbes, enzymes, and biopolymers. They also utilize the structural advantages of biopolymers and bio-derived compounds. Such systems provide naturally occurring reducing agents and stabilizers as shown in Fig. 5.2. These allow the controlled formation of nanoparticles under mild conditions without the use of toxic chemicals (Shah et al. 2025).

5.3.2.1 Plant-Mediated Synthesis or Phyto-Nanotechnology

Plant-mediated synthesis is often referred to as phyto-nanotechnology. It uses aqueous extracts derived from leaves, stems, roots, bark, or fruits, which are natural sources of reducing and stabilizing agents (Noah and Ndangili 2022). These plant extracts are rich in secondary metabolites and include flavonoids, tannins, saponins, alkaloids, and terpenoids. They facilitate the bioreduction of metal ions and stabilization of the formed nanoparticles (El-Seedi et al. 2019). Various plants such as *Azadirachta indica* (neem), *Ocimum sanctum* (tulsi), and *Camellia sinensis* (green tea) have been successfully employed for synthesis of silver, gold, and zinc oxide nanoparticles. ZnO nanoparticles have been prepared with a great variety of plant leaf extracts such as *Coriandrum sativum* (coriander), *Calotropis gigantean* (crown flower), *Acalypha indica* (copper leaf), *Hibiscus rosa-sinensis* (China rose), and

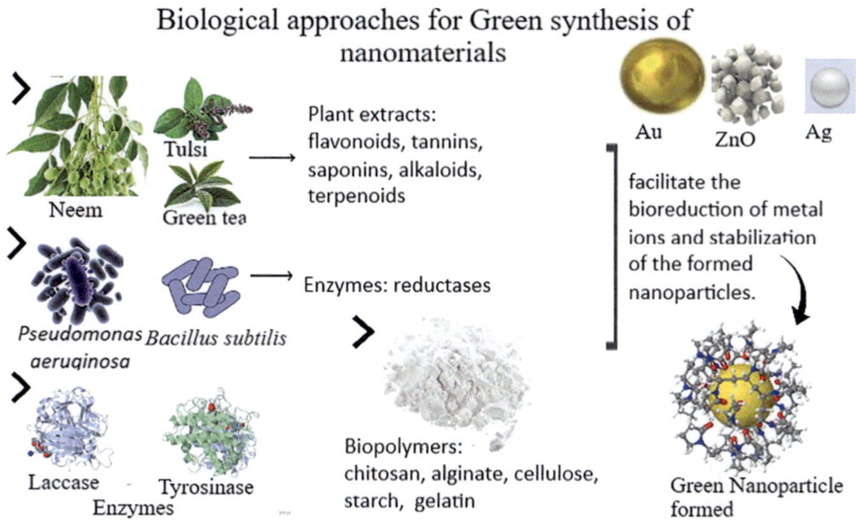

Fig. 5.2 Biological sources of reducing agents and stabilizers for green synthesis of nanomaterials

Camellia sinensis (green tea) (Singh et al. 2018). This method is useful for its simplicity, cost effectiveness, and environmental friendly nature. The reason is that it avoids high temperatures, pressures, or toxic reagents. However, challenges also remain, which include variability in phytochemical composition and the need for mechanistic understanding (Jiang et al. 2021).

5.3.2.2 Microbial Synthesis

Microbial synthesis is another promising approach in green nanotechnology. It utilizes the inherent ability of microorganisms to reduce metal ions either intracellularly or extracellularly. Bacterial species such as *Pseudomonas aeruginosa*, *Bacillus subtilis*, and *Escherichia coli* have demonstrated nanoparticle synthesis through enzyme-mediated reduction mechanisms. These mechanisms involve reductases and electron shuttling compounds (Pan et al. 2025). Fungi like *Fusarium oxysporum* and *Penicillium chrysogenum* are particularly effective. They have high metal ion tolerance and the ability to secrete large quantities of extracellular enzymes. These enzymes facilitate nanoparticle formation. Fungi, particularly species belonging to Ascomycota and Basidiomycota, respond to a spectrum of metal and metalloid precursors, including organic compounds and salts. The uptake and transport of these precursors within fungal cells via different proteins and channels set the stage for subsequent biochemical transformations. In the case of gold and silver NP synthesis, fungi often encounter precursor compounds such as gold chloride or silver nitrate. These ions are assimilated into the fungal biomass, triggering a cascade of biochemical reactions (Rami et al. 2024). Actinomycetes, especially those belonging to the genus *Streptomyces*, are capable of producing metallic and metal oxide nanoparticles (Manimaran and Kannabiran 2017). These microbes support ecofriendly synthesis and also control the shape and size control of nanoparticles. This is achieved by optimizing the culture media and process parameters, thereby finetuning the synthesis process to produce nanoparticles with desired characteristics. This makes them ideal for further genetic engineering to enhance biosynthesis (Adeleke et al. 2024).

5.3.2.3 Enzyme-Mediated Synthesis and Biocatalysis

Enzyme-mediated nanoparticle synthesis involves the use of isolated enzymes. These enzymes include nitrate reductase, laccase, tyrosinase, and peroxidase. They catalyze the reduction of metal ions into their nanoparticulate forms (Talekar et al. 2016) and enable synthesis under ambient temperature and pressure. They also offer a high degree of specificity and yield nanoparticles with uniform morphology. As biocatalysts, enzymes minimize unwanted side reactions and also reduce energy input requirements (Sen et al. 2024). Although enzyme-based synthesis methods are clean and efficient, they face certain limitations—the methods are expensive and have a risk of denaturation or loss of activity over time (Seth and Meena 2024).

5.3.2.4 Use of Biopolymers, Natural Extracts, and Renewable Feedstock

Biopolymers and renewable feedstock offer a biologically compatible and sustainable platform for synthesis of green nanoparticle. Natural polymers such as chitosan, alginate, cellulose, starch, and gelatin are biodegradable and nontoxic. They act as stabilizing and reducing agents (Fakayode et al. 2016) and help prevent nanoparticle agglomeration. They also improve dispersion and enhance surface properties for biomedical applications (Kučuk et al. 2023). Additionally, renewable raw materials are increasingly being utilized as cost-effective and sustainable feedstock for nanoparticle production. These raw materials include agricultural waste, food processing by-products, and microbial biomass. Their use contributes to the principles of the circular economy as they convert waste into high-value functional nanomaterials (Kadhim et al. 2025).

5.3.2.5 Role of Synthetic Biology and Genetic Engineering

The application of synthetic biology and genetic engineering in green nanotechnology enables the design of microorganisms with enhanced capabilities of nanoparticle synthesis. Genetic modifications can lead to overexpression of specific metal-binding peptides. They can modify biosynthetic enzymes or metabolic pathways that facilitate nanoparticle formation (Ji et al. 2024). For example, engineered strains of *Escherichia coli* can express phyto chelatin synthase, which have been used to produce cadmium-based quantum dots. Also genetically modified *Pseudomonas putida* strains have increased resistance to silver ions. So, they have improved nanoparticle production efficiency. These biotechnological tools are paving way for customized microbial factories that are capable of producing nanomaterials with tailored properties. They have diverse applications in medicine, agriculture, and environmental remediation (Valenzuela-Ibaceta et al. 2025).

5.3.2.6 Advantages of Bio-based Methods

Bio-based synthesis methods offer numerous advantages over traditional physical and chemical-based methods. They operate under mild reaction conditions and eliminate the need for toxic reagents. They produce nanoparticles that are often functionalized with biological molecules enhancing their stability and activity. These methods are ecofriendly and cost effective. They are also highly scalable, which makes them attractive for industrial applications. These bio-synthesized nanoparticles are compatible with living systems (Table 5.1), which further increases their potential for use in biomedical, pharmaceutical, and agricultural fields. It also aligns with the broader goals of green and sustainable development (Shah et al. 2025).

Table 5.1 Biological Sources for Nanoparticle Synthesis: Examples and Mechanisms

Biological source	examples	Mechanism of synthesis of nanoparticles	Type of nanoparticles synthesized	References
Plants	*Azadirachta indica, Camellia sinensis*	Phytochemicals act as reducing and stabilizing agents	Ag, Au, ZnO, CuO	Halder et al. (2025), Luque et al. (2021)
Bacteria	*Bacillus subtilis, Pseudomonas aeruginosa*	Enzyme-mediated reduction (nitrate reductase, etc.)	Ag, Pd, Fe_3O_4, Se	Alsamhary (2020)
Fungi	*Fusarium oxysporum, Penicillium chrysogenum*	Extracellular enzymes and proteins mediate synthesis	Au, Ag, TiO_2	Bakhtyar et al. (2024)
Actinomycetes	*Streptomyces* spp.	Secretion of bioactive compounds and metal ion reducers	Ag, ZnO, Au	Karthik et al. (2014)
Enzymes	Nitrate reductase, laccase, tyrosinase	Direct catalytic reduction of metal ions	Au, Ag, CuO	Sanket and Das (2021)
Biopolymers	Chitosan, starch, alginate	Structural matrix and stabilizing function	Various (used as support/capping agents)	Kučuk et al. (2023)
Synthetic biology	Genetically modified *E. coli, Pseudomonas*	Engineered proteins or peptides for controlled synthesis	Tailored nanoparticles (e.g., quantum dots)	Jewett and Patolsky (2013)

5.4 Applications of Green Nanotechnology

Green nanotechnology has opened new frontiers across a wide range of industries. It integrates sustainability and nanoscale innovation. Its interdisciplinary nature allows ecofriendly nanomaterials to be applied in various fields ranging from environmental protection and agriculture to medicine and renewable energy. These applications reduce ecological burdens. They also enhance efficiency, precision, and safety in resource intensive processes. This section highlights the key areas where green nanotechnology has a transformative impact (Lu and Ozcan 2015).

5.4.1 *Environmental Applications*

Environmental remediation is one of the most promising areas of green nanotechnology. Green-synthesized nanoparticles have shown excellent capabilities in wastewater treatment. They are used particularly for the removal of heavy metals, which include lead, cadmium, and mercury (Neeti et al. 2023). Nanoparticles synthesized from biological agents are effective adsorbents. For example, fungal or plant-based silver and iron oxide nanoparticles exhibit high adsorption capacities. It is due to their high surface area and reactive functional groups. Additionally, nanomaterials are being used in photocatalytic degradation of organic pollutants. They aid in the purification of water and air. Soil decontamination has also been benefited from nano-enabled solutions. These applications are shown in Fig. 5.3. Biofriendly nanoparticles immobilize or degrade harmful compounds without altering the

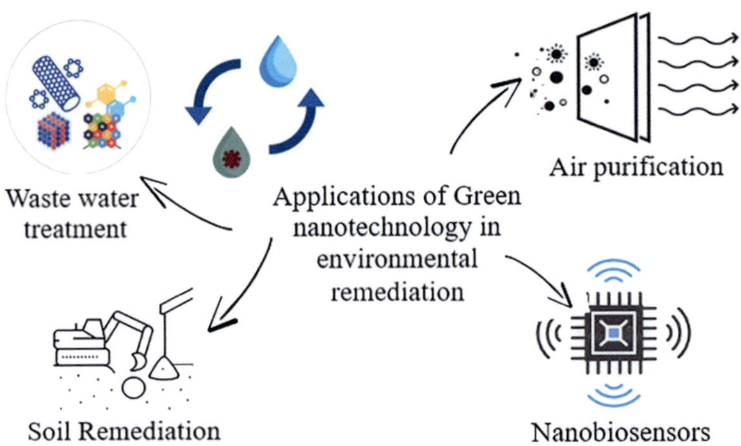

Fig. 5.3 Promising applications of green nanotechnology in environmental remediation

natural microbial flora (Roy et al. 2021). Furthermore, biosensors based on green nanomaterials have been developed for the real-time monitoring of environmental contaminants. These contaminants include pesticides, nitrates, and heavy metals. This enables early detection and informed decision-making, facilitating prompt action to mitigate environmental risks (Mohamed and Awad 2022).

5.4.2 Agricultural Applications

Green nanotechnology offers sustainable solutions for improving agricultural productivity and minimizes chemical load. Nanofertilizers and nanopesticides produced via biological synthesis methods provide enhanced nutrient delivery and targeted pest control and reduced environmental runoff and toxicity. These smart formulations ensure prolonged availability of active ingredients. They also decrease the frequency and quantity of applications needed (Kumar et al. 2025). Additionally, green nanocarriers are employed for controlled and stimuli responsive release of agrochemicals as shown in Fig. 5.4. They improve plant uptake and minimize losses. Certain nanoparticles, such as zinc oxide or silicon-based materials, can enhance soil structure, microbial health, and nutrient cycling. They contribute to long-term agricultural sustainability (Bansal et al. 2025). Moreover, bioengineered nanomaterials are being explored for seed coating and genetic transformation. They further optimizes crop traits without compromising ecological safety (Ijaz et al. 2023).

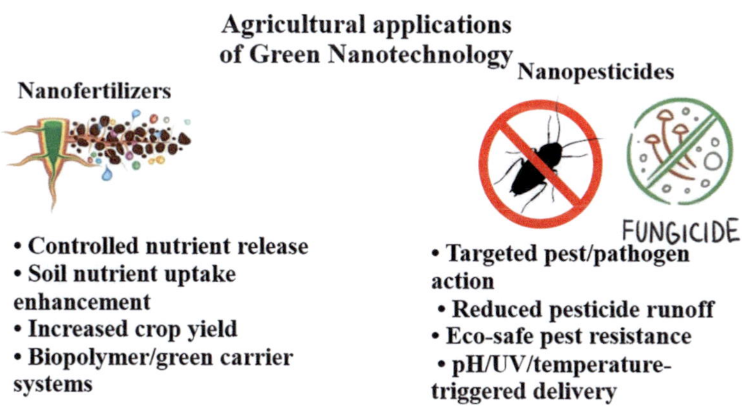

Fig. 5.4 Promising applications of green nanotechnology in agriculture domain. The advantages of nanofertilizers and nanopesticides are shown

5.4.3 Biomedical and Pharmaceutical Applications

Green-synthesized nanomaterials are biocompatible and they have minimum toxicity. It makes them ideal for medical and pharmaceutical applications. Plant- and microbial-derived nanoparticles have been extensively explored as drug and gene delivery carriers due to their surface modifiability, nanoscale size, and ability to cross biological barriers (Karnwal et al. 2024). Silver, gold, and iron oxide nanoparticles synthesized through green methods show antimicrobial, antioxidant, and anticancer properties. Their surface can be functionalized with biomolecules such as proteins, antibodies, or peptides, which is done to achieve targeted therapy. So, it reduces side effects and enhances therapeutic efficiency (Laib et al. 2025). Green nanomaterials are also being applied in tissue engineering. They serve as scaffolds that support cellular growth and regeneration. Additionally, diagnostic platforms incorporate green nanotechnology. For example, biosensors are being developed for early disease detection. They are especially useful in infectious and cancerous conditions (Hemdan et al. 2025).

5.4.4 Renewable Energy and Industrial Applications

Green nanotechnology contributes significantly to the energy and industrial sectors. It helps in the development of clean and efficient systems. Biosynthesized nanomaterials have been successfully integrated into photovoltaic devices such as dye-sensitized solar cells, which improves light absorption and energy conversion efficiency (Zhu et al. 2015). In hydrogen production, biogenic metal and metal oxide nanoparticles act as catalysts in photoelectrochemical and thermocatalytic reactions. They facilitate sustainable fuel generation (Bakranova and Nagel 2023). Similarly, bio-based nanocatalysts are employed in the transesterification of oils for biodiesel production (Fig. 5.5). They reduce the reliance on fossil fuels (Shaik and Kumari 2025). In the realm of energy storage, green nanomaterials are being used to develop high-performance electrodes for supercapacitors and lithium ion batteries. This enhances charge and discharge cycles, while also reducing toxic byproducts (Asamoah et al. 2024). Moreover, industries benefit from ecofriendly nanocatalysts in chemical synthesis, polymer production, and waste minimization. It aligns with goals of circular economy (Saxena et al. 2025).

5.5 Advantages of Green Nanotechnology

Green nanotechnology offers a transformative approach to material synthesis and application (Table 5.2) by prioritizing environmental safety and long-term sustainability. It is a sub-discipline grounded in the principles of green chemistry and

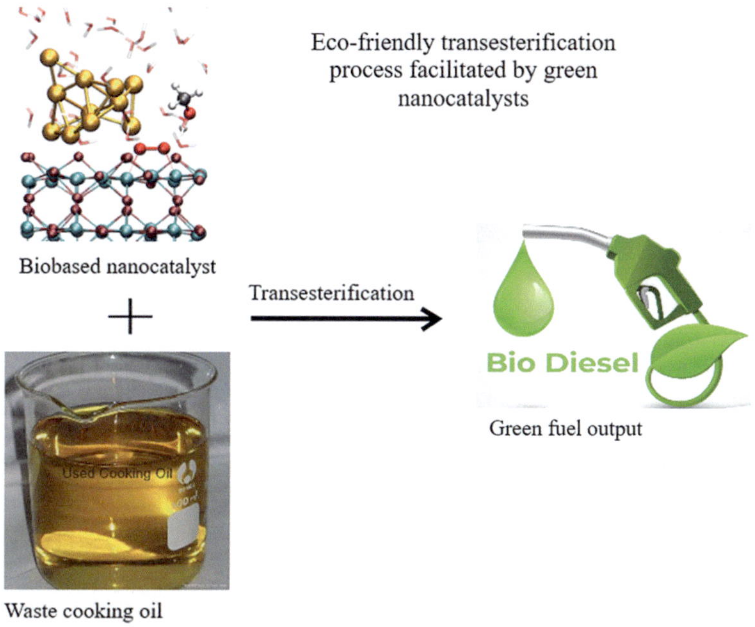

Fig. 5.5 Applications of green nanotechnology in energy sector; green nanomaterials are used to convert waste oil into biodiesel for sustainable fuel production

Table 5.2 Advantages of green nanotechnology with examples, aligned with Sustainable Development Goals (SDGs): SDGs are part of the 2030 Agenda for Sustainable Development

Advantage	Example	Aligned SDGs	References
Reduced toxicity and environmental impact	Silver nanoparticles synthesized from *Azadirachta indica* (neem) extract reduce cytotoxicity and maintain antimicrobial activity	SDG 3 (good health), SDG 15 (life on land)	Asimuddin et al. (2020)
Energy efficient and sustainable production	Gold nanoparticles synthesized using *Pseudomonas aeruginosa*, under ambient conditions, reduce energy use	SDG 12 (responsible consumption), SDG 13 (climate action)	Husseiny et al. (2007)
Biodegradability and biocompatibility	Zinc oxide nanoparticles from *Aloe vera* extract, used for wound healing due to high biocompatibility	SDG 3 (good health), SDG 9 (industry, innovation)	Hou et al. (2021)
Cost effectiveness	Iron nanoparticles synthesized using banana peel extract offer low-cost water treatment options	SDG 6 (clean water), SDG 1 (no poverty)	Ondiek (2016)
Enhanced public and regulatory acceptance	Plant-derived nanoparticles in cosmetics and drug delivery are considered safer and more natural	SDG 12 (responsible consumption), SDG 3 (good health)	Karnwal et al. (2024)
Alignment with circular economy and SDGs	Nanocellulose from agrowaste, used in biodegradable packaging and water filtration systems	SDG 12, SDG 6, SDG 13	Padhi et al. (2023)

engineering. This approach promotes the design and development of nanomaterials that are functional and effective and also environmentally benign and economically viable (Dhingra et al. 2010).

5.5.1 Reduced Toxicity and Environmental Impact

Green nanotechnology has the ability to minimize the ecological footprint of nanomaterial production and their use. It is a central advantage of green nanotechnology (Khan 2019). Traditional chemical methods of nanomaterial synthesis often involve toxic solvents and require high energy inputs. They also produce hazardous by-products that can persist in the environment and pose risks to ecosystems and human health (Hachem et al. 2022). In contrast, green synthesis uses biological agents such as plant extracts, bacteria, fungi, and enzymes. They are used to produce nanoparticles in aqueous systems and at ambient conditions (Malik et al. 2023). For instance, silver nanoparticles are synthesized using *Azadirachta indica* (neem) extracts. They have demonstrated antimicrobial efficacy. They exhibit significantly lower cytotoxicity compared to chemically synthesized counterparts. Such reductions in toxicity benefit environmental systems. They also enhance the safety profile of nanomaterials used in biomedical, agricultural, and environmental applications (Chinnasamy et al. 2021).

5.5.2 Energy Efficient and Sustainable Production

Green nanotechnology focuses on low-energy and resource-conserving methods of synthesis. This makes it more sustainable approach. Biological processes generally operate under mild temperatures and neutral pH, which eliminates the need for high-temperature furnaces, vacuum systems, or corrosive chemicals. This result in lower energy demand and reduces greenhouse gas emission (Rathod et al. 2024). For example, gold nanoparticles are synthesized using *Pseudomonas aeruginosa*, which requires minimal energy input compared to conventional chemical reduction methods. Those methods often necessitate strong reducing agents and thermal treatment. This energy efficiency is particularly important in the context of global climate goals. It is also important in the context of carbon neutral technologies (Moradi et al. 2025).

5.5.3 Biodegradability and Biocompatibility

Nanomaterials synthesized through green methods exhibit enhanced biodegradability and biocompatibility. These attributes are critical for biomedical and environmental applications (Hosseingholian et al. 2023). Biogenic nanoparticles often carry organic surface capping agents. These are derived from their biological source. Examples of organic capping agents include polyphenols, proteins, or polysaccharides. These can facilitate their breakdown in natural systems and reduce bioaccumulation. Furthermore, such surface modifications improve compatibility with living tissues. It makes green nanomaterials highly suitable for drug delivery, tissue engineering, and wound healing (Sidhu et al. 2022). For example, zinc oxide nanoparticles synthesized using *Aloe vera* extracts have shown promise as wound healing agents. It is due to their favorable interaction with skin cells and accelerated tissue regeneration properties. ZnO nanoparticles also show antibacterial capabilities. These nanostructures have been more harmful to bacteria and less reactive toward human cells. This is because they have an excellent antibacterial mechanism based on the production of reactive oxygen species (ROS) that kills bacterial cells (Al-darwesh et al. 2024; El-Saadony et al. 2024).

5.5.4 Cost-Effectiveness

Green nanotechnology holds promise for resource constrained settings. It is due to its cost-effective synthesis routes. It utilizes locally available biomass, agricultural waste, or microbial cultures. This eliminates the need for expensive laboratory reagents and infrastructure (Bahrulolum et al. 2021). For instance, banana peel and tea waste have been successfully used in the synthesis of iron and gold nanoparticles, respectively. It offers a low-cost alternative to traditional methods. This aspect of nanotechnology can empower developing regions to produce their own nano-enabled materials. They can use them in water purification, crop protection, and medical diagnostics. It reduces the need for heavy financial investment (Ashok et al. 2010; Ismael et al. 2023).

5.5.5 Enhanced Public and Regulatory Acceptance

There is growing public concern over the safety of nanomaterials. It has created a need for technologies that align with society and regulatory expectations. Green nanotechnology prioritizes safety and sustainability. It often faces fewer barriers to acceptance and commercialization (Khatoon and Velidandi 2025). Biologically synthesized nanoparticles are produced using food-grade organisms or plant extracts. So, they are perceived as more "natural" and less hazardous. That's why they are gaining easier entry into markets such as pharmaceuticals, cosmetics, and agriculture (Kumar 2022). Moreover, there are regulatory frameworks such as REACH (Registration, Evaluation, Authorization and Restriction of Chemicals) in the EU and EPA guidelines in the USA. These favor processes that demonstrate a reduced environmental impact. Green nanotechnology often fulfills these criteria (Naryal et al. 2025; Williams et al. 2009).

5.5.6 *Alignment with Circular Economy and Sustainable Development Goals (SDGs)*

Green nanotechnology aligns with the principles of circular economy and the United Nations Sustainable Development Goals (SDGs). It is one of the most strategic advantages of green nanotechnology (Rajpal et al. 2025). Green nanotechnology supports a closed loop production system. It does so by employing renewable resources. It also reduces waste and enables recycling or biodegradation (Narayanan and Bhaskar 2024). For example, nanocellulose can be derived from agricultural residues and can be used to create biodegradable packaging materials. It reduces reliance on plastics and contributes to SDG 12 (Responsible Consumption and Production) and SDG 13 (Climate Action). Similarly, green nanomaterials used in water purification align with SDG 6 (Clean Water and Sanitation). Applications in precision agriculture and energy storage also support SDGs related to food security and clean energy (Elzein 2024).

5.6 Limitations and Challenges

Green nanotechnology presents advantages over conventional approaches. But it also has certain limitations and challenges (Fig. 5.6). It is a relatively new and evolving field and faces scientific, technical, regulatory, and interdisciplinary hurdles. These must be addressed to unlock its full potential. Understanding these limitations is critical for researchers, policymakers, and industries. It is necessary to transition them toward sustainable nanomaterial production and application (Khan 2019).

5.6.1 *Lack of Standardization and Reproducibility*

Lack of standardization in synthesis protocols is one of the most prominent limitations of green nanotechnology. Biological systems used in nanoparticle synthesis such as plant extracts, microbial cultures, or enzymes are variable. This is due to differences in growth conditions, species, age, and extraction processes (Ying et al. 2022). For example, the phytochemical composition of a plant extract used in gold nanoparticle synthesis may differ significantly between batches. It can lead to variations in nanoparticle size, shape, and surface chemistry. This inconsistency hinders reproducibility, which makes it difficult to compare results across studies (Hosseingholian et al. 2023). It is also difficult to scale up production reliably. Standardized methodologies are needed to establish a reproducible and validated knowledge base. It is also important to develop more rigorous reporting of biological sources and synthesis parameters (Manzoor et al. 2025).

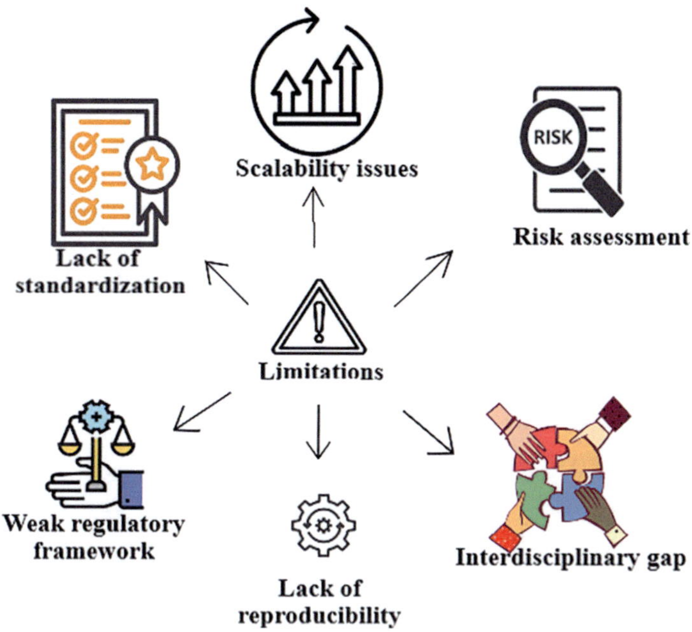

Fig. 5.6 Limitation of green nanotechnology that need to be addressed

5.6.2 Difficulties in Scaling up Bio-based Synthesis

Laboratory-scale biosynthesis of nanomaterials has shown promise. But scaling up these processes for commercial or industrial use remains a major challenge (Manzoor et al. 2025). Biological systems are often sensitive to environmental conditions. They may produce nanoparticles at lower yields or with lower uniformity compared to chemical methods (Stark 2011). Additionally, it can be technically demanding and cost intensive to maintain sterile conditions. It is also difficult to optimize bioreactor designs and ensure consistent biological activity on a larger scale (Mukherjee and Patra 2017). For example, *Fusarium oxysporum* has been used successfully to produce silver nanoparticles in lab settings. But it is difficult to scale up such microbial systems. There is need for biotechnological infrastructure and expertise. These may not be readily available in all regions or sectors (Nasr Azadani et al. 2024).

5.6.3 Long-Term Eco-toxicological Effects

Green nanotechnology follows ecofriendly synthesis routes. But many green nanomaterials still possess unknown long-term effects on ecosystems and human health (Hosseingholian et al. 2023). Nanoparticles may undergo transformation after they are released into the environment. These transformations may be oxidation, agglomeration, or degradation. This alters their reactivity and toxicity. Even biologically synthesized nanoparticles can accumulate in aquatic or terrestrial systems. They can potentially affect microbial communities, plant growth, or food chains (Khan et al. 2021). For example, silver nanoparticles whether chemically or biologically produced have demonstrated antibacterial properties. These could disrupt beneficial soil microbes if not properly assessed. Comprehensive ecotoxicological studies are essential to evaluate the true sustainability of these materials. These include chronic exposure scenarios and life cycle assessments (LCA) (Ameen et al. 2021).

5.6.4 *Limited Regulatory Frameworks and Risk Assessment Tools*

There is rapid pace of innovation in green nanotechnology. It has surpassed the development of regulatory policies and risk assessment tools. Most current regulatory frameworks are not tailored specifically to nanomaterials. Very few account for the unique features of biologically synthesized nanoparticles. This regulatory ambiguity can create uncertainty for developers and industries. So, it becomes difficult to bring green nanomaterials to market. Moreover, there is lack of standardized toxicity testing protocols and lifecycle-based evaluation models. This makes it difficult to assess the safety and compliance of green nanomaterials (Subhan and Subhan 2022). For example, a green-synthesized zinc oxide nanoparticle used in sunscreen might evade regulatory scrutiny despite potential risks. It will occur if no specific guidelines exist for evaluating nanoscale active ingredients (Sharma et al. 2020).

5.6.5 *Interdisciplinary Gaps in Knowledge and Technology Transfer*

Green nanotechnology is inherently interdisciplinary. It requires integration of knowledge from multiple disciplines. These disciplines include biology, chemistry, materials science, engineering, environmental science, and policy (Bensaude-Vincent 2016). However, effective communication and collaboration across these fields remain limited. Many biologists and microbiologists lack training in nanomaterial characterization techniques. These include transmission electron microscopy (TEM) or zeta potential analysis (Aziz et al. 2025). Similarly, materials scientists may not fully grasp the complexities of biological systems used in synthesis (Acharya and Owens 2024). This gap slows down the development of robust and scalable technologies. It also slows the translation of laboratory innovations into real-world applications. Furthermore, knowledge transfer between academia and industry is often constrained by intellectual property concerns. It is also due to lack of funding and insufficient collaboration networks (Wang and Grainger 2014).

5.7 Conclusion

Green nanotechnology has emerged as a transformative and interdisciplinary field. It aligns advanced nanoscience with the principles of environmental sustainability, biological safety, and circular economy. It utilizes biological resources such as plants, microbes, enzymes, and biopolymers. For this reason, the synthesis of nanomaterials can be achieved through eco-friendly and energy-efficient routes in contrast to conventional chemical and physical methods. The applications of green nanotechnology are expanding across multiple sectors. These include agriculture, environmental remediation, biomedicine, and renewable energy. There are also some notable limitations. These include lack of standardization, difficulties in scaling up bio-based synthesis, and incomplete understanding of long-term eco-toxicological effects. These challenges hinder the widespread adoption of green nanotechnologies. Additionally, regulatory frameworks, interdisciplinary gaps, and technological integration require further development to ensure responsible and safe implementation. Despite these challenges, the future of green nanotechnology is promising. Continued research, collaboration, and innovation will be essential in advancing this field toward practical, scalable, and globally accepted solutions.

References

Abady MM, Mohammed DM, Soliman TN, Shalaby RA, Sakr FA (2025) Sustainable synthesis of nanomaterials using different renewable sources. Bull Natl Res Cent 49(1):24

Acharya R, Owens RM (2024) Reaching across the divide: materials scientists interfacing with biologists. Mater Horiz 11(21):5141–5146

Adeleke BS, Olowe OM, Ayilara MS, Fasusi OA, Omotayo OP, Fadiji AE, Onwudiwe DC, Babalola OO (2024) Biosynthesis of nanoparticles using microorganisms: a focus on endophytic fungi. Heliyon 10(21):e39636

Afonso IS, Cardoso B, Nobrega G, Minas G, Ribeiro JE, Lima RA (2024) Green synthesis of nanoparticles from olive oil waste for environmental and health applications: a review. J Environ Chem Eng 12(5):114022

Ahluwalia V, Kidwai M (2004) Basic principles of green chemistry. In: New trends in green chemistry. Springer, Berlin/Heidelberg, pp 5–14

Al-darwesh MY, Ibrahim SS, Mohammed MA (2024) A review on plant extract mediated green synthesis of zinc oxide nanoparticles and their biomedical applications. Results Chem 7:101368

Ali Q, Zia MA, Kamran M, Shabaan M, Zulfiqar U, Ahmad M, Iqbal R, Maqsood MF (2023) Nanoremediation for heavy metal contamination: a review. Hybrid Adv 4:100091

Alqarni LS, Alghamdi MD, Alshahrani AA, Nassar AM (2022) Green nanotechnology: recent research on bioresource-based nanoparticle synthesis and applications. J Chem 2022(1): 4030999

Alsamhary KI (2020) Eco-friendly synthesis of silver nanoparticles by *Bacillus subtilis* and their antibacterial activity. Saudi J Biol Sci 27(8):2185–2191

Ameen F, Alsamhary K, Alabdullatif JA, ALNadhari, S. (2021) A review on metal-based nanoparticles and their toxicity to beneficial soil bacteria and fungi. Ecotoxicol Environ Saf 213:112027

Asamoah GA, Korsah M, Jeyasundar PGSA, Ahmed M, Lau SY, Danquah MK (2024) Nanotechnology-based lithium-ion battery energy storage systems. Sustainability 16(21):9231

Ashok B, Bhagyashree J, Ameeta R, Smita Z (2010) Banana peel extract mediated synthesis of gold nanoparticles. Colloids Surf B Biointerfaces 80:45–50

Asimuddin M, Shaik MR, Adil SF, Siddiqui MRH, Alwarthan A, Jamil K, Khan M (2020) *Azadirachta indica* based biosynthesis of silver nanoparticles and evaluation of their antibacterial and cytotoxic effects. J King Saud Univ-Sci 32(1):648–656

Aziz A, Shaikh H, Abbas A, Zehra KE, Javed B (2025) Microscopic techniques for nanomaterials characterization: a concise review. Microsc Res Tech 88(5):1599–1614

Bahrulolum H, Nooraei S, Javanshir N, Tarrahimofrad H, Mirbagheri VS, Easton AJ, Ahmadian G (2021) Green synthesis of metal nanoparticles using microorganisms and their application in the agrifood sector. J Nanobiotechnol 19(1):86

Bakhtyar R, Tofiq R, Hamzah H, Qurbani K (2024) Fabricated Fusarium species-mediated nanoparticles against Gram-negative pathogen. World Acad Sci J 7(1):1

Bakranova D, Nagel D (2023) ZnO for photoelectrochemical hydrogen generation. Clean Technol 5(4):1248–1268

Bansal M, Jyoti N, Bharti A, Sharma S, Thakur M, Chauhan V, Kumar R (2025) Green synthesis of nanomaterials used as nano-fertilizer for sustainability in crop production: a overview on recent advancements and future perspectives. Plant Nano Biol 11:100143

Bayda S, Adeel M, Tuccinardi T, Cordani M, Rizzolio F (2019) The history of nanoscience and nanotechnology: from chemical–physical applications to nanomedicine. Molecules 25(1):112

Bensaude-Vincent B (2016) Building multidisciplinary research fields: the cases of materials science, nanotechnology and synthetic biology. In: The local configuration of new research fields: on regional and national diversity. Springer, Cham, pp 45–60

Chinnasamy G, Chandrasekharan S, Koh TW, Bhatnagar S (2021) Synthesis, characterization, antibacterial and wound healing efficacy of silver nanoparticles from *Azadirachta indica*. Front Microbiol 12:611560

Dhingra R, Naidu S, Upreti G, Sawhney R (2010) Sustainable nanotechnology: through green methods and life-cycle thinking. Sustainability 2(10):3323–3338

Elezz MA, Aboeleneen N, Abd-ElMonem NM, Sorour FH (2025) A comprehensive review: functional nanomaterials for renewable energy: innovations, applications, and sustainable strategies. Next Mater 9:101001

El-Saadony MT, Fang G, Yan S, Alkafaas SS, El Nasharty MA, Khedr SA, Hussien AM, Ghosh S, Dladla M, Elkafas SS (2024) Green synthesis of zinc oxide nanoparticles: preparation, characterization, and biomedical applications—a review. Int J Nanomedicine 19:12889–12937

El-Seedi HR, El-Shabasy RM, Khalifa SA, Saeed A, Shah A, Shah R, Iftikhar FJ, Abdel-Daim MM, Omri A, Hajrahand NH (2019) Metal nanoparticles fabricated by green chemistry using natural extracts: biosynthesis, mechanisms, and applications. RSC Adv 9(42):24539–24559

Elzein B (2024) Nano Revolution:"Tiny tech, big impact: how nanotechnology is driving SDGs progress". Heliyon 10(10):e31393

Fakayode OJ, Oladipo AO, Oluwafemi OS, Songca SP (2016) Biopolymer-mediated green synthesis of Noble metal. In: Recent advances in biopolymers, vol 10. IntechOpen, London, p 19

García-Serna J, Pérez-Barrigón L, Cocero M (2007) New trends for design towards sustainability in chemical engineering: green engineering. Chem Eng J 133(1–3):7–30

Hachem K, Ansari MJ, Saleh RO, Kzar HH, Al-Gazally ME, Altimari US, Hussein SA, Mohammed HT, Hammid AT, Kianfar E (2022) Methods of chemical synthesis in the synthesis of nanomaterial and nanoparticles by the chemical deposition method: a review. BioNanoScience 12(3):1032–1057

Halder A, Mohan GR, Matheshwaran S, Jha SK (2025) Green synthesis of neem (*Azadirachta indica*) functionalized zinc oxide with enhanced antimicrobial properties. Next Mater 8:100725

Haleem A, Javaid M, Singh RP, Rab S, Suman R (2023) Applications of nanotechnology in medical field: a brief review. Glob Health J 7(2):70–77

Hemdan M, Abuelhaded K, Shaker AA, Ashour MM, Abdelaziz MM, Dahab MI, Nassar YA, Sarguos AM, Zakaria PS, Fahmy HA (2025) Recent advances in nano-enhanced biosensors: innovations in design, applications in healthcare, environmental monitoring, and food safety, and emerging research challenges. Sens Bio-Sens Res 48:100783

Hosseingholian A, Gohari S, Feirahi F, Moammeri F, Mesbahian G, Moghaddam Z, Ren Q (2023) Recent advances in green synthesized nanoparticles: from production to application. Mater Today Sustain 24:100500

Hou Y, Gao Y, Wang X, Zhang Y, Li J, Zhang H, Li X (2021) Alginate-aloe vera film contains zinc oxide nanoparticles with high degradability and biocompatibility on post-cesarean wounds. J Drug Delivery Sci Technol 66:102631

Husseiny M, Abd El-Aziz M, Badr Y, Mahmoud M (2007) Biosynthesis of gold nanoparticles using Pseudomonas aeruginosa. Spectrochim Acta A Mol Biomol Spectrosc 67(3–4): 1003–1006

Ijaz M, Khan F, Ahmed T, Noman M, Zulfiqar F, Rizwan M, Chen J, Siddique KH, Li B (2023) Nanobiotechnology to advance stress resilience in plants: current opportunities and challenges. Mater Today Bio 22:100759

Ismael AF, Ahmed NM, Ibrahim KH, AL-Kubaisi AA (2023) Tea plant leaves for green synthesis of metallic nanoparticles. Macromol Symp 407:2100377

Jewett MC, Patolsky F (2013) Nanobiotechnology: synthetic biology meets materials science. Curr Opin Biotechnol 24:551–554

Ji J, Li L, Guo W, Zhang J, Yao Y, Chen H, Liao F, Jin Z, Liu L, Ouyang J (2024) Nanoscale synthetic biology with innovative medicinal applications. Fundam Res 5(5):1889–1910

Jiang M, Song Y, Kanwar MK, Ahammed GJ, Shao S, Zhou J (2021) Phytonanotechnology applications in modern agriculture. J Nanobiotechnol 19(1):430

Kadhim SA, Alslami SA, Sadaka MWM, Mohammadalizadeh M, Kashi MA (2025) Green synthesis of sucrose-mediated copper oxide nanoparticles: a comprehensive study on photocatalytic efficiency, antioxidant potential, and antimicrobial properties. BioNanoScience 15(1):14

Karnwal A, Jassim AY, Mohammed AA, Sharma V, Al-Tawaha ARMS, Sivanesan I (2024) Nanotechnology for healthcare: plant-derived nanoparticles in disease treatment and regenerative medicine. Pharmaceuticals 17(12):1711

Karthik L, Kumar G, Kirthi AV, Rahuman A, Bhaskara Rao K (2014) Streptomyces sp. LK3 mediated synthesis of silver nanoparticles and its biomedical application. Bioprocess Biosyst Eng 37(2):261–267

Ke M, Zhang K, Hicks AL, Wu F, You J (2025) A life cycle risk assessment of nanopesticides in freshwater. Environ Sci Ecotechnol 25:100565

Kekeli MA, Wang Q, Rui Y (2025) The role of nano-fertilizers in sustainable agriculture: boosting crop yields and enhancing quality. Plants 14(4):554

Khafaga DS, El-Khawaga AM, Elfattah Mohammed RA, Abdelhakim HK (2023) Green synthesis of nano-based drug delivery systems developed for hepatocellular carcinoma treatment: a review. Mol Biol Rep 50(12):10351–10364

Khan SH (2019) Green nanotechnology for the environment and sustainable development. In: Green materials for wastewater treatment. Springer, Cham, pp 13–46

Khan M, Khan MSA, Borah KK, Goswami Y, Hakeem KR, Chakrabartty I (2021) The potential exposure and hazards of metal-based nanoparticles on plants and environment, with special emphasis on ZnO NPs, TiO2 NPs, and AgNPs: a review. Environ Adv 6:100128

Khatoon UT, Velidandi A (2025) An overview on the role of government initiatives in nanotechnology innovation for sustainable economic development and research progress. Sustainability 17(3):1250

Kučuk N, Primožič M, Knez Ž, Leitgeb M (2023) Sustainable biodegradable biopolymer-based nanoparticles for healthcare applications. Int J Mol Sci 24(4):3188

Kumar P (2022) Nanomaterial's synthesis, types and their use in bioremediation and agriculture. Nat Resour Hum Health 2(3):349–365

Kumar S, Sharma Y, Khandelwal V, Rawat K, Patil A (2025) Applications of nanotechnology in fertilizers: a review study. Environment 10:100247

Laib I, Gheraissa N, Benaissa A, Benkhira L, Azzi M, Benaissa Y, Abdelaziz AG, Tian F, Walsh M, Bechelany M (2025) Tailoring innovative silver nanoparticles for modern medicine: the importance of size and shape control and functional modifications. Mater Today Bio 33:102071

Lu Y, Ozcan S (2015) Green nanomaterials: on track for a sustainable future. Nano Today 10(4): 417–420

Luque P, Chinchillas-Chinchillas M, Nava O, Lugo-Medina E, Martínez-Rosas M, Carrillo-Castillo A, Vilchis-Nestor A, Madrigal-Muñoz L, Garrafa-Gálvez H (2021) Green synthesis of tin dioxide nanoparticles using *Camellia sinensis* and its application in photocatalytic degradation of textile dyes. Optik 229:166259

Maksimović M, Omanović-Mikličanin E (2017) Towards green nanotechnology: maximizing benefits and minimizing harm. In: CMBEBIH 2017: proceedings of the international conference on medical and biological engineering 2017. Springer Singapore, Singapore

Malik AQ, Mir TUG, Kumar D, Mir IA, Rashid A, Ayoub M, Shukla S (2023) A review on the green synthesis of nanoparticles, their biological applications, and photocatalytic efficiency against environmental toxins. Environ Sci Pollut Res 30(27):69796–69823

Manimaran M, Kannabiran K (2017) Actinomycetes-mediated biogenic synthesis of metal and metal oxide nanoparticles: progress and challenges. Lett Appl Microbiol 64(6):401–408

Manzoor U, Masood S, Nazir S, Younis L, Waheed A, Qurashi SU, Sheikh FA, Majeed S (2025) Limitations and concerns of nanotechnology in obtaining the desirable products. In: Nanotechnology based microbicides and immune stimulators. Springer, pp 217–236

Mohamed EF, Awad G (2022) Development of nano-sensor and biosensor as an air pollution detection technique for the foreseeable future. In: Comprehensive analytical chemistry, vol 99. Elsevier, pp 163–188

Moradi M, Rouhani M, Hekmati M, Veisi H, Esmaeli D (2025) Green synthesis of gold nanoparticles supported over the *Myrtus communis* L extract modified magnetic iron oxide as a novel antibacterial agent. Resul Chem 15:102323

Mukherjee S, Patra CR (2017) Biologically synthesized metal nanoparticles: recent advancement and future perspectives in cancer theranostics. Future Sci OA 3:FSO203

Naikoo GA, Mustaqeem M, Hassan IU, Awan T, Arshad F, Salim H, Qurashi A (2021) Bioinspired and green synthesis of nanoparticles from plant extracts with antiviral and antimicrobial properties: a critical review. J Saudi Chem Soc 25(9):101304

Nain R, Patel H, Chahar M, Kumar S, Rohilla D, Pal M (2025) Biosynthesized metallic nanoparticles for sustainable environmental remediation: mechanisms, applications, and future perspectives. Discov Chem 2(1):124

Narayanan KB, Bhaskar R (2024) Green nanotechnology: paving the way for environmental sustainability. Sustainability 16(14):6262

Naryal S, Rana M, Sharma N, Ahmed B, Sharma P, Kujur S, Jaiswal S, Gupta SD, Patel SMB, Kaushal S (2025) Navigating sustainable and healthy future: green nanotechnology, regulatory priorities, and challenges. In: Role of science and technology for sustainable future: volume 2—Applied sciences and technologies. Springer, pp 353–369

Nasr Azadani F, Madani M, Karimi J, Sepahvand S (2024) Green synthesis of silver nanoparticles by *Fusarium oxysporum* and its function against *Aspergillus* and *Fusarium fungi*. Indian J Microbiol 64(1):213–224

Neeti K, Singh R, Ahmad S (2023) The role of green nanomaterials as effective adsorbents and applications in wastewater treatment. Mater Today Proc 77:269–276

Noah NM, Ndangili PM (2022) Green synthesis of nanomaterials from sustainable materials for biosensors and drug delivery. Sens Int 3:100166

Ondiek JK (2016) Synthesis and characterization of iron nanoparticles using banana peels extracts and their application in aptasensor. University of Nairobi

Padhi S, Singh A, Routray W (2023) Nanocellulose from agro-waste: a comprehensive review of extraction methods and applications. Rev Environ Sci Biotechnol 22(1):1–27

Pan J, Qian H, Sun Y, Miao Y, Zhang J, Li Y (2025) Microbially synthesized nanomaterials: advances and applications in biomedicine. Precis Med Eng 2:100019
Pati P, McGinnis S, Vikesland PJ (2014) Life cycle assessment of "green" nanoparticle synthesis methods. Environ Eng Sci 31(7):410–420
Rajpal VR, Nongthongbam B, Bhatia M, Singh A, Raina SN, Minkina T, Rajput VD, Zahra N, Husen A (2025) The nano-paradox: addressing nanotoxicity for sustainable agriculture, circular economy and SDGs. J Nanobiotechnol 23(1):314
Rami MR, Meskini M, Sharafabad BE (2024) Fungal-mediated nanoparticles for industrial applications: synthesis and mechanism of action. J Infect Public Health 17(10):102536
Rathod S, Preetam S, Pandey C, Bera SP (2024) Exploring synthesis and applications of green nanoparticles and the role of nanotechnology in wastewater treatment. Biotechnol Rep 41: e00830
Roy A, Sharma A, Yadav S, Jule LT, Krishnaraj R (2021) Nanomaterials for remediation of environmental pollutants. Bioinorg Chem Appl 2021(1):1764647
Sanket S, Das SK (2021) Role of enzymes in synthesis of nanoparticles. In: Bioprospecting of enzymes in industry, healthcare and sustainable environment. Springer, pp 139–153
Saxena S, Moharil M, Jadhav P, Ghodake B, Deshmukh R, Ingle AP (2025) Transforming waste into wealth: leveraging nanotechnology for recycling agricultural byproducts into value-added products. Plant Nano Biol 11:100127
Schmutz M, Borges O, Jesus S, Borchard G, Perale G, Zinn M, Sips ÄAM, Soeteman-Hernandez LG, Wick P, Som C (2020) A methodological safe-by-design approach for the development of nanomedicines. Front Bioeng Biotechnol 8:258
Sen A, Oswalia J, Yadav S, Vachher M, Nigam A (2024) Recent trends in nanozyme research and their potential therapeutic applications. Curr Res Biotechnol 7:100205
Seth R, Meena A (2024) Enzymes-based nanomaterial synthesis: an eco-friendly and green synthesis approach. Clean Techn Environ Policy 27:1–24
Shah ST, Sari IP, Yanto DHY, Chowdhury ZZ, Bashir MN, Badruddin IA, Hussien M, Lee JS (2025) Nature's nanofactories: biogenic synthesis of metal nanoparticles for sustainable technologies. Green Chem Lett Rev 18(1):2448171
Shaik B, Kumari AM (2025) A review on nanoparticles as a catalyst for biodiesel production. Resul Chem 16:102426
Shanker U, Jassal V, Rani M, Kaith BS (2016) Towards green synthesis of nanoparticles: from bio-assisted sources to benign solvents. A review. Int J Environ Anal Chem 96(9):801–835
Sharma R, Garg R, Kumari A (2020) A review on biogenic synthesis, applications and toxicity aspects of zinc oxide nanoparticles. EXCLI J 19:1325
Sidhu AK, Verma N, Kaushal P (2022) Role of biogenic capping agents in the synthesis of metallic nanoparticles and evaluation of their therapeutic potential. Front Nanotechnol 3:801620
Singh J, Dutta T, Kim K-H, Rawat M, Samddar P, Kumar P (2018) 'Green' synthesis of metals and their oxide nanoparticles: applications for environmental remediation. J Nanobiotechnol 16(1): 84
Singh H, Desimone MF, Pandya S, Jasani S, George N, Adnan M, Aldarhami A, Bazaid AS, Alderhami SA (2023) Revisiting the green synthesis of nanoparticles: uncovering influences of plant extracts as reducing agents for enhanced synthesis efficiency and its biomedical applications. Int J Nanomedicine 18:4727–4750
Soni AK, Jha RK, Jha RK (2024) Nanotechnology's applications and potential in various fields. Cureus 16(4):e59234
Srivastava S, Bhargava A (2021) Green nanotechnology: an overview. In: Green nanoparticles: the future of nanobiotechnology. Springer, pp 1–13
Stark WJ (2011) Nanoparticles in biological systems. Angew Chem Int Ed 50(6):1242–1258
Subhan MA, Subhan T (2022) Safety and global regulations for application of nanomaterials. In: Nanomaterials recycling. Elsevier, pp 83–107

Talekar S, Joshi A, Chougle R, Nakhe A, Bhojwani R (2016) Immobilized enzyme mediated synthesis of silver nanoparticles using cross-linked enzyme aggregates (CLEAs) of NADH-dependent nitrate reductase. Nano-Struct Nano-Obj 6:23–33

Valenzuela-Ibaceta F, Álvarez SA, Pérez-Donoso JM (2025) Production of minicell-like structures by *Escherichia coli* biosynthesizing cadmium fluorescent nanoparticles: a novel response to heavy metal exposure. J Nanobiotechnol 23(1):111

Vinukonda A, Bolledla N, Jadi RK, Chinthala R, Devadasu VR (2025) Synthesis of nanoparticles using advanced techniques. Next Nanotechnol 8:100169

Wang Y, Grainger DW (2014) Barriers to advancing nanotechnology to better improve and translate nanomedicines. Front Chem Sci Eng 8(3):265–275

Williams ES, Panko J, Paustenbach DJ (2009) The European Union's REACH regulation: a review of its history and requirements. Crit Rev Toxicol 39(7):553–575

Ying S, Guan Z, Ofoegbu PC, Clubb P, Rico C, He F, Hong J (2022) Green synthesis of nanoparticles: current developments and limitations. Environ Technol Innovation 26:102336

Zhu G, Wang H, Zhang Q, Zhang L (2015) Enhanced photovoltaic performance of dye-sensitized solar cells based on NaYF4: Yb3+, Er3+-incorporated nanocrystalline TiO2 electrodes. J Colloid Interface Sci 451:15–20

Chapter 6
Nanotechnology: Properties and Applications

Muhammad Islam, Muhammad Zubair Ali, Muhammad Sajjad Ahmad, Hafiz Abdullah Shakir, Muhammad Khan, Ozod Sunnatalievich Turaev, Marcelo Franco, and Muhammad Irfan

Abstract Nanotechnology involves manipulating matter at the nanoscale for various industrial applications. Highly featured electrochemical, optical, and mechanical properties of nanomaterials promote their application in the fields of agriculture, medicine, drug discovery, environment, and renewable energy production. The development of nanocatalysts has redefined the landscapes of biofuels production from green biomass. There have been approximately 764,279 experimental studies conducted in the field of nanotechnology during the last 15 years. The current status of nanotechnology reflects its ability to meet increasing global demands. Reports indicate that since 2001, the incorporation of nanotechnology across seven major economic sectors generated an estimated $1 trillion in profits by 2015, of which $284 billion was attributed to the United States. The US had approximately 33% ($110 billion) of global nanotechnology revenue in 2010, and maintained 33% in 2011, 28% in 2013, and 25% ($750 billion) of global nanotechnology revenue of almost 3 trillion in 2020. In addition to the United States, numerous nanotech companies from different states including Japan, China, Switzerland, and France are also playing pivotal roles in the manufacture of nanoscale-based products. There is a need for global cooperation to harness the actual benefits of nanotechnology applications.

Keywords Nanomaterials · Current status · Sustainable development · Challenges · Future directions

---

M. Islam · M. Z. Ali · M. S. Ahmad · M. Irfan (✉)
Department of Biotechnology, University of Sargodha, Sargodha, Pakistan
e-mail: irfan.ashraf@uos.edu.pk

H. A. Shakir · M. Khan
Institute of Zoology, University of the Punjab New Campus, Lahore, Pakistan

O. S. Turaev
Research Institute of Plant Genetic Resources, Tashkent, Uzbekistan

M. Franco
Department of Exact Science, State University of Santa Cruz, Ilheus, Brazil

N. Srivastava (ed.), *Sustainable Nanostrategies for Biorefineries*, Clean Energy Production Technologies, https://doi.org/10.1007/978-981-95-5875-9_6

6.1 Introduction

Nanotechnology has emerged as a revolutionary field because it operates at the nanoscale, enabling unique and transformative properties. Nanotechnology involves manipulating matter at the atomic and molecular levels to create devices and structures ranging from 1 to 100 nm in size. These nanomaterials show significant potential for a wide range of industrial and agricultural applications. Due to their highly distinct mechanical, electrochemical, and optical properties, these materials are widely used in various applications in the agriculture sector. Nanomaterials are utilized for the targeted delivery of fertilizers, pesticides, and genetic material in crops (Santos et al. 2025). Currently, nanotechnology has revolutionized industrial sectors and reshaped their landscapes with their its nanoscale properties. Between 2009 and 2024, approximately 764,279 experimental studies on nanotechnology were undertaken. These research studies involved the development of nanoscale particles and their applications among different sectors, as shown in Fig. 6.1. Their scope was not limited to agriculture and industry; rather, it has become an emerging tool in biomedical applications. Nanotechnology has made disease diagnoses easier (Alfei and Zuccari 2025) and has enabled efficient, controlled, and targeted release of therapeutics. The development of nanocarriers such as dendrimers, liposomes, inorganic nanomaterials, and polymeric nanoparticles has led towards improved bioavailability, enhanced therapeutic efficiency, and reduced side effects. Currently, different research is being conducted on how to use nanoparticles for the treatment of chronic disorders, including neurodegenerative diseases, cancer, and infectious

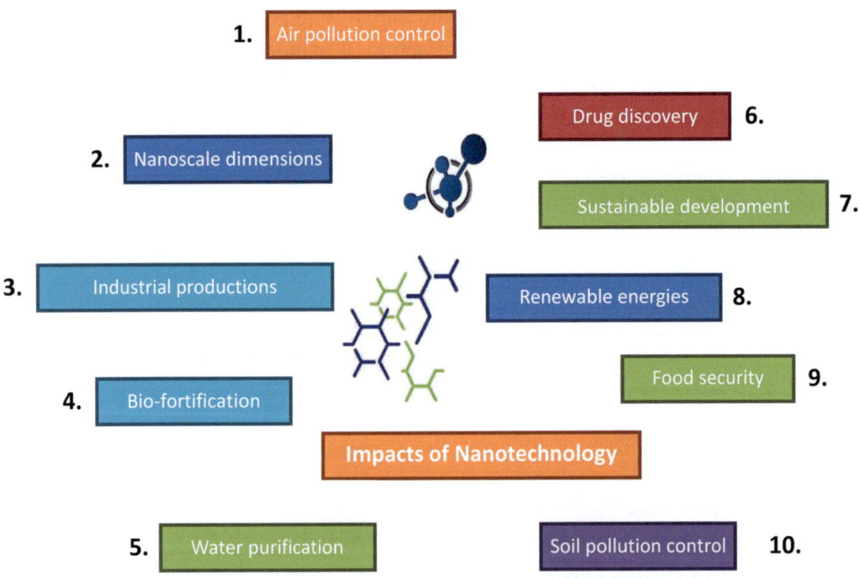

Fig. 6.1 Impacts of nanotechnology on sustainable development and other fields

diseases (Bhargavi et al. 2025). In addition, nanomaterials can be used for solving environmental pollution problems. Nanomaterials such as carbon-based structures (carbon nanotubes, graphene), polymeric nanoparticles, and metal oxides are widely utilized for water purification. These nanoparticles possess a significant capability to degrade, neutralize, and adsorb environmental pollutants. They can also remove harmful chemicals from water and make it suitable for drinking. Similarly, nanotechnology uses nanoparticles to control air pollution by breaking down poisonous gases such as volatile organic compounds and nitrogen oxides. They facilitate the development of clean air technologies to cope with air pollution. Similarly, it mitigates soil pollution to a greater level. Nanoparticles play a vital role in the removal of harmful pollutants from soil by breaking down and immobilizing petroleum residues, pesticides, and other hazardous contaminants. The development of nanosensors enabled early detection and real-time monitoring of environmental pollutants. In addition, nanoparticles are used for the production of green biofuels and can stimulate sustainable development by reducing dependence on fossil fuels. Nanoparticles are also used to enhance photocell efficiency and solar energy production (Prasad et al. 2025). Since its inception in 2000, there has been swift advancement in the field of nanotechnology. As of 2024, the total federal investment in nanotechnology exceeds approximately $43 billion (Khatoon and Velidandi 2025). Since 2015, federal funding for nanotechnology in the United States has totaled nearly $1.5 billion. (Khatoon and Velidandi 2025). Furthermore, nanotechnology is emerging as a key technology to help achieve 17 sustainable development goals by 2030 (Raman et al. 2024).

6.2 Nanotechnology—Current Status

Nanotechnology is not only a regional step but also involves global cooperation for research and development at nanoscale. Globally, a number of states have announced their national research and development programs (R&D), including Japan (2001), the European community (March 2002), Germany (May 2002), South Korea (July 2001), Taiwan (September 2002), and China (2002). It became evident when 80 countries collectively were working on nanotechnology by 2005 (Roco 2023). It not only promoted nano-based products all over the world but also created a number of job opportunities (Elzein 2024). It was analysis of emerging trends in the field of nanotechnology. According to the research analysts nanotechnology applications were estimated to reach 1 trillion USD by 2015 (Elzein 2024). Out of $1 trillion, $284 billion was held in the United States. Collectively, the United States had approximately 33% ($110 billion) of global revenue totaling 335 USD in 2010, 33% in 2011, 28% in 2013, and 25% ($750 billion) of global revenue totaling almost 3 trillion USD in 2020 (Elzein 2024). By 2021, nanotechnology was projected to add over $3 trillion to the global economy, supporting 6 million jobs worldwide in nanomaterials-based industries (Roco and Bainbridge 2005). Companies across the world are adopting nano-enabled products to boost enhanced market

Table 6.1 Top manufacturing companies in the field of nanotechnology (Khatoon and Velidandi 2025)

Company	Total products	Country	Area
Amphenol Corporation	206	USA	Electronics
Intel	189	USA	Electronics
Nano world	197	Switzerland	Electronics
Samsung Electronics Co. Ltd.	138	Republic of Korea	Home appliances, environment
OMRON corporation	135	Japan	Electronics
Shanghai Huzheng Nanotechnology Co. Ltd.	50	China	Construction, medicine, textiles
Beiersdorf AG	38	Germany	Cosmetics, medicine
NANOSKIN Car Care	38	USA	Automotive
Pfizer Inc.	30	USA	Medicine, cosmetics
Hitachi Ltd.	38	Japan	Renewable energies

competitiveness (Bandyopadhyay et al. 2023). Top-most companies in the field of nanotechnology are shown in Table 6.1 (Ferrari 2005).

6.3 Nanotechnology Applications

6.3.1 Drug Discovery

Development of nanocarriers fostered applications of nanotechnology in the field of drug discovery. The success of nanotechnology in drug discovery relies mainly on the composition, functionality, and design of nanocarriers. These nanocarriers enable highly precise delivery of drugs to target locations, significantly improving drug efficiency and lowering side effects to a significant level. Different nanocarriers have been discovered for biomedical applications (Gianchandani and Meng 2012). Liposomes are important nanocarriers in drug delivery to target sites. These are spherical vesicles covered with phospholipid bilayer that make them suitable targets for drug delivery applications (Gianchandani and Meng 2012). Both lipophilic and hydrophilic drugs can be encapsulated in them to improve drug solubility, enable controlled drug release, and optimize pharmacokinetics (Bhargavi et al. 2025). Another nanocarrier includes dendrimers with a tree-shaped branched nanoscale structure (Bhargavi et al. 2025). Likewise, polymeric nanoparticles are colloidal, solid drug carrier molecules composed of biocompatible and biodegradable polymers such as PLGA (poly(lactic-co-glycolic acid)), polycaprolactone, and chitosan (Manoukian et al. 2019). Other nanocarriers include inorganic nanomaterials such as gold nanoparticles, silica nanoparticles, magnetic nanoparticles, and quantum dots (Bhargavi et al. 2025). Likewise, nanotechnology holds significant potential applications across various sectors, as illustraed in Fig. 6.2.

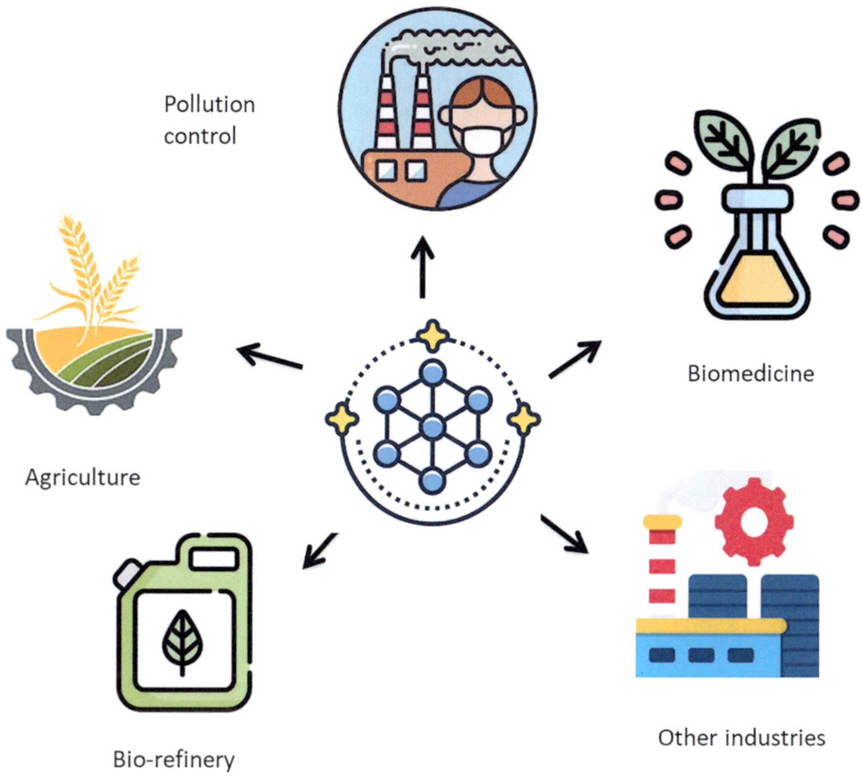

Fig. 6.2 Applications of nanotechnology among agriculture, bio-refinery and various industries

6.3.2 Food and Agriculture

Nano fertilizers are an important group of fertilizers that consist of nanoparticles and possess significant nutrient potential for crops and plant growth. These nanoparticles can be applied individually as well as with fertilizers and enhance productivity (Ramkumar et al. 2024). Nanoparticles can also be used for bio-fortification applications, especially for staple crops like rice and wheat (Johnson 2022; Prasad 2012). Nanotechnology plays a pivotal role in the mitigation of hunger and food insecurity all over the world. Hidden hunger is a global problem directly affecting 800 million people and causing micronutrient deficiency in more than 2 billion people from various regions in the world (Kapoor et al. 2022; Wessells and Brown 2012). Biofortification can also be attained via foliar sprays or soil applications (Marzouk et al. 2019; Mondal 2022). Common nanofertilizers include nanoZn, nanoFe, and nanoMn, which can enhance crop yields by approximately 30% to 56% (Hu et al. 2017). Research studies have found that Fe2o3 nanoparticles improved iron content in bread wheat (*Triticum aestivum L.*) and pomelo (*Citrus maxima*) (Du et al. 2019;

Sundaria et al. 2019). Nano-chelated iron fertilizers enhanced Fe content in white rice and enhanced uptake of micro- and macronutrients like phosphorus (43%), nitrogen (46%), zinc (50%) and and potassium (41%) (Mahdieh et al. 2018; Moghaddasi et al. 2017; Taheri et al. 2016). Other essential nutrients, including Fe, K, and Mn, are fortified across numerous fruits and crops; boron content in potatoes is also fortified by the use of nanoparticles (El-Hak El et al. 2019; Hussein and Abou-Baker 2018; Mahmoud et al. 2019).

6.3.3 Environmental Applications

Environmental pollution has become a global challenge and causes severe climate change problems, infectious diseases, and loss of biodiversity; water pollution, particularly, is a category of environmental pollution that affects both humans and aquatic life. Advancement in the field of nanotechnology has revolutionized polluted water treatment to bring water security all over the world by efficiently removing impurities from water, making it fit to drink. Nanotubes, nanoparticles, and nanofibers are widely applied in water purification due to their ability to remove heavy metals, organic polymers, and other contaminants. This approach has gained significance because of its low operational costs, increased safety, and improved efficiency. Common nanomaterials utilized for water treatment include carbon-based materials, graphene, silver, titanium, magnetic nanoparticles, magnetic graphene oxide, cerium oxide, zinc oxide, and alumina (Baskaran et al. 2008; McCollough 1997). In addition, nanoparticles are widely used for eliminating gaseous impurities and for soil remediation, targeting emerging contaminants in soil (Prasad et al. 2025).

6.3.4 Renewable Energy Production

A type of energy produced from biomass that can are be replenished naturally over time, such as sunlight, wind, water, and biomass. Biomass includes forestry, agricultural waste, animal waste, municipal waste, food and feed industry waste, and other organic industrial waste. This biomass can be used to produce green biofuels such as bio-ethanol, bio-butanol, biodiesel, and biogas. The use of these fuels reduces dependence on fossil-based energy sources, which are rapidly depleting due to overuse. Moreover, burning fossil fuels causes serious environmental problems, including climate change, global warming, and pollution. The development of nanocatalysts supports the sustainable production of biofuels from biomass. Overall, nanotechnology has paved the way for optimization of biofuel production processes from biomass (Prasad et al. 2025).

6.4 Types of Nanoparticles

6.4.1 Gold Nanoparticles

There exist numerous types of nanoparticles, as noted in Fig. 6.3. Because of their remarkable stability, controlled size and shape, and unique optical features, gold nanoparticles (AuNPs) are among the most researched and used forms of metallic nanoparticles. As a particle size increases, its surface plasmon resonance (SPR) behavior causes size-dependent coloration that ranges from ruby red to purple or blue (Zhang et al. 2023). Because of their ease of surface functionalization and biocompatibility, gold nanoparticles exhibit exceptional diversity in biomedical applications. Through sophisticated techniques like optical vortex-induced forward transfer, they may be accurately assembled in two-dimensional patterns, allowing for the production of nanoscale metallic structures with previously unheard-of definition

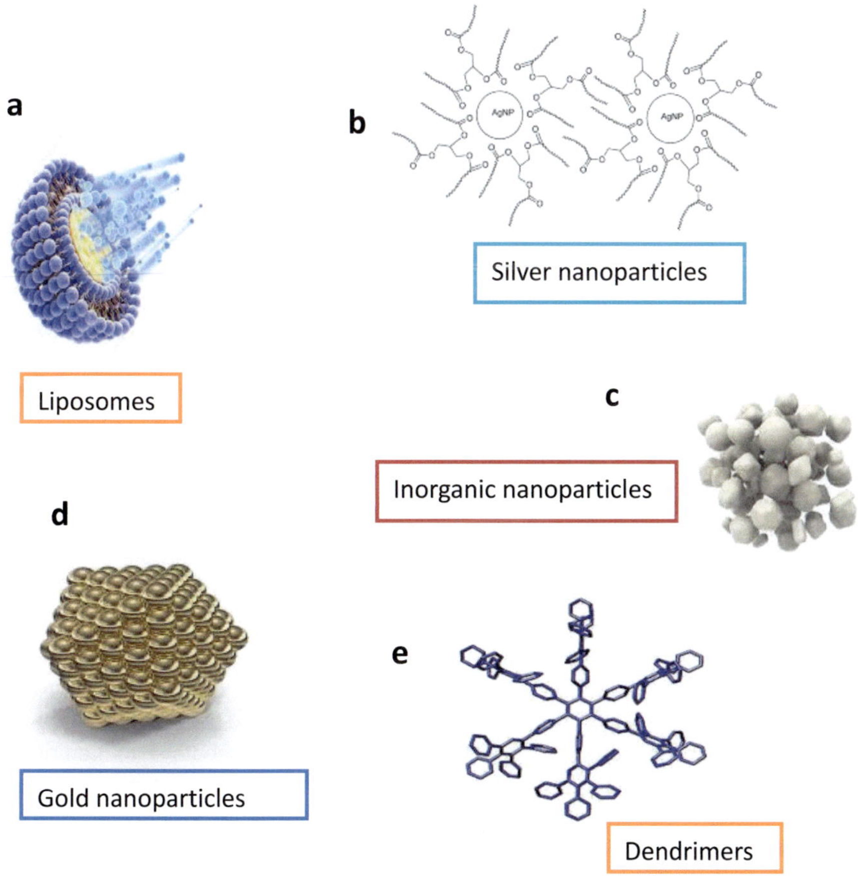

Fig. 6.3 Types of different nanoparticles used for numerous industrial applications

(Al-Dulimia et al. 2022; Kawaguchi et al. 2023). The development of electronic components, catalysts, and high-precision sensors will be greatly impacted by this capability.

Thiol-gold bonding makes it easy to alter the surface chemistry of gold nanoparticles, enabling the attachment of different biomolecules such as DNA, antibodies, and medicinal substances. Because of their adaptability, gold nanoparticles are now considered to be useful platforms for photothermal treatment, biosensing, targeted drug administration, and diagnostic imaging (Kawaguchi et al. 2023; Taha et al. 2024).

6.4.2 *Silver Nanoparticles*

Known for their remarkable antibacterial qualities, silver nanoparticles (AgNPs) are used in various commercial goods, from textiles and personal care products to wound dressings. Through a variety of processes, such as the breakdown of cell membranes and the production of reactive oxygen species, these nanoparticles demonstrate broad-spectrum antibacterial action against bacteria, fungi, and certain viruses (Aljezani and Ali 2022; Bukhari et al. 2021). Because of their special physicochemical characteristics, silver nanoparticles show promise in catalysis, electronics, and optical sensing in addition to antibacterial uses. They can be used in surface-enhanced Raman spectroscopy (SERS) and other sensing applications because of their surface plasmon resonance properties. Plant extract-based green synthesis techniques have become popular as eco-friendly ways to create silver nanoparticles with regulated characteristics for a range of uses (Bukhari et al. 2021; Yahya et al. 2024).

6.4.3 *Zinc Oxide Nanoparticles*

Zinc oxide nanoparticles (ZnO NPs) are a significant family of semiconductor nanoparticles with unique electrical, optical, and antibacterial characteristics. These multifunctional nanoparticles are useful for applications in electronics, optoelectronics, and photocatalysis because of their substantial excitation binding energy and wide bandgap (3.37 eV) (Bukhari et al. 2021; Mohammed et al. 2024). ZnO nanoparticles' antibacterial and anticancer properties have drawn interest in the biomedical domain. Their medicinal potential is enhanced by their capacity to produce reactive oxygen species when photoactivated. In contrast to traditional chemical synthesis techniques, green synthesis methods employing plant extracts have effectively generated zinc oxide nanoparticles with improved biocompatibility and a lower environmental impact (Mohammed and Alwan 2024). Because of their ability to resist UV rays, zinc oxide nanoparticles are also used in cosmetics, especially sunscreens. They can also be used as photocatalysts in wastewater treatment procedures to break down organic contaminants, demonstrating their versatility in environmental remediation.

6.4.4 Iron Nanoparticles

Due to their special magnetic characteristics, iron and iron oxide nanoparticles can be used in a number of industries such as data storage, environmental cleanup, and biomedicine. Pure iron (Fe), magnetite (Fe3O4), and hematite (Fe2O3) are among several types of nanoparticles that may be manufactured; each has unique properties that make it suitable for a particular use (Bukhari et al. 2021). The utilization of iron oxide nanoparticles as "nanozymes", which have catalytic properties similar to those of natural peroxidases, is an especially intriguing application. This characteristic has been used to create catalytic systems and sensitive biosensing platforms. The magnetic characteristics of iron nanoparticles allow for their manipulation with external magnetic fields in biomedical applications as well as their recovery and reuse in environmental applications (Zandieh and Liu 2024).

Superparamagnetic iron oxide nanoparticles (SPIONs) are used in medicine as therapeutic agents in magnetic hyperthermia for curing cancer; they are also used in magnetic resonance imaging (MRI). By reducing heavy metals and halogenated chemicals, zero-valent iron nanoparticles have proven to be efficient in environmental remediation, especially for groundwater decontamination.

6.5 Green Nanoparticle Synthesis Techniques

An environmentally responsible and sustainable substitute for traditional synthetic techniques is the green synthesis of nanoparticles. With benefits including decreased toxicity, enhanced biocompatibility, and less environmental effect, this method produces nanoparticles utilizing natural agents like proteins, microbes, and plant extracts (Anselmo and Mitragotri 2016). Green synthesis techniques can be roughly categorized into two primary groups: top-down and bottom-up procedures. The formation of nanoparticles and the methods used to regulate their size, shape, and characteristics vary among these approaches.

6.5.1 Top-Down Techniques

Top-down methods convert heavy materials into nanosize particles using mechanical or physical methods. These techniques are commonly used in classic industrial sectors and have been modified to produce nanoparticles. Several key top-down strategies consist of the following steps, as shown in the Fig. 6.4.

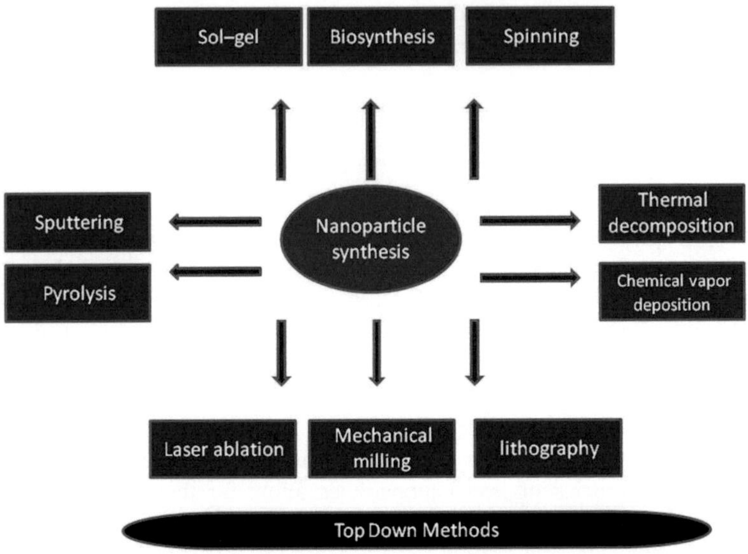

Fig. 6.4 Top down methods for green nanoparticle synthesis

6.5.1.1 Mechanical Milling

Mechanical energy is used in mechanical milling to crush heavy materials into tiny nanoparticles. Ball milling, which uses steel or ceramic balls, is a common technique for crushing larger particles. This method is highly effective when large amounts of metal and alloy nanoparticles are required. Although the technique is scalable, it often results in an unequal distribution of particle sizes and can show faults such as surface roughness or milling media contamination (Mba and Nweze 2021).

6.5.1.2 Lithography

Nanoscale patterns are created on substrates using lithography techniques such as photolithography, electron beam lithography, and nano imprint lithography, especially in the manufacture of semiconductors and nanoelectronics. These methods, which are widely used to synthesize complex nanoparticles with accurate measurements, allow for high precision. However, because of its high cost, complexity, and commonly limited range of materials that can be handled using these methods, lithography is less ideal for creating nanoparticles on a large scale (Zhang et al. 2019).

6.5.1.3 Laser Ablation

By employing powerful laser pulses to remove material from a solid target, laser ablation produces nanoparticles. High-purity metal, ceramic, and composite nanoparticles can be produced with this approach since the laser allows bulk materials to be directly ablated into the vapor phase, followed by quick cooling and nanoparticle formation. Despite producing high-quality nanoparticles, laser ablation is not practical for large-scale production because it requires sophisticated equipment (Lingamurthy et al. 2023).

6.5.1.4 Etching

Both chemical and physical etching techniques are used to remove material from a substrate in order to create nanostructures. Wet etching often involves chemical interactions with a solution, whereas dry etching uses ionized gases or plasmas to erode surfaces. Etching is widely used in photonics, microelectronics, and the production of nanostructured surfaces. It is a labor-intensive method that needs costly apparatus and may not be appropriate for large-amount nanoparticle production, despite offering exceptional precision (Tamil Selvan 2024).

6.5.2 Bottom-up Techniques

Bottom-up approaches build nanoparticles from atomic or molecular antecedents. These techniques allow for the creation of very uniform and customized particles by precisely controlling the size, shape, and content of nanoparticles (Huang et al. 2021). The following are the main techniques for bottom-up green synthesis as shown in Fig. 6.5.

6.5.2.1 Chemical Vapor Deposition (CVD)

Chemical Vapor Deposition (CVD) is a technique that dissolves gaseous precursors on a heated substrate to produce nanoparticles. It is often used to create graphene, carbon nanotubes, and other nanostructures. CVD is a preferred option for large-scale manufacturing due to its exceptional purity, uniformity, and scalability. However, it requires carefully controlled conditions and high temperatures—typically above 800°C—to ensure effective deposition and nanoparticle formation. Compared to other green synthesis methods, the procedure may use more energy, which could reduce its environmental friendliness (Hyder et al. 2023).

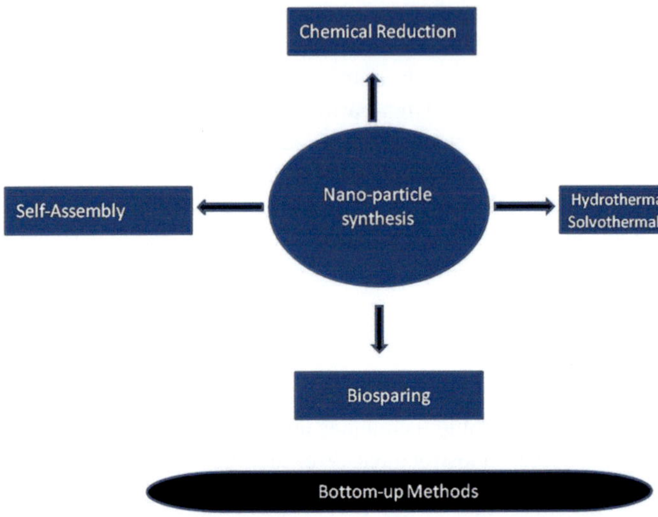

Fig. 6.5 Bottom-up methods Green Nanoparticle Synthesis

6.5.2.2 Sol-Gel Synthesis

The sol-gel process turns precursor materials into a gel-like phase, which is subsequently dried and calcined to create nanoparticles. Ceramic and metal oxide nanoparticles, including silica (SiO_2), titanium dioxide (TiO_2), and zinc oxide (ZnO), are frequently produced using sol-gel synthesis (Paulami et al. 2023). Because of its great versatility, this method can produce nanoparticles with exact control over their size and composition. To get the best results from sol-gel synthesis, reaction parameters such as temperature, pH, and precursor concentration must be carefully controlled (Wahab et al. 2023).

6.5.2.3 Chemical Reduction

Chemical reduction is a straightforward and affordable technique for creating metallic nanoparticles, such as copper, silver, and gold nanoparticles. Using reducing chemicals like hydrogen, sodium citrate, and plant-based agents, this method decreases metal salts (Saraswat et al. 2023). Since this approach doesn't require complicated equipment or high temperatures, it is very popular due to its simplicity and versatility. To guarantee consistent size and superior nanoparticles, it is crucial to regulate the quantity of the reducing agent and other reaction parameters (Hheidari et al. 2024).

6.5.2.4 Self-Assembly

The process by which molecules spontaneously arrange themselves into nanoscale structures without outside help is known as self-assembly. Van der Waals forces, hydrogen bonds, and electrostatic interactions are examples of intermolecular interactions that can direct this process (Li et al. 2023). Self-assembly is used to arrange molecules into micelles, quantum dots, and other nanoscale structures while making nanoparticles. Self-assembly provides a high level of accuracy, but it is slow, material-specific, and challenging to scale for large-scale manufacturing. The materials chosen and the conditions of the reaction determine its efficiency. (Nie et al. 2023).

6.5.2.5 Hydrothermal and Solvothermal Synthesis

High temperatures and high pressures are used in the hydrothermal and solvothermal synthesis processes to create nanoparticles. Hydrothermal synthesis uses water as the solvent, whereas solvothermal synthesis uses organic solvents. These techniques are quiet successful for the production of metal oxide, carbon-based, and composite nanoparticles. (Raza et al. 2024). They require specialized equipment that can survive the difficult conditions involved, but they also provide amazing control over particle size and purity. Although these processes can yield high-quality nanoparticles, they frequently require a lot of energy and might not be as eco-friendly as other green synthesis methods (Nazir et al. 2024).

6.6 Future Directions

Because of their remarkable physicochemical and antibacterial qualities, ZnO nanoparticles are garnering a lot of interest and have a lot of potential for use in active packaging and food preservation. In addition to preserving the functionality of the nanoparticles, green synthesis processes, which have major environmental benefits over conventional procedures, also lessen their ecological imprint. ZnO nanoparticles improve mechanical strength, shelf life, and antibacterial qualities when added to food packaging. To guarantee their safe and efficient use, however, issues such nanoparticle migration, aggregation, and environmental safety need to be resolved.

With a number of recent advancements that could impact their future development, the future of metal oxide nanoparticles in the context of environmental cleanup and dye degradation seems bright (Sarathi et al. 2023). Interestingly, a lot of focus is on creating nanoparticles with improved properties, such as higher photocatalytic activity and selectivity for particular dye impurities. This calls for advanced methods of producing nanoparticles, like surface modifications and controlled doping, in order to tailor their reactivity and selectivity (Kalsoom et al. 2022).

Artificial intelligence (AI) and machine learning (ML) are being used more frequently to improve dye degradation processes. To optimize the efficiency of metal oxide nanoparticles, artificial intelligence (AI) and machine learning (ML) systems can evaluate complicated data, forecast ideal operating conditions, and modify the treatment procedure in real time (Meneceur et al. 2024). This fits in with the larger movement toward self-sufficient and intelligent environmental remediation systems. Concerns regarding environmental and regulatory issues will develop as metal oxide nanoparticle use is predicted to expand (Meneceur et al. 2024). Future research will likely concentrate on creating thorough risk assessment procedures and setting guidelines for the appropriate use of nanoparticles in environmental applications. Ensuring their safety and avoiding unintended repercussions will be crucial to winning over the public and obtaining regulatory approval (Kalsoom et al. 2023).

In dentistry, nanoparticles are receiving a lot of attention, but further evidence-based research is needed to properly explore their potential. Innovative nanoparticles for dental applications are being actively developed by researchers with industry and clinical backing. According to certain studies, smart nanoparticles—such as electroactive and stimuli-responsive nanoparticles—may provide effective remedies for dental conditions.

Numerous developments in environmentally friendly nanoparticle production employing fungi, bacteria, and plants provide viable substitutes that reduce environmental hazards. By avoiding the use of hazardous chemicals, green synthesis techniques frequently produce harmless, biodegradable nanoparticles that are more in line with health and environmental regulations. As the market for sustainable products grows, it is anticipated that these environmentally friendly synthesis techniques will gain popularity (Vakili et al. 2025). Additionally, green synthesis creates new opportunities for the creation of multipurpose nanoparticles. Nanoparticles can be utilized in agriculture to develop controlled-release fertilizer and pesticide systems that reduce environmental harm and increase crop productivity. Similarly, developments in targeted drug delivery and nanotheranostics hold promise for improving the treatment of illnesses like cancer by offering more accurate therapies with fewer adverse effects (Jayakodi et al. 2024).

6.7 Conclusion

Nanotechnology utilizes nano scale materials for various industrial applications. Due to nano scale dimension, it has gained tremendous significance across the fields of agriculture, biomedical application, and industrial productions. Between 2009 and 2024, there were 764,279 experimental studies conducted to explore nanomaterials and their properties. The current status of nanotechnology seems to meet growing demands. It was analysis of emerging trends in the field of nanotechnology.

According to the research analysts nanotechnology applications were estimated to reach 1 trillion USD by 2015. Out of $1 trillion, $284 billion was held in the United States. Collectively, the United States had approximately 33% ($110 billion) of global revenue of 335 USD in 2010, 33% in 2011, 28% in 2013 and 25% ($750 billion) of global revenue of almost three trillion USD in 2020. Despite advancements in the field of nanotechnology, there exist various challenges that need to be addressed and resolved to reap actual benefits from this technology. There is a need for global collaboration to promote sustainable development practices all over the world.

References

Al-Dulimia AG, Hasan AF, Al-Mogadamy OA (2022) Anti-tumor activity of gold nanoparticles by use of high-content screening technique (HCS). J Med Life Sci 4(3):27–40

Alfei S, Zuccari G (2025) Last fifteen years of nanotechnology application with our contribute. Nano 15(4):265

Aljezani HA, Ali IH (2022) Cytotoxicity assessment of silver nanoparticles on liver and spleen of female rats. J Med Life Sci 4(3):53–60

Anselmo AC, Mitragotri S (2016) Nanoparticles in the clinic. Bioeng Transl Med 1(1):10–29

Bandyopadhyay S, Dorbala HC, Mandal S (2023) An overview of the impact of nanotechnology on economy and business. In: Biological applications of nanoparticles, pp 201–216

Baskaran R, Selvakumaran T, Rodrigues G, Kanjilal D, Roy A (2008) Measurements and analysis of bremsstrahlung x-ray spectrum obtained in NANOGAN electron cyclotron resonance ion source. Rev Sci Instrum 79(2):02A324

Bhargavi DM, Kavyasri C, Jyothi VJ (2025) Nanotechnology: pioneering drug discovery and delivery innovations. In: Convergence of technology & biology— transforming life sciences, pp 255–262

Bukhari A, Ijaz I, Gilani E, Nazir A, Zain H, Saeed R et al (2021) Green synthesis of metal and metal oxide nanoparticles using different plants' parts for antimicrobial activity and anticancer activity: a review article. Coatings 11(11):1374

Du W, Yang J, Peng Q, Liang X, Mao H (2019) Comparison study of zinc nanoparticles and zinc sulphate on wheat growth: from toxicity and zinc biofortification. Chemosphere 227:109–116

El-Hak El R, El-Aty El SA, El-Gazzar AAE-F, Shaaban EAE-A, Saleh MMS (2019) Efficiency of nano-zinc foliar spray on growth, yield and fruit quality of Flame seedless grape. J Appl Sci 19(6):612–617

Elzein B (2024) Nano revolution: Tiny tech, big impact: how nanotechnology is driving SDGs progress. Heliyon 10(10):e31393

Ferrari M (2005) Cancer nanotechnology: opportunities and challenges. Nat Rev Cancer 5(3): 161–171

Gianchandani Y, Meng E (2012) Emerging micro-and nanotechnologies at the interface of engineering, science, and medicine for the development of novel drug delivery devices and systems. (Preface). Adv Drug Deliv Rev 64(14):1545–1546

Hheidari A, Mohammadi J, Ghodousi M, Mahmoodi M, Ebrahimi S, Pishbin E, Rahdar A (2024) Metal-based nanoparticle in cancer treatment: lessons learned and challenges. Frontiers in Bioengineering and Biotechnology 12:1436297

Hu J, Guo H, Li J, Wang Y, Xiao L, Xing B (2017) Interaction of γ-Fe2O3 nanoparticles with Citrus maxima leaves and the corresponding physiological effects via foliar application. J Nanobiotechnol 15(1):51

Huang D, Zhang G, Yi J, Cheng M, Lai C, Xu P et al (2021) Progress and challenges of metal-organic frameworks-based materials for SR-AOPs applications in water treatment. Chemosphere 263:127672

Hussein M, Abou-Baker N (2018) The contribution of nano-zinc to alleviate salinity stress on cotton plants. R Soc Open Sci 5(8):171809

Hyder S, Ul-Nisa M, Shahid H, Gohar F, Gondal AS, Riaz N et al (2023) Recent trends and perspectives in the application of metal and metal oxide nanomaterials for sustainable agriculture. Plant Physiol Biochem 202:107960

Jayakodi S, Sriramulu M, Govindasamy M, Pandian E, Shanmugam R, Ameen F (2024) Visible light photocatalytic degradation and biological activities of Aegle marmelos-mediated cerium oxide nanoparticles. Green Proc Synth 13(1):20240060

Johnson R (2022) Examining energy intake and select macronutrients and micronutrients of formula fed WIC infants as per recommendations of the dietary guidelines. Rutgers The State University of New Jersey, School of Graduate Studies

Kalsoom U, Ahsan Z, Bhatti HN, Amin F, Nadeem R, Aftab K, Bilal M (2022) Iron oxide nanoparticles immobilized aspergillus flavus manganese peroxidase with improved biocatalytic, kinetic, thermodynamic, and dye degradation potentialities. Process Biochem 117:117–133

Kalsoom U, Khalid N, Ibrahim A, Ashraf SS, Bhatti HN, Ahsan Z et al (2023) Biocatalytic degradation of reactive blue 221 and direct blue 297 dyes by horseradish peroxidase immobilized on iron oxide nanoparticles with improved kinetic and thermodynamic characteristics. Chemosphere 312:137095

Kapoor P, Dhaka RK, Sihag P, Mehla S, Sagwal V, Singh Y et al (2022) Nanotechnology-enabled biofortification strategies for micronutrients enrichment of food crops: current understanding and future scope. NanoImpact 26:100407

Kawaguchi H, Takahashi K, Wei R, Yamane K, Morita R, Yuyama K-I et al (2023) High-definition microprint of metal-nanoparticles by optical vortex induced forward transfer. Paper presented at the 2023 Conference on Lasers and Electro-Optics (CLEO)

Khatoon UT, Velidandi A (2025) An overview on the role of government initiatives in nanotechnology innovation for sustainable economic development and research progress. Sustainability 17(3):1250

Li Y, Zhang P, Li M, Shakoor N, Adeel M, Zhou P et al (2023) Application and mechanisms of metal-based nanoparticles in the control of bacterial and fungal crop diseases. Pest Manag Sci 79(1):21–36

Lingamurthy DB, Hiregoudar S, Nidoni U, Ganachari SV, Patil VB, Hiremath V (2023) Adsorptive removal of organic pollutants from milk-processing industry effluents through chitosan-titanium dioxide nanoadsorbent-coated sand. Environ Sci Pollut Res 30(9):24101–24119

Mahdieh M, Sangi MR, Bamdad F, Ghanem A (2018) Effect of seed and foliar application of nano-zinc oxide, zinc chelate, and zinc sulphate rates on yield and growth of pinto bean (Phaseolus vulgaris) cultivars. J Plant Nutr 41(18):2401–2412

Mahmoud AWM, Abdeldaym EA, Abdelaziz SM, El-Sawy MB, Mottaleb SA (2019) Synergetic effects of zinc, boron, silicon, and zeolite nanoparticles on confer tolerance in potato plants subjected to salinity. Agronomy 10(1):19

Manoukian OS, Arul MR, Rudraiah S, Kalajzic I, Kumbar SG (2019) Aligned microchannel polymer-nanotube composites for peripheral nerve regeneration: small molecule drug delivery. J Control Release 296:54–67

Marzouk NM, Abd-Alrahman HA, El-Tanahy AMM, Mahmoud SH (2019) Impact of foliar spraying of nano micronutrient fertilizers on the growth, yield, physical quality, and nutritional value of two snap bean cultivars in sandy soils. Bull Natl Res Cent 43(1):1–9

Mba IE, Nweze EI (2021) Nanoparticles as therapeutic options for treating multidrug-resistant bacteria: research progress, challenges, and prospects. World J Microbiol Biotechnol 37(6):108

McCollough CH (1997) The AAPM/RSNA physics tutorial for residents. X-ray production. Radiographics 17(4):967–984

Meneceur S, Hemmami H, Bouafia A, Laouini SE, Tedjani ML, Berra D, Mahboub MS (2024) Photocatalytic activity of iron oxide nanoparticles synthesized by different plant extracts for the degradation of diazo dyes Evans blue and Congo red. Biomass Convers Biorefinery 14(4): 5357–5372

Moghaddasi S, Fotovat A, Karimzadeh F, Khazaei HR, Khorassani R, Lakzian A (2017) Effects of coated and non-coated ZnO nano particles on cucumber seedlings grown in gel chamber. Arch Agron Soil Sci 63(8):1108–1120

Mohammed HF, Alwan LH (2024) Preparation and identification of nanomaterials from myrtle and its use in treating lung cancer. J Biosci Appl Res 10(5):1–17

Mohammed IS, Nafea MH, Jawad RS, Abd DA, Aljoubory HM (2024) Evaluation of the effect of green synthesis of zinc oxide nanoparticles on Candida albicans. J Biosci Appl Res 10(3): 518–529

Mondal S (2022) A review: Nanofertilizers for biofortification and sustainable agriculture. Vidyodaya J Sci 25(2):6–17

Nazir S, Zhang J-M, Junaid M, Saleem S, Ali A, Ullah A, Khan S (2024) Metal-based nanoparticles: basics, types, fabrications and their electronic applications. Z Phys Chem 238(6):965–995

Nie D, Li J, Xie Q, Ai L, Zhu C, Wu Y et al (2023) Nanoparticles: a potential and effective method to control insect-borne diseases. Bioinorg Chem Appl 2023(1):5898160

Paulami D, Paret ML, Mondal R, Mandal AK (2023) Advancement of noble metallic nanoparticles in agriculture: a promising future. Pedosphere 33(1):116–128

Prasad AS (2012) Discovery of human zinc deficiency: 50 years later. J Trace Elem Med Biol 26(2–3):66–69

Prasad RD, Prasad NR, Prasad N, Prasad RS, Prasad SR, Shrivastav OP et al (2025) A review on aspects of nanotechnology in environmental science and engineering. ES General 7:1397

Raman R, Singhania M, Nedungadi P (2024) Advancing the United Nations sustainable development goals through digital health research: 25 years of contributions from the Journal of Medical Internet Research. J Med Internet Res 26:e60025

Ramkumar S, Seethalaxmi M, Rekha T, Abiyoga M, Baskar V, Sureshkumar S (2024) Biofortification of crops with nutrients by the application of nanofertilizers for effective agriculture. Plant Physiol Biochem 212:108772

Raza A, Khandelwal K, Pandit S, Singh M, Kumar S, Rustagi S et al (2024) Exploring the potential of metallic and metal oxide nanoparticles for reinforced disease management in agricultural systems: a comprehensive review. Environ Nanotechnol Monitor Manag 22:100998

Roco MC (2023) National Nanotechnology Initiative at 20 years: enabling new horizons. J Nanopart Res 25(10):197

Roco MC, Bainbridge WS (2005) Societal implications of nanoscience and nanotechnology: maximizing human benefit. J Nanopart Res 7(1):1–13

Santos PA, Biraku X, Nielsen E, Ozketen AC, Ozketen AA, Hakki EE (2025) Agricultural nanotechnology for a safe and sustainable future: current status, challenges, and beyond. J Sci Food Agric 105(6):3159–3169

Saraswat P, Singh S, Prasad M, Misra R, Rajput VD, Ranjan R (2023) Applications of bio-based nanomaterials in environment and agriculture: a review on recent progresses. Hybrid Adv 4: 100097

Sarathi R, Devi LR, Sheeba N, Esakki ES, Sundar SM (2023) Photocatalytic degradation of malachite green dye by metal oxide nanoparticles-mini review. J Chem Rev 5(1):15–30

Sundaria N, Singh M, Upreti P, Chauhan RP, Jaiswal J, Kumar A (2019) Seed priming with iron oxide nanoparticles triggers iron acquisition and biofortification in wheat (Triticum aestivum L.) grains. J Plant Growth Regul 38(1):122–131

Taha YA, Elrasheed FE, Ibrahim AI, Abdoon YH, Elhussien EA (2024) Meta-analysis of radio sensitivity enhancement by gold nanoparticles in tumor-bearing mice. J Biosci Appl Res 10(3): 536–544

Taheri M, Qarache HA, Qarache AA, Yoosefi M (2016) The effects of zinc-oxide nanoparticles on growth parameters of corn (SC704). STEM Fellowship J 1(2):17–20
Tamil Selvan S (2024) Sustainable bioremediation of cassava waste effluent using Nannochloropsis Salina TSD06: an eco-technological approach. Discov Appl Sci 6(12):642
Vakili M, Ebadi T, Hajbabaie M (2025) A systematic analysis of research trends on the permeable reactive barrier in groundwater remediation. Int J Environ Sci Technol 22(1):503–520
Wahab A, Munir A, Saleem MH, AbdulRaheem MI, Aziz H, Mfarrej MFB, Abdi G (2023) Interactions of metal-based engineered nanoparticles with plants: an overview of the state of current knowledge, research progress, and prospects. J Plant Growth Regul 42(9):5396–5416
Wessells KR, Brown KH (2012) Estimating the global prevalence of zinc deficiency: results based on zinc availability in national food supplies and the prevalence of stunting. PLoS One 7(11): e50568
Yahya A, Adil Obaid W, Mohammed Hameed O, Hasan AF (2024) Histopathological and immunohistochemical studies on the effects of silver oxide nanoparticles (AgNPs) on male rats' liver. J Biosci Appl Res 10(3):392–398
Zandieh M, Liu J (2024) Nanozymes: definition, activity, and mechanisms. Adv Mater 36(10): 2211041
Zhang M, Yang J, Cai Z, Feng Y, Wang Y, Zhang D, Pan X (2019) Detection of engineered nanoparticles in aquatic environments: current status and challenges in enrichment, separation, and analysis. Environ Sci Nano 6(3):709–735
Zhang R, Kiessling F, Lammers T, Pallares RM (2023) Clinical translation of gold nanoparticles. Drug Deliv Transl Res 13(2):378–385

Chapter 7
Sustainability Assessment of Green Nanotechnology Applications in Bioenergy Area: Promises and Feasibility Analysis

Anjali Pandit and Sonia Sethi

Abstract The demand for a steady supply of clean and green energy sources has made the biofuel sector a crucial area of study. It has been recognized that biofuels, such as bioethanol, biodiesel, biogas, biohydrogen, and microbial fuel cells, are sustainable energy substitutes for fossil fuels. Nanotechnology presents a revolutionary way forward for the biofuel sector as it looks for sustainable substitutes for fossil fuels. The incorporation of nanoparticles into biofuel production has shown considerable advantages over time owing to their nanoscale dimensions and unique structural characteristics. Their application improves operating efficiency, boosts yield, and expedites the transformation of biomass into biofuels. It bears the possibility to bring about change in biofuel sector by tackling major challenges and interpreting emerging opportunities for boosting the production of biofuel, feedstock consumption, and sustainability. This chapter emphasizes on the significant potential of bioenergy, including enhanced energy production, resource utilization, and alignment with sustainable development goals.

Keywords Green nanotechnology · Bioenergy · Biofuel · Biomass Conversion · Sustainable energy

7.1 Introduction

The fast depletion of fossil fuel reserves, elevating fuel prices, and the pollutants released during fuel combustion threatens the ecological balance. Along with this, the continuous utilization of non-renewable energy undermines the stability of both, the environment and the economy, which has made it critical to transit towards renewable energy sources (Chandel et al. 2024). Climate change has become increasingly evident to the scientific community as a result of environmental issues

A. Pandit · S. Sethi (✉)
Dr. B Lal Institute of Biotechnology, Malviya Industrial Area, Jaipur, Rajasthan, India
e-mail: soniakaura@drblal.com

N. Srivastava (ed.), *Sustainable Nanostrategies for Biorefineries*, Clean Energy Production Technologies, https://doi.org/10.1007/978-981-95-5875-9_7

such as greenhouse gas emissions, particularly those caused by CO_2 produced by the combustion of fossil fuels (Rai and Da Silva 2017). This serious issue of climate change along with the increasing population necessitates a drastic change in the way we generate and use energy. With biomass being converted into a wide range of solid, liquid, and gaseous fuels, bioenergy becomes an intriguing alternative in the global energy transition (Blay-Roger et al. 2024).

Biofuels are seen as a substitute for fossil fuels and have attracted global interest recently because of their unique characteristics (Arya et al. 2021). Oil extracts from the seeds or kernels of inedible crops are among the possible feedstocks used to make biodiesel (Ahmmed et al. 2019). The growing global reliance on biofuels as a substitute for fossil fuels has prompted significant study into enhancing their production methods. Nanotechnology has emerged as a transformative instrument for improving biofuel yield, purity, and efficiency, presenting substantial commercial prospects (Goyal et al. 2025). The application of nanotechnology and nanomaterials has emerged as an effective tool for offering economical methods to raise production quality (Dikshit et al. 2021). The size of nanoparticles (NPs) and their special qualities, like their high surface area to volume ratio and a notable degree of crystallinity, catalytic properties, adsorption capacity, and stability, make them superior to other sources for the synthesis of biofuel (Saoud 2018). For the production of biofuels, nanotechnology has shown promise when combined with other procedures such as gasification, pyrolysis, hydrogenation, and anaerobic digestion (Ali et al. 2020).

7.2 Biofuels

Biofuels are produced from renewable sources, thereby safeguarding the environment and addressing the issue of fossil fuel depletion, and are regarded as an alternative to fossil fuels. The technology of biofuels has advanced significantly over numerous generations. Depending on the feedstock or biosynthetic platform, biofuels are classified as first, second, or fourth generation (Liu et al. 2021). Biofuels are divided into four generations, first, second, third, and fourth generation, on the basis of sources and production (Fig. 7.1).

7.2.1 First-Generation Biofuels

This generation of biofuels is derived from food-based sources such as starch, sugar, vegetable oil, etc. In countries like USA and Brazil, corn and sugarcane are the commonly used feedstocks. Bioalcohols like bioethanol and biomethanol are produced through the fermentation of starch and sugar with the help of microbes and their enzymes (Deora et al. 2022). Biodiesel is synthesized through transesterification process using animal fats and vegetable oil resulting in fatty

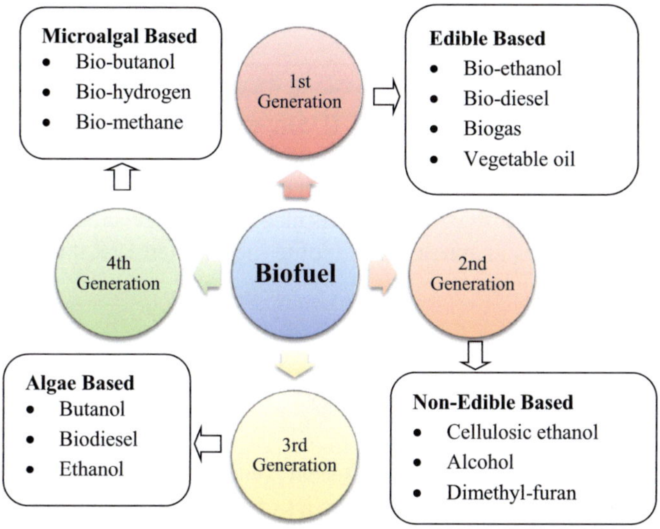

Fig. 7.1 Biofuels and its generations

acid esters. Vegetable oils can themselves be used as fuel directly or after some modifications. Fermentation or anaerobic digestion of organic wastes like agricultural, municipal, and industrial residues leads to the generation of biogas (mainly yielding methane) (Obergruber et al. 2021).

7.2.2 *Second-Generation Biofuels*

These are advanced biofuels that are made from a variety of agricultural wastes, plants, including grass, bushes, and trees. They come in a variety of forms depending on the methods used to manufacture them. Cellulosic ethanol is extracted from fermented sugar, which is made from lignocellulosic compounds. Algae can also be used as biofuel that can be developed in open and closed systems. An advanced trait algae is to transform into various biofuels, including hydrogen, biogas, and biodiesel. The methods utilized for biomass extraction and concentration include centrifugation, flotation, aggregation, flocculation, and purification (Ho et al. 2011). With a gravimetric oxygen concentration of 17%, dimethyl-furan is an oxygenated hydrocarbon and it functions as a possible addition for butanol and ethanol in diesel fuels (Xu and Wang 2016).

7.2.3 Third-Generation Biofuels

This generation of fuel utilizes energy-rich oil from algae, known as algal oil. These can easily replace starch-based biofuels (Deora et al. 2022). Photosynthetic organisms such as diatoms, euglena, and cyanobacteria serve as the raw material for third-generation biofuels (Darzins et al. 2010). Due to its high investment and operating costs, algae is more expensive per unit mass than other biofuel crops; nevertheless, it produces 10–100 times as much oil or biofuel per unit area (Kafarov and Rosso-Cerón 2021).

7.2.4 Fourth-Generation Biofuels

Fourth-generation biofuels, the most recent generation of biofuels, use genetic engineering to enhance desired characteristics of organisms used in biofuel production (Cavelius et al. 2023). Two distinct strategies are currently being used: the engineering of native producer pathways and the reconstruction of natural producer pathways in more genetically reachable biological models. The native producer pathway engineering can be achieved through growth rate optimization, utilization of various carbon sources, redirection of metabolic flux toward the production of biofuel, and increased production titers (Moravvej et al. 2019).

7.3 Green Nanotechnology in Bioenergy

For resilient or renewable energy, future eco clean energy sources are crucial as they remain a carbon-neutral with proper management. Bioenergy leverages the realistic carbon cycle to offset carbon dioxide emanations from burning with carbon captured during biomass development. Bioenergy is intriguing in combating climate change because of its sustainability (Rony 2023). Bioenergy can be altered introducing modern research and technology by the use of nano-catalysts and enzymatic reactions. The throughput and quality of biodiesel production can be enhanced by harnessing nanocatalysts with metal oxides and optimizing operating conditions (Zhang et al. 2021).

Future research should focus on making highly energetic, particular, and customized nanocatalysts that can suit the needs of generation of biofuel at industrial level. Because nano enzymes have high surface area-to-volume ratios, they can improve the enzymatic processes and bioethanol and biodiesel can be produced from the feedstocks. Therefore, the outcome of bioenergy generation is greener and more sustainable for the use in automobile and electricity. Nowadays nano-enzymes provide more sustainable production of bioenergy by reducing greenhouse gas emissions and therefore resulting in acceleration of the transition to a more sustainable energy landscape. Therefore, recent advances in nanoenzymes through innovative research and technology have enhanced the bioenergy generation by breaking down complicated organic molecules into energy-rich chemicals. Hence, nowadays more research is focused on biofuel or bioenergy generation including biodiesel, bioethanol, biogas, etc. Biofuels are environmentally friendly, cheap, easily accessible and can be easily obtained from non-edible feedstock (Sekoai et al. 2019).

7.3.1 Nanotechnology in Bioethanol Production

Among all liquid biofuels, bioethanol is most commonly used globally in spark ignition engines as fuel or fuel additive. Also, it has applications for cogeneration of fuel, fuel cells, in personal care products, explosives, paints, cosmetics, and beverages. Due to variety of applications, sustainable and monetary benefits, the production has been considerably augmented over years and also it is predicted that in future there will be sustained rapid increase in production. As the demand increases, production increases, so too research, for the optimization of production process and reduced production costs (Awogbemi and Kallon 2024).

The bioethanol production from edible crops is easy, but it led to debate because of different social concerns such as food security and rivalry. The crop plants that contain more sugar including maize, sugarcane, sorghum, or lignocellulosic biomass are commonly used. And other crops, like corn, wheat, grains, waste straw, willow, sawdust, reed canary grass, cordgrasses, miscanthus, and sorghum plants, are used exceptionally for energy usage (Silva et al. 2011).

In order to achieve sustainable biofuel production, it is crucial to select cost-effective feedstock and for these only lignocellulosic materials are considered to be alluring and best feedstock due to its eco-friendly and monetary aspects. Utilizing lignocellulosic residues resulting from crops that are not edible and agricultural residues holds special potential because it supports a number of Sustainable Development Goals (SDGs), such as SDG7 (cheap and clean energy) and SDG12 (responsible consumption and production).

The production of bioethanol occurs through different technological pathways, on the basis of which it is categorized into three generations: first-generation (1G), second-generation (2G), and third-generation (3G) biofuels. First generation obtained from edible feedstocks, second generation from lignocellulosic biomass, third-generation bioethanol refers to ethanol derived from algae (Yadav 2021). The conversion of various feedstocks into bioethanol generally involves four major steps: pretreatment, hydrolysis, fermentation, and ethanol recovery.

Ethanol-blended fuels are able to reduce net greenhouse gas emissions by up to 37.1%, resulting in substantial environmental benefit; also, gases from ethanol combustion due to more efficient burning are cleaner and carbon-neutral. This means that the gases released can be easily absorbed by the crops and therefore helps reduce ozone formation, thereby lowering the potential for ozone damage.

Table 7.1 Nanocatalysts in production of various biofuels

Nanoparticle	Example
Metal/non-metal oxide	TiO_2, Fe_3O_4, ZnO, CaO, SiO_2, Al_2O_3
Carbon-based	Carbon nanotubes, multi-walled carbon nanotubes, fullerene graphene
Metallic	Ni, Ag, Au, Pt
Core–shell	Ag/SiO_2, Au/SiO_2, Ni/SiO_2, Fe_3O_4/ SiO_2
Hybrid	Pd/PRGO/ Ce-MOF

However, there are various different challenges like feedstock's complex nature, high cost, increased concentrations of NOx, while levels of CO and HC are reduced. Besides these challenges, nanotechnology presents a promising solution, potentially improving feedstock properties and boosting engine performance (Amirabedia et al. 2019).

Among such catalysts, the nanoparticles are being given a great deal of attention as it has the capability of dissolving in a variety of solvents and capability to modify surface functionality and/or activity, and considering the discreet sizes, shapes, surface reactivity, and adsorption capacity, we can obtain this either directly or indirectly; such catalysts could be utilized for the biofuel production by improving the metabolic process. These nanoparticles like nanocrystals, etc., can be used in gasoline having diesel blend so as to increase the performance of the same.

Furthermore, they serve as catalysts for improving the anaerobic microbial communities, promoting electron transfer, and mitigating the inhibition of fermentation in the process of biofuel production. These nanoparticles have given new directions to research resulting in more efficient production process of bioethanol. There are various challenges in biofuel production like high costs, etc., so the possible solution for these economic and environmental challenges is nanotechnology. The application of nanoparticles at each and every step of production results in increase in efficiency of production by affecting their enzymatic activity (Table 7.1).

Various nanoparticles can enhance bioethanol production to 166.1% like methyl functionalized silica nanoparticles and 213.5% by methyl functionalized magnetic nanoparticle. Even the nanoparticles can be reused and properly recovered, which make the entire process highly economically adaptable (Kim and Lee 2016). There is various other potential application of nanoparticles in production including carbon nanotubes, immobilized nanoparticles, zinc oxide nanoparticles, magnetic nanoparticles, nickel oxide nanoparticles, and Mg-Zn ferrite nanoparticles (Sanusi et al. 2020), etc., which act as catalysts, resulting in expedition of chemical reactions which break complex sugars more efficiently, increasing yield.

Nanocomposites and nanomaterials also play a pivotal role during key stages of bioethanol production starting from biomass fermentation to distillation and separation. Other ways of the utilization of nanoparticles are in the form of nanosensors leveraged for monitoring of process and the analysis of different environmental and sustainability considerations which provide a broader outlook on the use of nanocomposites in bioethanol production (de Oliveira et al. 2025).

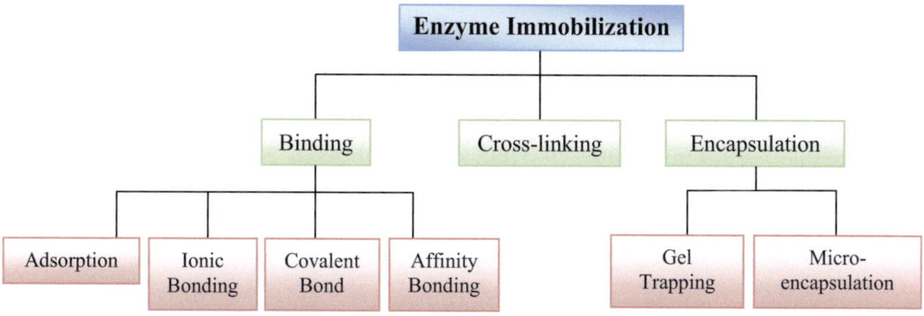

Fig. 7.2 Enzyme immobilization

Also, the first step, which involves pretreatment of biomass for production, seems to be costly but cannot be avoided and can be significantly improved by the application of nanoparticles in conjunction with other approaches, making the process more robust. Applying nanoparticles during pretreatment improves the chemistry at molecular level making it more specific and targeted. For these, metal nanoparticles can be efficiently used because it can easily penetrate the cell wall of raw biomass, resulting in interaction of biomolecules easily and thus releasing carbohydrates used in bioethanol production (Razack et al. 2016).

In a study, biolignification of elephant grass by using nickel oxide nanoparticles in combination with recombinant *Trichoderma reesei* (*T. reesei*) resulted in 54% lignin removal. The reason behind this mechanism is the ability of nanoparticle to penetrate the cell membrane of biomass and act as cofactors for enzymes responsible for breaking the biomass, while also providing the structural support for enhancement of biodelignification process. In one more study, the ability of chitosan-chitin nanofiber hybrid with hydrogel beads (C-CNBs) releases 93.7% glucose and 91.8% xylose from sugarcane bagasse hydrolysates (Sun et al. 2022).

Another method for improving enzyme catalysis process in production is immobilization of enzymes which is now being recognized as a viable option. Immobilizing various biocatalysts on support materials such as nanoparticles, resulting in formation of nano biocatalysts, boosts the enzymatic efficiency and its stability. There is significant interest to immobilize enzymes because the immobilized enzymes can be used repeatedly, can sequester enzymes, are more stable at different temperatures and pH, and remain active for a longer time; enzyme immobilization can be achieved by trapping the enzyme within a porous support. The enzyme's substrates and products must be relatively free to diffuse in and out of the immobilized portion. The carrier must however be biocompatible and insoluble and must have a high surface area with functional groups to enable effective enzyme attachment (Naqash and Naik 2022). There are different types of nanoparticles made up of metal oxide including zinc oxide (ZnO), Titanium dioxide (TiO_2), iron oxide, or magnetic oxide and others like polymer-based nanoparticles, fullerenes, graphene, and carbon nanotubes which act as a lattice for enzyme immobilization (Fig. 7.2).

Among all nanoparticles, magnetic nanoparticles (MNPs) provide better advantage over others in biomass pretreatment because of easy recovery of catalysts allowing repeated leverage in process (Prasad and Ingle 2023). Cellulase enzyme obtained from *Penicillium* sp. when impregnated with magnetic gold silica nanostructures resulted in enhancement of catalytic performance. This setup attained 76% enzyme immobilization proficiency and stability, which represents a sustainable roadmap for continuous enzyme utilization even protecting from various inhibitory aspects during process.

7.3.2 Nanotechnology in Biodiesel Production

Another industry, which has beheld substantial breakthrough in recent years, is biodiesel production through the unification of nanotechnology. These tiny nanoparticles, of nanoscale, with distinctive characteristics, can be leveraged to improve and increase the production process, contributing to a more sustainable future. Therefore, nanoparticles have provided a viable instrument by giving streamlined mechanism for better production with improved quality at rational costs. Nanoparticles provide better alternative to be used to manufacture biodiesel because of their easy recovery, e.g., carbon nanotubes and metal oxide nanoparticles (Kumar et al. 2021).

For biodiesel production, microalgae is auspicious feedstock as we all know; therefore, for enhancement of effectiveness of process, nanoparticles can be used and incorporated for cell harvest, disruption, and extraction at low costs and also improve the conversion efficiency of feedstock into product (Verma and Kuila 2020). Feedstock like vegetable oils and fats are generally used as an alternative to fossil fuels because of reduced gas emissions, but there are certain limitations like lower energy content, poor cold flow properties, and oxidative stability issues, so nanoparticles use these limitations and helps in optimizing the process of diesel production.

The nanoparticles application in biodiesel production lies in catalyst design, i.e., nanocatalysts. These have demonstrated exceptional catalytic activity and stability, resulting in more ecofriendly and efficient process, e.g., platinum, palladium, gold nanoparticles. Also, these nanoparticles are used to improve the characteristics of biodiesel, lubricity, anti-wear properties, demonstrating the increased fuel efficiency and prolonged engine life, e.g., molybdenum disulfide and tungsten disulfide.

Another area where nanoparticles have contributed efficient strides is in improving the cold flow properties of biodiesel where they act as flow improvers by changing their crystallization behavior, thereby preventing large crystal formation and improving low temperature performance. Also, nanotechnology has enabled the maturation of innovative fuel additives for biodiesel.

Another method of modifying the combustion properties and emissions of less pollutant is the formation of nanoemulsion fuels consisting of small droplets. This can be done by increasing surface area to volume ratio, resulting in better mixing of air and fuel promoting complete combustion. For example, graphene oxide and TiO_2 nanoparticles when integrated with biodiesel increase brake thermal efficiency by

5% and reduce brake-specific fuel consumption (BSFC) by 6%. Also, they cause reduction of wall wetting effects by improving the evaporation characteristics (Venu and Madhavan 2016).

The application of nanoparticles with biodiesel also significantly causes reduction in emissions of harmful gases like carbon monoxide, Nox, etc. For example, ZnO nanoparticles reduce CO emissions by up to 8% and NOx emissions by approximately 11%, and cerium oxide nanoparticles result in mitigation of thermal NOx formation. The amalgam of graphene oxide and carbon nanotubes results in the formation of hybrid nanoparticles that enable high combustion efficiency, reduced CO and HC emissions, thermal, conductivity, high oxidation rate, increased BTE by 20.66%, demonstrating the effectiveness of nanoparticles with biodiesel.

Nanoparticles result in improvement of the thermal and oxidative stability of biodiesel, which is one of its limitations, for example, silica-coated nanoparticles, palladium nanoparticles, ZnO and Al_2O_3 nanoparticles, etc. Nanoparticles help preserve biodiesel quality by preventing the formation of peroxides, which are a key cause of oxidative instability. By slowing down this degradation process, they help the fuel retain its original properties, promote smoother engine performance, and lower the need for frequent maintenance (Gohain and Paul 2022).

Leverage of nanoenzymes as catalysts provides a promising substitute for traditional catalysts for biodiesel production because of their exceptional remarkable productivity and reusability. Molybdenum (VI) complexes tethered with silica-coated magnetic nanoparticles have been applied for transesterification, yielding above 96% of biodiesel. These Fe_3O_4 @SiO_2-APTES-MoO_2 catalysts demonstrated their cost effectivity and long-term utility across various feedstocks for biodiesel production (Majedi and Safaei 2023).

Another example of nanoenzymes used to achieve 95% biodiesel yield by single-step transesterification of rice bran oil is calcium oxide nanoparticles. Further, ZnO nanoparticles also exhibit strong catalytic activity, for biodiesel production by yielding 97.8%, with high thermal stability and structural integrity. Therefore, these research findings have demonstrated the growing role of nanoparticle for biodiesel production with improved efficiency, stability, and reusability.

7.3.3 Nanotechnology in Biogas Production

By the use of available organic feedstock, various types of biofuels can be produced and there are various protocols available for this energy conversion, but among all anaerobic digestion is most preferred especially for biogas production (Feng et al. 2021). For the decomposition of organic matter in presence of microorganisms into methane and other by products, oxygen absence provides beneficial ecosystem and this system finds its implications for treatment of waste into biogas.

The field of bioenergy generation can be updated to modern styles by the refinement of catalytic transitions and by increasing efficiency of catalysis process. Huge amount of literature is available from the latest implications of nanomaterials in bioenergy generation. Few literature reviews highlighted the relevance of various operating parameters for biogas production and to understand the significance of additional technologies for its optimization (Wu et al. 2021).

Various studies have reported the fact that the amalgation of nanoparticles is one of the promising option for the enhanced biogas production. However, there are various challenges like high cost of investment, high energy demand, and high risk of disposal, which need to be minimized. These challenges can be controlled or reduced by the implications of nanoparticles. Especially, the nanoparticles composed of metals are the most favorable and famous nanoparticles, which are found to have more potential to improve the biogas production (Hassanein et al. 2021).

The mechanism behind the enhanced biogas production by the nanoparticles incorporation with different inorganic wastes is the interaction of nanoparticles with the cell membrane of waste, leading to changes in structural integrity making permeable for the entry of the microorganisms. Therefore, the microorganisms are able to find their way to attack, resulting in enhanced biogas production.

7.3.3.1 Use of Metal Oxide Nanoparticle in Anaerobic Digestion

The metal oxides nanoparticles have better surface structure; strong catalytic activity and other unique properties cause an impact on anaerobic digestion of biogas production. There are literatures available which show that the integration or addition of NPs during digestion results in increased biogas production like Fe_3O_4 nanoparticles with poultry litter (Hassanein et al. 2019) increased production by 25.6%; Fe_3O_4 nanoparticles with cattle dung slurry increased by 1.7 and 2.0 times (Abdelsalam et al. (2016); the addition of 100 mg/g-TSS Fe_3O_4 nanoparticles (Wang et al. 2016) increases CH_4 production by 117%; the addition of Fe_3O_4 nanoparticles and TiO_2 with cattle manure increases by 1.19 times. There are results which suggest that the addition of 500 mg/L TiO_2 NPs had the highest specific biogas and CH_4 production by 336.25 and 192.31 mg/g VS, respectively.

7.3.3.2 Use of NP for Process Stability

There are several indicators used for monitoring the process of anaerobic digestion and ensuring its stability like total solids (TS), volatile solids (VS), volatile fatty acids (VFAs), and COD. Therefore, the incorporation of NPs during this process results in decomposition of these indicators and in turn enhances the biogas production. For example, the use of Co, Ni, Fe, and Fe_3O_4 NPs showed that the highest decomposition of TS and VS by 5.08%, 4.95%, 4.36%, and 4.15%, respectively (Abdelsalam et al. 2016); addition of 75 mg/L Fe_3O_4 NPs decreases the TS and VS by 31.10% and 36.15%, respectively (Ali et al. 2017); also, the addition of metal oxide NPs in AD not only affects TS and VS decomposition but also influences microbial activation for each stage of the AD process.

7.3.3.3 Impact on Biogas Yield

Among all NP, zero-valent iron nanoparticles (nZVI) showed tremendous effect on biogas production from the AD of Municipal sludge (Amen et al. 2017). Different concentrations of nZVI addition increase biogas volume by 1.19, 1.31, 1.73, and 2.05 times as compared with control. Like wise1 000 mg/L nZVI addition increased the cumulative biogas production by 18.11% (Jia et al. 2017; Suanon et al. 2017; Zhou et al. 2019). Other NP like graphene oxide NPs (GO NPs) showed a positive impact on biogas and CH_4 yield from mixture substrate of cow dung, food waste, and wheat straw increasing the yield by 2.1 times (Kaushal and Baitha 2019). Multiwall carbon nanotubes and fullerenes addition at different concentrations significantly increased accumulative CH_4 yield by 46.8% and 33.5%, respectively (Hao et al. 2019).

7.4 Advantages and Functional Benefits

Nanotechnology has application in energy conservation by developing energy saving and technologies for renewable energy. The utilization of nanotechnology results in production of stable, cost effective, and energy efficient products. So these help in stabilizing future demands and supply of renewable resources. Nanoparticles or nanocatalysts, due to their high surface area, active sites, high conversion rate, result in improvement of overall efficiency and selectivity and biomass conversion leading to higher quality and yield.

7.5 Challenges and Future Prospects

The need for a consistent supply of green and clean energy sources has made research into the biofuel industry essential. As the biofuel industry seeks sustainable alternatives to fossil fuels, nanotechnology offers a significant path forward (Markandan and Chai 2022). It has been demonstrated that employing nanomaterials improves performance in ways like increasing catalytic and microbial activity and decreasing inhibition formation (Prabu 2018). Implementing nanotechnology can be very beneficial, especially when designing catalysts at the level of particles or atoms to improve specificity, effectiveness, spectrum of operational temperature, and various other crucial catalyst features. However, even though using nanotechnology in this field has many benefits, further research is needed to fully understand how it interacts with biomass because nanomaterials have unique properties. The conversion of macroscale metal oxides to nanoparticles is the obvious problem that is typically faced in the case of metal oxides, and it requires the right infrastructure, labor, experience, and other needs. Energy, time, and other resources are needed for the recovery, regeneration, alterations, and deployment of nanoparticles. These raise the cost of the catalyst and, eventually, the cost of producing bioenergy (Awogbemi and Kallon 2024). Despite the benefits that nanomaterials provide for the production of bioenergy, a number of problems currently prevent their widespread application. In terms of cost, nanoparticles are little costly, which prevents their widespread use on an industrial scale. Significant work needs to be done to further lower the cost of additional nanomaterials. The recoverability problem may soon be fixed with more investigation. Furthermore, not much is known about the toxicity caused by nanoparticles. Therefore, when using nanomaterials for the production of bioenergy, safety evaluation is crucial to reducing these concerns.

7.6 Conclusion

Because of global pursuit of clean, cost effective, eco-friendly energy, many efforts have been taken to transit from traditional or conventional fossil fuels to environment friendly alternatives. Therefore, nanotechnology transformative potential for the improvement of biofuel production including bioethanol, biodiesel, biogas, etc., by better conversion of feedstock, and therefore, enhancing the yield. All types of nanoparticles play a crucial role at all process steps including catalysis, purification, and process efficiency for increased yield.

References

Abdelsalam E, Samer M, Attia YA, Abdel-Hadi MA, Hassan HE, Badr Y (2016) Comparison of nanoparticles effects on biogas and methane production from anaerobic digestion of cattle dung slurry. Renew Energy 87:592–598. https://doi.org/10.1016/j.renene.2015.10.053

Ahmmed B, Samaddar OU, Kibria KQ (2019) Production of biodiesel from used vegetable oils production of biodiesel from used vegetable oils. Int J Sci Res Sci Technol 6

Ali A, Mahar RB, Soomro RA, Sherazi STH (2017) Fe_3O_4 nanoparticles facilitated anaerobic digestion of organic fraction of municipal solid waste for enhancement of methane production. Energy Sour Part A Recov Utilizat Environ Effects 39(16):1815–1822. https://doi.org/10.1080/15567036.2017.1384866

Ali S, Shafique O, Mahmood S, Mahmood T, Khan BA, Ahmad I (2020) Biofuels production from weed biomass using Nanocatalyst technology. Biomass Bioenergy 139:105595

Amen TW, Eljamal O, Khalil AM, Matsunaga N (2017) Biochemical methane potential enhancement of domestic sludge digestion by adding pristine iron nanoparticles and iron nanoparticles coated zeolite compositions. J Environ Chem Eng 5(5):5002–5013

Amirabedia M, Jafarmadar S, Khalilarya S, Kheyrollahi J (2019) Experimental comparison the effect of Mn_2O_3 and Co_3O_4 nano additives on the performance and emission of SI gasoline fueled with mixture of ethanol and gasoline. Int J Eng Trans 32(5):769–776

Arya I, Poona A, Dikshit PK, Pandit S, Kumar J, Singh HN, Jha NK, Rudayni HA, Chaudhary AA, Kumar S (2021) Current trends and future prospects of nanotechnology in biofuel production. Catalysts 11(11):1308

Awogbemi O, Kallon DVV (2024) Recent advances in the application of nanomaterials for improved biodiesel, biogas, biohydrogen, and bioethanol production. Fuel 358:Article 130261

Blay-Roger R, Saif M, Bobadilla LF, Ramirez-Reina T, Nawaz MA, Odriozola JA (2024) Embracing the sustainable horizons through bioenergy innovations: a path to a sustainable energy future. Front Chem 12:1416102

Cavelius P, Engelhart-Straub S, Mehlmer N, Lercher J, Awad D, Brück T (2023) The potential of biofuels from first to fourth generation. PLoS Biol 21(3):e3002063

Chandel H, Kumar P, Chandel AK, Verma ML (2024) Biotechnological advances in biomass pretreatment for bio-renewable production through nanotechnological intervention. Biomass Convers Biorefinery 14(3):2959–2981

Darzins A, Pienkos P, Edye L (2010) Current status and potential for algal biofuels production. Rep IEA Bioenerg Task 39(13):403–412

de Oliveira JAR et al (2025) Nanocomposites/nanomaterials for bioethanol production. In: Dave V, Kuila A (eds) Nanomaterials as a catalyst for biofuel production. Clean energy production technologies. Springer, Singapore. https://doi.org/10.1007/978-981-96-1706-7_4

Deora PS, Verma Y, Muhal RA, Goswami C, Singh T (2022) Biofuels: an alternative to conventional fuel and energy source. Mater Today Proc 48:1178–1184

Dikshit PK, Kumar J, Das AK, Sadhu S, Sharma S, Singh S, Gupta PK, Kim BS (2021) Green synthesis of metallic nanoparticles: applications and limitations. Catalysts 11:902

Feng R, Li Q, Zaidi AA, Peng H, Shi Y (2021) Effect of autoclave pretreatment on biogas production through anaerobic digestion of green algae. Period Polytech Chem Eng 65:483–492. https://doi.org/10.3311/ppch.18064

Gohain PP, Paul S (2022) Transition metal oxide nanoparticles for biofuel production. J Nano-Electron Phys 14:03005

Goyal A, Meena PK, Shelare S (2025) Nanotechnology in biofuel production: enhancing efficiency and sustainability through nanomaterials. J Clust Sci 36(3):87

Hao Y, Wang Y, Ma C, White JC, Zhao Z, Duan C, Zhang Y, Adeel M, Rui Y, Li G (2019) Carbon nanomaterials induce residue degradation and increase methane production from livestock manure in an anaerobic digestion system. J Clean Prod 240:118257

Hassanein A, Lansing S, Tikekar R (2019) Impact of metal nanoparticles on biogas production from poultry litter. Bioresour Technol 275:200–206

Hassanein A, Kumar AN, Lansing S (2021) Impact of electro-conductive nanoparticles additives on anaerobic digestion performance-a review. Bioresour Technol 342:126023. https://doi.org/10.1016/j.biortech.2021.126023

Ho SH, Chen CY, Lee DJ, Chang JS (2011) Perspectives on microalgal CO2-emission mitigation systems—a review. Biotechnol Adv 29(2):189–198

Jia T, Wang Z, Shan H, Liu Y, Gong L (2017) Effect of nanoscale zero-valent iron on sludge anaerobic digestion. Resour Conserv Recycl 127:190–195. https://doi.org/10.1016/j.resconrec.2017.09.007

Kafarov V, Rosso-Cerón AM (2021) Biomass as a source for heat, power and chemicals. In: Advances in carbon management technologies. CRC Press, pp 3–36

Kaushal R, Baitha R (2019) Biogas and methane yield enhancement using graphene oxide nanoparticles and Ca $(OH)_2$ pre-treatment in anaerobic digestion. Int J Ambient Energ 5:1–8

Kim YK, Lee H (2016) Use of magnetic nanoparticles to enhance bioethanol production in syngas fermentation. Bioresour Technol 204:139–144

Kumar Y, Yogeshwar P, Bajpai S (2021) Nanomaterials: stimulants for biofuels and renewables, yield and energy optimization. Mater Adv 2:5318–5343

Liu Y, Cruz-Morales P, Zargar A, Belcher MS, Pang B, Englund E, Dan Q, Yin K, Keasling JD (2021) Biofuels for a sustainable future. Cell 184(6):1636–1647

Majedi M, Safaei E (2023) Molybdenum (VI) complex of resorcinol-based ligand immobilized on silica-coated magnetic nanoparticles for biodiesel production. Appl Organomet Chem 37:e7216

Markandan K, Chai WS (2022) Perspectives on nanomaterials and nanotechnology for sustainable bioenergy generation. Materials 15(21):7769

Moravvej Z, Makarem MA, Rahimpour MR (2019) The fourth generation of biofuel. In Second and third generation of feedstocks (pp. 557–597). Elsevier.
Naqash S, Naik H (2022) Peach wastes and by-products: chemistry, processing, and utilization. Hand Fruit Wastes By-Products Chem Process Technol Util 2022:161–174. https://doi.org/10.1201/9781003164463-11
Obergruber M, Hönig V, Procházka P, Kučerová V, Kotek M, Bouček J, Mařík J (2021) Physico-chemical properties of biobutanol as an advanced biofuel. Materials 14(4):914
Prabu A (2018) Nanoparticles as additive in biodiesel on the working characteristics of a DI diesel engine. Ain Shams Eng J 9(4):2343–2349
Prasad S, Ingle AP (2023) The emerging role of nanotechnology in ethanol production. In: Nanotechnol Biorefinery, pp 235–256. https://doi.org/10.1016/B978-0-323-95965-0.00001-9
Rai M, Da Silva SS (2017) Nanotechnology for bioenergy and biofuel production. Springer, Dordrecht
Razack SA, Duraiarasan S, Mani V (2016) Biosynthesis of silver nanoparticle and its application in cell wall disruption to release carbohydrate and lipid from C. vulgaris for biofuel production. Biotechnol Rep 11:70–76
Rony ZI (2023) Alternative fuels to reduce greenhouse gas emissions from marine transport and promote UN sustainable development goals. Fuel 338:127220
Sanusi IA, Suinyuy TN, Lateef A, Kana GE (2020) Effect of nickel oxide nanoparticles on bioethanol production: process optimization, kinetic and metabolic studies. Process Biochem. 92:386–400
Saoud K (2018) Nanocatalyst for biofuel production: a review. In: Srivastava N, Srivastava M, Pandey H, Mishra PK, Ramteke PW (eds) Green nanotechnology for biofuel production. Springer, Cham, Switzerland, pp 39–62
Sekoai PT, Ouma CNM, Preez SPD, Modisha P, Engelbrecht N, Bessarabov DG (2019) Application of nanoparticles in biofuels: an overview. Fuel 237:380–397
Silva LEE, Escobar PJC, Rocha MH, Grillo RML, Venturini OJ, Almanza DOO (2011) Issues to consider, existing tools and constraints in biofuels sustainability assessments. Energy 36(4): 2097e110
Suanon F, Sun Q, Li M, Cai X, Zhang Y, Yan Y, Yu CP (2017) Application of nanoscale zero valent iron and iron powder during sludge anaerobic digestion: impact on methane yield and pharmaceutical and personal care products degradation. J Hazard Mater 321:47–53
Sun H, Liu L, Liu W, Liu Q, Zheng Z, Fan Y (2022) Removal of inhibitory furan aldehydes in lignocellulosic hydrolysates via chitosan-chitin nanofiber hybrid hydrogel beads. Bioresour Technol 346:Article 126563
Venu H, Madhavan VR (2016) Effect of Al_2O_3 nanoparticles in biodiesel-diesel-ethanol blends at various injection strategies: performance, combustion and emission characteristics. Fuel 186: 176–189
Verma S, Kuila A (2020) Involvement of green technology in microalgal biodiesel production. Rev Environ Health 35:173–188
Wang T, Zhang D, Dai L, Chen Y, Dai X (2016) Effects of metal nanoparticles on methane production from waste-activated sludge and microorganism community shift in anaerobic granular sludge. Sci Rep 6:25857. https://doi.org/10.1038/srep25857
Wu D, Peng X, Li L, Yang P, Peng Y, Liu H (2021) Commercial biogas plants: review on operational parameters and guide for performance optimization. Fuel 303:121282. https://doi.org/10.1016/j.fuel.2021.121282
Xu H, Wang C (2016) A comprehensive review of 2, 5-Dimethylfuran as a biofuel candidate. In: Biofuels from lignocellulosic biomass: innovations beyond bioethanol. Wiley-VCH, Weinheim, pp 105–129

Yadav KK (2021) Review on evaluation of renewable bioenergy potential for sustainable development: bright future in energy practice in India. ACS Sustain Chem Eng 9(48):16007–16030

Zhang J, Sewell CD, Huang H, Lin Z (2021) Closing the anthropogenic chemical carboncycle toward a sustainable future via CO2 valorization. Adv Energy Mater 11(47):2102767

Zhou J, You X, Jia T, Niu B, Gong L, Yang X, Zhou Y (2019) Effect of nanoscale zero-valent iron on the change of sludge anaerobic digestion process. Environ Technol 6:1–11

Chapter 8
Nanomaterials for Industrial and Environmental Sustainability of Bioenergy: Economic and Social Impact

Sakshi Kabra Malpani, Annu Yadav, Mahesh Chand, Pooja Devi, Deepti Goyal, and Rajesh Kumar Meena

Abstract Nanomaterials like carbon nanomaterials, metal nanoparticles offer promising solutions to improve the efficacy, sustainability, and economic viability of bioenergy resources on an industrial scale. Their tunable features like high surface area, small size, enhanced stability, good catalytic activity, and adsorption capacity enable advanced biomass conversion, biofuel production, hydrogen generation, and hence, waste valorization. Looking through industrial and environmental lenses, nanomaterials support green chemical, cleaner, and scalable processes with higher yields, reducing carbon footprints, energy use, and waste. Economically speaking, nanomaterials lower production costs, create new market avenues, and socially, contribute to job creation, rural energy access, and advancement of the Sustainable Development Goals. This chapter explores the industrial applications, environmental benefits, and wider socio-economic impacts of nanomaterials in production of bioenergy, while addressing challenges and outlining future opportunities in this area.

Keywords Nanomaterials · Bioenergy · Environmental sustainability · Biomass conversion · Biofuel production · Industrial scale

---

S. K. Malpani
Independent Researcher and Consultant, Issaquah, WA, USA

A. Yadav · R. K. Meena (✉)
Department of Chemistry, Kalindi College, University of Delhi, Delhi, India

M. Chand
Department of Chemistry, Guru Gobind Singh IP University Delhi, Delhi, India

P. Devi
Department of Physics, Kalindi College, University of Delhi, Delhi, India

D. Goyal
Department of Chemistry, Gautam Buddha University, Greater Noida, India

N. Srivastava (ed.), *Sustainable Nanostrategies for Biorefineries*, Clean Energy Production Technologies, https://doi.org/10.1007/978-981-95-5875-9_8

8.1 Introduction

Bioenergy, renewable energy produced from plant and animal waste collectively known as biomass, is helping us in meeting our increased energy demands. It helps with significant cutting down industrial carbon emissions, climate change, global warming, and promoting ecological sustainability. It is a reliable alternative to non-renewable, fossil fuel-based energy resources. About 80% of our energy needs are fulfilled by different fossil fuels like oil, coal, and natural gas (Dadi et al. 2025). Biomass is converted into bioenergy using various processes like thermochemical, physicochemical, biochemical processes, pyrolysis, combustion, anaerobic digestion. Comprehensive fuels including solid, liquid, and gaseous biofuels are transformed into bioenergy using the above-mentioned processes (Rather et al. 2021). Utilizing bioenergy in industrial processes is often considered as zero carbon or carbon neutral processes as the carbon dioxide released during biomass conversion is balanced by the carbon dioxide absorbed during the growth cycle of biomass. Recent reports showed that carbon dioxide emission will decrease by 95% in bioenergy generation (Munawar et al. 2021). They also reported that at present, biomass generates 5×10^{19} kJ bioenergy every year which contributes about 10% of global energy consumption and it should increase up to 150×10^{19} kJ by 2050 (Alper et al. 2020). Solid biofuels are directly converted into bioenergy. Liquid and gaseous biofuels are biomass derivatives, also known as first, second, and third generation biofuels. For example, if we decompose sugar and starch, bioethanol is produced and on transesterification of seed-based oils, biodiesel is formed. These liquid fuels are considered as first-generation biofuels. Biogas generated from burning animal and plant wastes is second generation biofuel. Third-generation biofuels are less common; they are derived from oil-producing microorganisms. Apart from these resources, municipal solid waste, agro-waste, industrial waste like sawmill dust, and paper sludge are also used to produce bioenergy. In Europe, many countries like Poland, Sweden, Denmark are stepping towards clean, green energy and their 50% of energy needs are fulfilled by bioenergy production (Saleem 2022).

Nanomaterials are small-sized materials having as a minimum one dimension in the nano range, i.e., less than 100 nm. They have greater surface area and show different size-dependent properties like higher adsorption, superior mechanical, optical, electrical properties, optimum porosity, good reactivity, superparamagnetism, greater adhering with functional groups. Their surface can be conveniently modified using thermal, physical, and chemical processes (Malpani and Goyal 2023). Due to these peculiar properties, nanomaterials are extensively used in many applications like photocatalysis, adsorption, wastewater treatment, drug delivery, energy storage, and many more (Feliczak-Guzik 2023; Liu et al. 2022; Naser et al. 2021; Zhang et al. 2023a; Farooq et al. 2024; Malpani et al. 2022). Nanomaterials improve bioenergy production by enhancing the potential of processes involved in bioenergy generation like anaerobic digestion, hydrolysis, pyrolysis, and pretreatment of biomass. In these processes, nanomaterials act as catalysts, additives, and adsorbents. They also help in improving efficiency and properties of fuel, immobilizing enzymes, thus enhancing the cost efficiency, storage capacity, and potency of bioenergy production along with minimizing chemical and energy consumption. Usually, carbon nanotubes (CNTs), graphene, metal nanoparticles, nano quantum dots have been seen in bioenergy-related applications (Markandan and Chai 2022).

Besides environmental applications, reduction in carbon footprint, greenhouse gas emissions, bioenergy generation also impart economic and social benefits, which include creating new job opportunities, societal and rural development, and meeting global energy demands. Bioenergy generation requires manpower while production phase, processing, transportation, supply chain, and distribution. If biomass and agricultural residues are utilized in bioenergy creation, it helps in rural development this way by gaining additional income. It further increases savings, purchasing power, and income diversification. In another aspect, bioenergy generation reduces our reliance on fossil fuels for our energy requirements, which decreases oil and gas prices globally and thus, reduces overall pricing of raw materials, consumer goods, and industrial supplies. It also helps in creating new avenues for market and industrial growth. Government and corporates can spend more money on public welfare. This results in community empowerment and improvements in healthcare, education, skill training, and social upliftment. Reduction in fossil fuel burning decreases air pollution and thus improves public health by moderating cardiovascular, respiratory, and dermatological disorders. Some reports have projected these socio-economic impacts of bioenergy generation, analyzed the outcomes using quantification methods, and described the limitations of ongoing practices (Brinkman et al. 2019).

In this chapter, we will provide a perception about different types of nanomaterials which are usually used for bioenergy production. Several research works discussed about synthesis of nanomaterials and their different applications. The economic and social aspects of bioenergy production are also summarized. Prospects and challenges are elucidated. This chapter also highlights that nanotechnology research can become a golden opportunity for large-scale use of green nanomaterials and it can also be helpful in the principles of circular economy and sustainable development.

8.2 Nanomaterials in Bioenergy Production

All types of nanomaterials, including zero-dimensional (like graphene, fullerenes, magnetic nanomaterials, quantum dots), one-dimensional (like carbon nanotubes, nanofibers/nanowires), two-dimensional (like carbon nanosheets, nanoflakes), and three-dimensional (like nanograins) have been utilized for bioenergy production. Usually inorganic nanomaterials like metal/metal oxide based, carbon-based, magnetic nanomaterials are used in bioenergy sector. But recently some hybrid nanomaterials which are a combination of both inorganic and organic nanomaterials also show their potential in biomass conversion (Kumar et al. 2021a). In this section,

Fig. 8.1 Different types of nanomaterials used in bioenergy production

we will discuss the different nanomaterials that have been used for bioenergy production, which is summarized in Fig. 8.1.

8.2.1 Carbon-Based Nanomaterials

Carbon-based nanomaterials like fullerenes, graphite nano oxides, graphene quantum dots, carbon nanotubes (CNTs) are majorly used in bioenergy-related applications (Khan and Rashid 2024). Their excellent biocompatibility helps in enzyme immobilization. Specifically, multi-walled CNTs are more compatible with immobilization of enzymes as compared to single-walled CNTs (Mubarak et al. 2014). Enzyme immobilization is a key step in the production of biogas, bioethanol, and biodiesel. Researchers also reported that some other types of carbon nanoparticles like sulfonated CNTs, carbon nano catalysts, acid and base-activated carbon catalysts have also shown their potential in biofuel production. Apart from catalysis, carbon-based nanomaterials can also convert biomass like agri-waste, algae into biofuels. Agri-wastes like rice husk ash, fruit peels, coffee husk are utilized in this process. Multi-walled CNTs, N-Ni-carbon nanofibers, and graphite nano granules

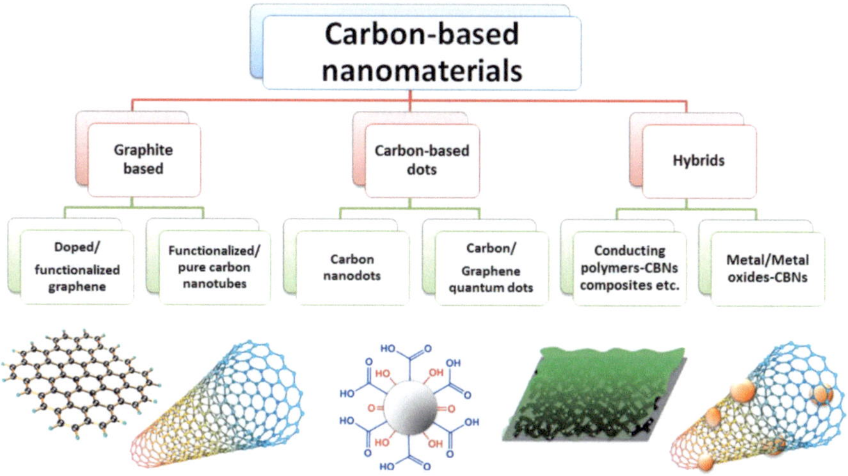

Fig. 8.2 Carbon-based nanomaterials in microbial fuel cell applications (Hoa et al. 2017)

have shown great potential as anode and cathode materials in microbial fuel cells (Hoa et al. 2017). Some of these materials are shown in Fig. 8.2.

The higher electrical conductivity, larger surface area, and porosity of amorphous carbon materials make them suitable material for electrode preparation and remarkably decrease size and weight of electrodes. They increase active surface area, conductivity, reaction kinetics, energy conversion efficiency, mass transport of fuels, power density in fuel cells. They also act as catalysts in fuel cells. Monosaccharide fuel cells, bio-ethanol fuel cells, and bio-hydrogen fuel cells are different biomass-fed fuel cells. Fullerenes have also demonstrated their activity in solar cells and nano-electronic components (Paukov et al. 2023). Fullerenes are proficient electron acceptors, when combined with electron donor materials conveniently form solar panel construction. Power conversion efficiency of solar cells can be increased by using fullerene tailoring leading to strong absorption, enhanced solubility, and miscibility. Anaerobic digestion is a crucial step in bioenergy production; single-walled CNTs can enhance direct interspecies electron transfer (DIET) in anaerobic digestion. Researchers reported that using single-walled CNTs improves consumption of raw materials and production rates of methane (Zhang et al. 2023b). These conductive materials facilitate DIET process and methane production. Similarly multi-walled CNTs also support biogas expansion and rate of methane production. Carbon nanomaterials derived from municipal solid waste incineration ash have also been used as additives in biogas production (Lo et al. 2012). They accelerate % yield of biogas during a 90-day anaerobic digestion period at 35 °C. Owing to improved thermal stability and chemical reactivity, CNTs are efficiently used in biodiesel synthesis. They performed the role of a catalyst during the process and enhanced burning rate of fuel and their cetane number (Tomar and Kumar 2020;

Soudagar et al. 2019). It also reduces ignition time, ejects less carbon monoxide than ordinary fuels, and hence, diminishes air pollution.

These nanomaterials also act as photocatalysts in water splitting and hydrogen generation. Their excellent surface area, optimum morphology, chemical reactivity, and thermal durability help carbon nanomaterials in improving absorption capacity, number of active sites, band gap narrowing, electron acceptors, and photosensitization. Apart from catalysis, they are also used in semiconductors and energy storage devices (Cao and Yu 2016; Rentería et al. 2022). The surface of carbon nanomaterials can be modified by doping, functional group grafting, interface engineering, activation by acids, and bases.

8.2.2 Silica-Based Nanomaterials

Silica-based nanomaterials are greatly used in second-generation biofuel production. If agricultural residues are used as feedstock in bioethanol production, then an extra step of their pretreatment gets included in the whole procedure. Silica-based nanomaterials are often used as catalysts in the next step of hydrolysis. They are quite cost-efficient, clean, and reusable to certain extent. Such nanomaterials are efficiently used in immobilization of many enzymes and later used in bioenergy generation. Kumari et al. observed immobilization of two enzymes, xylanase and lichenase, on silica nanoparticles through covalent bonding and used those in saccharification of biomass rich in lignin and cellulose (Kumari et al. 2018). They transformed silica nanoparticles into magnetic nanoparticles to improve the saccharification process. It is a crucial step in bioethanol production. They have recycled these nanoparticles 13 times and all enzymes are sustainably retained on nanoparticles. In another saccharification study, alkali-pretreated sugarcane bagasse was hydrolyzed using iron oxide coated silica nanoparticles (Yadav et al. 2024). In this study, Cellic CTec2 enzyme was covalently bonded on iron oxide-coated silica nanoparticles. Box Behnken design was used to optimize this process. The catalyst was recovered using magnetic field and reused five times. The procedure is shown in Fig. 8.3.

Lee et al. encapsulated iron oxide on mesoporous silica nanoparticles and used them in conversion of biomass enriched in cellulose to fructose (Lee et al. 2014). They did a comparative study with iron oxide-enzyme immobilized mesoporous silica nanoparticles and found them more effective in catalyzing destruction of glycosidic bonds of cellulose in biomass. Silica in form of zeolites act as heterogeneous catalysts in bioenergy generation. Srinivasan et al. used active acidic sites of zeolitic catalysts in pyrolysis of cellulose content of wood biomass (Srinivasan et al. 2014). In another study, it was observed that if other metals like Co, Ni, Mn, Ga, Pt, or Ce are doped on ZSM-5 zeolites, then the acidic strength and density of zeolite catalysts can be enhanced and yield of desired products can be increased (Vichaphund et al. 2014). Jeon et al. used silica nanoparticles and methyl-functionalized silica nanoparticles in culture of *Chlorella vulgaris* and improved

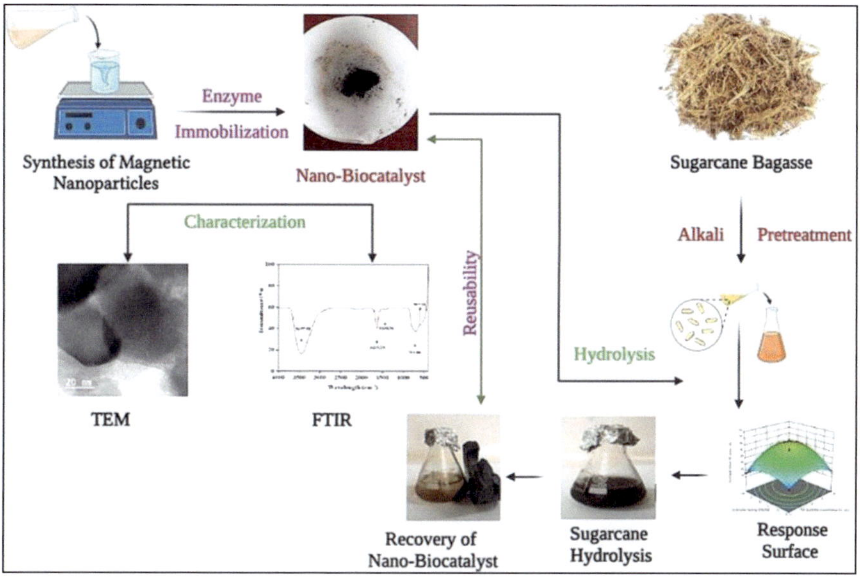

Fig. 8.3 Silica nanoparticles as nano-biocatalyst in hydrolysis of sugarcane bagasse (Yadav et al. 2024)

production of fatty acid methyl esters (Jeon et al. 2017). They concluded that methyl-functionalized silica nanoparticles are more effective in ester formation. Nanomaterials help in increasing biodiesel production by immobilizing enzymes which accelerate their stability, reusability, lifespan, and reactivity. Fatimah et al. also reviewed mesoporous silica nanoparticles as carriers for enzymes and their use in improving biodiesel yield (Fatimah et al. 2023). Such materials are reusable many times and can effectively retain enzyme activity on all cycles. These materials also help in separation of byproducts like glycerol during biodiesel synthesis, hence improve yield and purity, and lower the downstream processing costs (Ndaba et al. 2025; Kumar et al. 2021b).

8.2.3 *Metal/Metal Oxide Nanomaterials*

Metal, metal oxide nanoparticles, nanotubes, nanosheets, and nanofibers act as nanocatalysts on biofuel generation. Enzymes like hemicellulase, cellulase are magnetically immobilized on these nanomaterials and can be reused many times (Alftrén and Hobley 2013). Silver nanoparticles extracted from waste X-ray films, raw Camellia tea extracts, and oak galls have been used to produce biofuels from Jatropha oil. 85% biofuel yield and 90% Jatropha oil conversion with 76% gasoline selectivity were reported (Ganesan et al. 2020). Another work studied silver nanoparticles supported on alumina as solid catalyst in producing biodiesel from

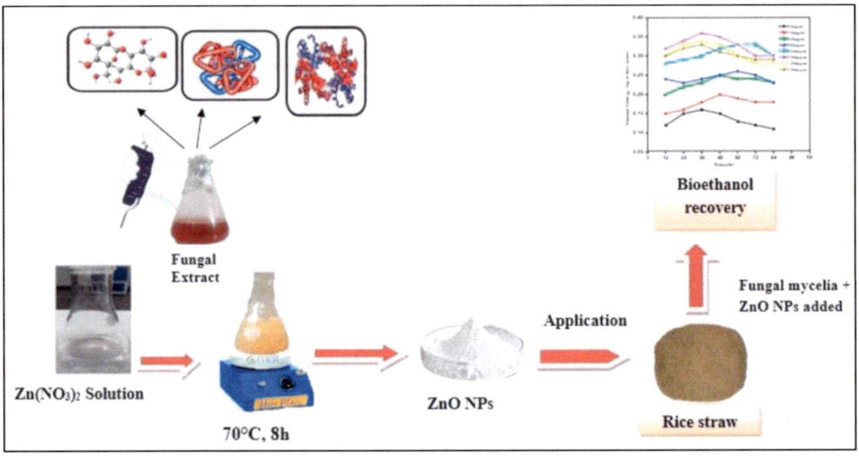

Fig. 8.4 ZnO nanoparticles in bioethanol synthesis from rice straw (Gupta and Chundawat 2020)

castor seeds. Silver nanoparticles were synthesized by chemical reduction method and then loaded on alumina support (Dankane et al. 2020). Silver nanoparticles have also been used in hydrogen production in fermentation process (Kumar et al. 2019). Gold nanoparticles with copper doping supported on spinel catalyzed aerobic oxidation of bioethanol under adequate hydrothermal conditions. Strong affinity between metal nanoparticles and support materials is responsible for their enhanced catalytic activity (Wang et al. 2013). Support materials stabilize metal nanoparticles and help in catalyst recovery and reusability. They also help in enhancing interactions with enzymes and microorganisms even at low concentration. Under anaerobic conditions, nanoparticles help bacteria in electron transfer to acceptor molecules and boost microbial activity in bioenergy production. Gold nanoparticles transfer electrons in fermentation process and produce hydrogen. Bacteria have hydrogenase enzymes in the cell which catalyze the reversible oxidation of molecular hydrogen (Engliman et al. 2017). In fuel cells, palladium nanoparticles catalyze electrochemical oxidation. Wu et al. used *Shewanella oneidensis* bacteria as a catalyst and substrate for palladium nanoparticles immobilization. These nanoparticles improve electron transmission and electrical conductance of fuel cells (Wu et al. 2018).

Biogenic ZnO nanoparticles have been used as a catalyst in ethanol production. Gupta and Chundaawat synthesized ZnO nanoparticles from a fungus, *Fusarium oxysporum*, and then used in bioethanol production by rice straw fermentation (Gupta and Chundawat 2020). This fungus also formed enzymes which decompose lignocellulosic content of rice straw to monosaccharide sugar. The sugar molecules were fermented using ZnO nanoparticles under microaerobic conditions to form ethanol. The whole process is shown in Fig. 8.4. Iron oxide nanoparticles also help in hydrogen generation through photocatalysis. They also play important role in immobilization of enzymes like hydrogenase. Fe^{2+} ions and magnetite nanoparticles concentration has demonstrated efficiency in dark fermentation studies of hydrogen production from sugarcane bagasse (Das et al. 2006). Ni-Fe and Fe-Fe hydrogenases are the common enzymes used in this process. Iron nanoparticles enhance the bioactivity of bacteria which helps in producing hydrogen. They catalyze the fermentation process, which results in higher and faster yield of hydrogen (Gadhe and Gupta 2006).

NiO nanoparticles act as biocatalysts in the saccharification and fermentation of potato peel and substantially improve the yield of bioethanol. Torija et al. showed the intervention of NiO nanoparticles in bioprocess of bioethanol formation (Torija 2003). Potato peel is a good source of starch and lignocellulose and has been known as significant biofuel raw material. Nowadays it has been gathering attention in biofuel generation. NiO and Fe_3O_4 nanoparticles can conveniently alter oxidation-reduction potential and contribute to glucose-nanoparticles electropositive interaction. These nanoparticles are strongly attracted to electrons due to their small size and redox potential values (Zhang and Shen 2006). The attraction between nanoparticles and bacteria grew stronger under anaerobic conditions. It ultimately leads to faster and better fermentation processes and higher yield of bioethanol and hydrogen production. Gadhe and co-workers studied the effect of NiO and Fe_3O_4 nanoparticles in hydrogen production under anaerobic conditions using industrial wastewater generated from dairy and molasses industry (Gadhe et al. 2015).

SrO nanoparticles have been suitably used as a heterogeneous catalyst in transesterification reactions producing biodiesel. SrO nanoparticles are highly basic and can be easily modified by activation techniques. Owing to their large number of basic sites, SrO nanoparticles proved to be a highly effective catalyst, converting a variety of lipid-enriched raw materials to biodiesel, giving approximately 100% yield (Islam et al. 2014). They are cost-effective catalysts, possess greater surface area and nano porous, and thus give better selectivity of desired products. In transesterification reactions, oil reacts with alcohol in the presence of strong acidic or basic catalysts. Similarly, microalgae can also be converted into biodiesel at slower speed. For example, *Nannochloropsis* microalgae has been turned 100% into biodiesel in around 3 min using SrO nanoparticles in microwave and ultrasound radiation (Koberg et al. 2010). Strontium titanate nanoparticles have been reported as potential heterogeneous photocatalysts. If SrO nanoparticles are supported on a stable matrix like silica or alumina, then their activity and reusability increase up to multiple folds. They increase photon absorption, minimize electron loss during excitation steps, and enhance yield of bioethanol under solar radiation, UV, or visible light. Researchers are now working on introducing dopants on SrO nanoparticles, forming a binary photocatalyst for water splitting applications, their morphology, functioning, and many more aspects (Phoon et al. 2019).

8.3 Economic Impacts of Bioenergy Production

Sustainable development refers to three pillars comprising economic, social, and environmental impacts. Sometimes if economic and social impacts are stronger, then they can balance the detrimental effects of sustainable development on environment. Bioenergy refers to Sustainable Development Goal No. 7 of United Nations, which means access to clean energy. In literature, several reports are documented focusing on environmental impacts of bioenergy production and climate change areas, but very less works have been reported assessing economic and social impacts. Economic aspects deal with the evaluation of costs and the revenues expected. Bioenergy production focuses on energy generation in a cost-efficient manner compared to other traditional methods of energy production. In most of the cases, cost of raw materials is less, so the overall procedure cost also decreases. Government also provides tax concessions and subsidies for bioenergy production. They also provide loans on less interest rates to set up plants for bioenergy generation. Some quantification methods and indicators are used to project and assess economic impacts of bioenergy production. Indicators study different variables for assessment (Van Eijck and Faaij 2014). These variables include cost and price volatility, the cost of carbon dioxide deduction, the financial viability of production, and the social, community benefits harnessed by the bioenergy production projects. The common indicators are per capita income, economic productivity, number of job opportunities created, financial viability, and gender equality. Other less common indicators include work hours, investment in R&D, land tenure, access to bioenergy, cost of education, etc. Some *ex ante* or forecast methods are studied on the potential positive and negative effects of bioenergy production and the equilibrium between both effects (Brinkman et al. 2019). All these results are used by government and policy makers to diminish the unwanted effects. Nowadays mathematical, statistical models, computation techniques, Artificial Intelligence, and Machine Learning methods have been used in projecting socio-economic aspects of bioenergy generation.

Using nanomaterials in generating bioenergy significantly reduces cost by process optimization. Traditional catalysts often require high energy inputs and frequent replacement. Catalysts based on nanomaterials, on the other hand, show better selectivity, thermal stability, and recyclability.

This leads to:

- Reduced consumption of raw materials
- Cost-efficiency and abundance of feedstock
- Lower energy expenditure
- Less downtime in industrial operations

For example, Au(III)-calcium oxide nanocatalyst synthesized from waste shells can catalyze biodiesel production at lower temperatures with minimal catalyst loading (Liu et al. 2020). Similarly, nanomaterial-assisted enzyme immobilization enhances enzyme stability and reusability, significantly reducing enzyme costs in bioethanol production (Gupta and Chundawat 2020). The optimization of process parameters through nanotechnology thus translates into quantifiable financial savings and higher production efficiency. Nanomaterials improve not only the economics but also the technical performance of bioenergy processes. Their ability to facilitate efficient energy conversion increases the overall yield of biofuels such as bioethanol, biodiesel, and biohydrogen. In photocatalytic hydrogen generation, water-splitting process, metal-doped nanostructures, and quantum dots enhance light absorption and electron transfer, leading to greater hydrogen output from the same biomass feedstock (Gadhe et al. 2015). This improvement in energy yield directly contributes to reduced costs per unit of energy produced, making bioenergy more competitive with fossil-based alternatives. Moreover, nanocatalysts ensure complete and selective conversion of feedstock, minimizing by-products and waste generation. This efficient resource utilization strengthens the economic case for bioenergy systems employing nanomaterials.

The combination of nanotechnology and bioenergy has catalyzed the formation of a new industrial niche, the nano bioenergy sector. This emerging market attracts cross-disciplinary investments from nanomaterial producers, bioenergy companies, and green technology startups. The economic implications include:

- Diversification of industrial portfolios through nanotechnology-driven bioenergy projects
- Increased export potential for biofuels produced via advanced catalytic routes
- Strengthening energy independence by reducing reliance on imported fossil fuels

As nanomaterial synthesis becomes more cost-effective with scale, the profit margins for bioenergy producers improve, stimulating industrial competitiveness and accelerating the transition toward sustainable energy economies. Nanomaterial-based bioenergy systems promote inclusive economic growth by creating employment opportunities and supporting rural development. Bioenergy plants established near agricultural regions utilize local biomass, generating income for farmers and reducing waste disposal costs. This decentralized energy production model fosters:

- Temporary/permanent job opportunities in feedstock collection, processing, and maintenance
- Skill development in advanced material handling and green engineering
- Infrastructure growth in rural and semi-urban areas

Thus, nanotechnology-enabled bioenergy not only contributes to energy transition but also advances social equity and regional development. The long-term economic success of nanomaterial-based bioenergy depends on investment and policy support. Governments can encourage this sector by offering incentives for renewable energy projects and funding for nanotechnology research. From an investor's perspective, nanomaterials improve the return on investment (ROI) by enhancing process efficiency, scalability, and reliability. Additionally, promoting safe and recyclable nanomaterials ensures environmental compliance and financial sustainability. The convergence of sound policy frameworks, technological innovation, and public-private partnerships will be essential for realizing the full economic potential of nanomaterial-assisted bioenergy systems.

8.4 Social Impacts of Bioenergy Production

The bioenergy sector powered by nanomaterials offers significant employment potential across multiple levels. From research laboratories to industrial production and rural biomass collection, a variety of jobs are being created. The integration of nanomaterials requires skilled personnel for catalyst synthesis, material handling, and process monitoring, encouraging skilled education and technical training. Such initiatives nurture local capacity building, particularly in developing countries. As communities engage in biomass and feedstock supply chains and decentralized energy systems, income diversification and rural employment rise. Workforce training programs in bioenergy and manufacturing of nanomaterials also foster a new generation of professionals equipped for the green economy.

Bioenergy production systems are often located near agricultural or forestry regions, making them vital for rural empowerment. With nanomaterials improving the conversion efficiency of biomass, small-scale bioenergy plants can generate affordable and reliable local power, reducing dependence on fossil fuels and centralized grids. This accessibility enhances quality of life through improved fundamental requirements like food, clothing, accommodation, electricity, as well as advanced ones, digital connectivity, and agricultural productivity. Rural communities benefit economically by selling agricultural residues and participating in energy cooperatives. The decentralized nature of bioenergy systems also encourages community ownership and participation, strengthening social cohesion and self-reliance.

Conventional energy sources contribute to air pollution, respiratory disorders, and environmental degradation. Nanomaterial-based bioenergy systems offer cleaner combustion and lower emissions, improving air quality and reducing public health risks (Khoo et al. 2020). In addition, nanomaterials used in bioenergy systems can play a dual role in wastewater purification and pollution control, reducing exposure to pollutants in both air and water. As a result, communities benefit from a healthier environment, lower medical costs, and increased well-being. However, the safe handling of nanomaterials is crucial to avoid accidental health risks. Public awareness and proper safety protocols are essential to mitigate any concerns about nanotoxicity.

The encouraging use of nanomaterials in generating bioenergy promotes education and research collaboration. Universities and research institutes are now offering advanced courses and certification programs in relatively new fields like green nanotechnology, development of innovative nanomaterials, clean and renewable energy in their curriculum, preparing students for future energy challenges. Public outreach programs, science communication initiatives, and community workshops on bioenergy and nanomaterials enhance scientific literacy and empower citizens to make informed choices. Exposure to innovative technologies also inspires young minds to pursue careers in science and engineering, strengthening the knowledge economy. Energy access and technological advancement have profound implications for gender equality and social inclusion. In many rural areas, women are primarily responsible for gathering conventional fuels. Access to bioenergy reduces this burden, allowing more time for education, entrepreneurship, and participation in community development. While nanomaterial-assisted bioenergy presents significant social benefits, it also raises ethical and regulatory challenges (Markandan and Chai 2022). Concerns about nanoparticle toxicity, waste management, and long-term environmental effects must be addressed through rigorous safety standards and transparent public communication. Governments and industries must adopt responsible innovation frameworks—ensuring legal sourcing, environmental protection, and reasonable access. Social acceptance is a key factor in the success of emerging energy technologies. Public perception of nanotechnology and use of nanomaterials is shaped by understanding, trust, and communication. Awareness campaigns that highlight the benefits and safety measures of nanomaterials in bioenergy can help dispel misconceptions and foster confidence. Community engagement, participatory pilot projects, and open discussions create mutual understanding between developers and citizens, ensuring that the technology is both socially acceptable and locally beneficial (Edrisi and Abhilash 2015).

8.5 Conclusion

This chapter outlines the concept of use of nanomaterials in producing bioenergy and the socio-economic impacts of bioenergy production using them. Some research works were discussed in the chapter related to the advancements in using different types of nanomaterials for producing biodiesel, hydrogen, and bioethanol through different ways. It sketches a picture of different classes of nanomaterials like carbon-based, silica-based, metal/metal oxide based, magnetic nanomaterials in the field of bioenergy production. The optimum features of nanomaterials like high surface area, smaller size, excellent thermal stability, chemical reactivity, catalytic activity, and reusability make generation of bioenergy more convenient. This work also demonstrates that bioenergy production using nanomaterials has impacted our society in various economic and social manners. As per industrial and environmental perspective, nanomaterials favor clean, green chemical and scalable processes with enhanced yields, diminishing carbon footprints, energy use, and waste. From an economic perspective, the use of nanomaterials in bioenergy production reduces production costs, creates new employment opportunities and also helps in achieving the sustainable development goals. This chapter reviews the types of nanomaterials and their use in bioenergy production, economic and social impacts of using nanomaterials in bioenergy production, while focusing on challenges and outlining future opportunities in this area.

References

Alftrén J, Hobley TJ (2013) Covalent immobilization of B-glucosidase on magnetic particles for lignocellulose hydrolysis. Appl Biochem Biotechnol 169(7):2076–2087. https://doi.org/10.1007/s12010-013-0122-5

Alper K, Tekin K, Karagöz S, Ragauskas AJ (2020) Sustainable energy and fuels from biomass: a review focusing on hydrothermal biomass processing. Sustain Energy Fuels 4(9):4390–4414. https://doi.org/10.1039/d0se00784f

Brinkman ML, Wicke B, Faaij AP, Van Der Hilst F (2019) Projecting socio-economic impacts of bioenergy: current status and limitations of ex-ante quantification methods. Renew Sust Energ Rev 115:109352. https://doi.org/10.1016/j.rser.2019.109352

Cao S, Yu J (2016) Carbon-based H2-production photocatalytic materials. J Photochem Photobiol C Photochem Rev 27:72–99. https://doi.org/10.1016/j.jphotochemrev.2016.04.002

Dadi M, Siwale W, Munalula F, Ravi SN, Mundike J, Ramasamy S, Mulenga C (2025) A comprehensive review of advances in bioenergy including emerging trends and future directions. Discov Energy 5(1):26. https://doi.org/10.1007/s43937-025-00095-3

Dankane I, Chapi S, Bello K (2020) The potential of silver nanoparticles as heterogeneous catalysts for generation of biodiesel from vegetable oils. AIP Conf Proc 2244:110017. https://doi.org/10.1063/5.0010272

Das D, Dutta T, Nath K, Kotay SM, Das AK, Veziroglu TN (2006) Role of Fe-hydrogenase in biological hydrogen production. Curr Sci 90(12):1627–1637. http://www.iisc.ernet.in/currsci/jun252006/1627.pdf

Edrisi SA, Abhilash PC (2015) Book review: socio-economic impacts of bioenergy production. Front Bioeng Biotechnol 3:310. https://doi.org/10.3389/fbioe.2015.00174

Engliman NS, Abdul PM, Wu S, Jahim JM (2017) Influence of iron (II) oxide nanoparticle on biohydrogen production in thermophilic mixed fermentation. Int J Hydrog Energy 42(45): 27482–27493. https://doi.org/10.1016/j.ijhydene.2017.05.224

Farooq N, Rehman ZU, Khan MI, Asghar S, Saleem M, Irshad R, Sheikh A, Shanableh A, Manzoor S, Khan ZU (2024) Nanomaterial-based energy conversion and energy storage devices: a comprehensive review. New J Chem 48(19):8933–8962. https://doi.org/10.1039/d3nj04846b

Fatimah I, Fadillah G, Sagadevan S, Oh W, Ameta KL (2023) Mesoporous silica-based catalysts for biodiesel production: a review. Chem Eng 7(3):56. https://doi.org/10.3390/chemengineering7030056

Feliczak-Guzik A (2023) Nanomaterials as photocatalysts-synthesis and their potential applications. Materials 16(1):193. https://doi.org/10.3390/ma16010193

Gadhe J, Gupta R (2006) Hydrogen production by methanol reforming in supercritical water: catalysis by in-situ-generated copper nanoparticles. Int J Hydrog Energy 32(13):2374–2381. https://doi.org/10.1016/j.ijhydene.2006.10.050

Gadhe A, Sonawane SS, Varma MN (2015) Enhancement effect of hematite and nickel nanoparticles on biohydrogen production from dairy wastewater. Int J Hydrog Energy 40(13): 4502–4511. https://doi.org/10.1016/j.ijhydene.2015.02.046

Ganesan R, Narasimhalu P, Joseph AIJ, Pugazhendhi A (2020) Synthesis of silver nanoparticle from X-ray film and its application in production of biofuel from jatropha oil. Int J Energy Res 45(12):17378–17388. https://doi.org/10.1002/er.6106

Gupta K, Chundawat TS (2020) Zinc oxide nanoparticles synthesized using Fusarium oxysporum to enhance bioethanol production from rice-straw. Biomass Bioenergy 143:105840. https://doi.org/10.1016/j.biombioe.2020.105840

Hoa L, Vestergaard M, Tamiya E (2017) Carbon-based nanomaterials in biomass-based fuel-fed fuel cells. Sensors (Basel) 17(11):2587. https://doi.org/10.3390/s17112587

Islam A, Taufiq-Yap YH, Chan E, Moniruzzaman M, Islam S, Nabi MN (2014) Advances in solid-catalytic and non-catalytic technologies for biodiesel production. Energy Convers Manag 88: 1200–1218. https://doi.org/10.1016/j.enconman.2014.04.037

Jeon H, Park SE, Ahn B, Kim Y (2017) Enhancement of biodiesel production in Chlorella vulgaris cultivation using silica nanoparticles. Biotechnol Bioprocess Eng 22(2):136–141. https://doi.org/10.1007/s12257-016-0657-8

Khan NR, Rashid AB (2024) Carbon-based nanomaterials: a paradigm shift in biofuel synthesis and processing for a sustainable energy future. Energy Convers Manage X 22:100590. https://doi.org/10.1016/j.ecmx.2024.100590

Khoo KS, Chia WY, Tang DYY, Show PL, Chew KW, Chen W (2020) Nanomaterials utilization in biomass for biofuel and bioenergy production. Energies 13(4):892. https://doi.org/10.3390/en13040892

Koberg M, Cohen M, Ben-Amotz A, Gedanken A (2010) Bio-diesel production directly from the microalgae biomass of nannochloropsis by microwave and ultrasound radiation. Bioresour Technol 102(5):4265–4269. https://doi.org/10.1016/j.biortech.2010.12.004

Kumar G, Mathimani T, Rene ER, Pugazhendhi A (2019) Application of nanotechnology in dark fermentation for enhanced biohydrogen production using inorganic nanoparticles. Int J Hydrog Energy 44(26):13106–13113. https://doi.org/10.1016/j.ijhydene.2019.03.131

Kumar P, Khan M, Hu J, Kibria MG (2021a) Single-atom catalysts for biomass-derived drop-in chemicals. In: Advanced catalysis for drop-in chemicals. Elsevier eBooks, pp 63–100. https://doi.org/10.1016/b978-0-12-823827-1.00009-2

Kumar Y, Yogeshwar P, Bajpai S, Jaiswal P, Yadav S, Pathak DP, Sonker M, Tiwary SK (2021b) Nanomaterials: stimulants for biofuels and renewables, yield and energy optimization. Mater Adv 2(16):5318–5343. https://doi.org/10.1039/d1ma00538c

Kumari A, Kaila P, Tiwari P, Singh V, Kaul S, Singhal N, Guptasarma P (2018) Multiple thermostable enzyme hydrolases on magnetic nanoparticles: an immobilized enzyme-mediated approach to saccharification through simultaneous xylanase, cellulase and amylolytic glucanotransferase action. Int J Biol Macromol 120:1650–1658. https://doi.org/10.1016/j.ijbiomac.2018.09.106

Lee Y, Dutta S, Wu KC (2014) Integrated, cascading enzyme-/chemocatalytic cellulose conversion using catalysts based on mesoporous silica nanoparticles. ChemSusChem 7(12):3241–3246. https://doi.org/10.1002/cssc.201402605

Liu J, Liu M, Chen S, Wang B, Chen J, Yang D, Zhang S, Du W (2020) Conversion of Au(III)-polluted waste eggshell into functional CaO/Au nanocatalyst for biodiesel production. Green Energy Environ 7(2):352–359. https://doi.org/10.1016/j.gee.2020.07.019

Liu W-X, Song S, Ye M-L, Zhu Y, Zhao Y-G, Lu Y (2022) Nanomaterials with excellent adsorption characteristics for sample pretreatment: a review. Nanomaterials 12(11):1845. https://doi.org/10.3390/nano12111845

Lo H, Chiu H, Lo S, Lo F (2012) Effects of micro-nano and non micro-nano MSWI ashes addition on MSW anaerobic digestion. Bioresour Technol 114:90–94. https://doi.org/10.1016/j.biortech.2012.03.002

Malpani SK, Goyal D (2023) Synthesis, analysis, and multi-faceted applications of solid wastes-derived silica nanoparticles: a comprehensive review (2010–2022). Environ Sci Pollut Res Int 30(11):28321–28343. https://doi.org/10.1007/s11356-022-23873-1

Malpani SK, Meena RK, Katara S, Goyal D (2022) Agro-industrial wastes-derived carbon nanomaterials: synthesis, characterization, and applications. In: Innovations in green nanoscience and nanotechnology, 1st edn. CRC Press, p 28. https://doi.org/10.1201/9781003319153

Markandan K, Chai WS (2022) Perspectives on nanomaterials and nanotechnology for sustainable bioenergy generation. Materials 15(21):7769. https://doi.org/10.3390/ma15217769

Mubarak N, Wong J, Tan K, Sahu J, Abdullah E, Jayakumar N, Ganesan P (2014) Immobilization of cellulase enzyme on functionalized multiwall carbon nanotubes. J Mol Catal B Enzym 107:124–131. https://doi.org/10.1016/j.molcatb.2014.06.002

Munawar MA, Khoja AH, Naqvi SR, Mehran MT, Hassan M, Liaquat R, Dawood UF (2021) Challenges and opportunities in biomass ash management and its utilization in novel applications. Renew Sust Energ Rev 150:111451. https://doi.org/10.1016/j.rser.2021.111451

Naser JA, Ahmed ZW, Ali EH (2021) Nanomaterials usage as adsorbents for the pollutants removal from wastewater; a review. Mater Today Proc 42:2590–2595. https://doi.org/10.1016/j.matpr.2020.12.584

Ndaba B, Rama H, Bingwa N, Roopnarain A (2025) Unravelling the role of nanoparticles during bioethanol production: a review on pretreatment, hydrolysis and fermentation. Fuel 396: 135336. https://doi.org/10.1016/j.fuel.2025.135336

Paukov M, Kramberger C, Begichev I, Kharlamova M, Burdanova M (2023) Functionalized fullerenes and their applications in electrochemistry, solar cells, and nanoelectronics. Materials 16(3):1276. https://doi.org/10.3390/ma16031276

Phoon BL, Lai CW, Juan JC, Show P, Pan G (2019) Recent developments of strontium titanate for photocatalytic water splitting application. Int J Hydrog Energy 44(28):14316–14340. https://doi.org/10.1016/j.ijhydene.2019.01.166

Rather RA, Wani AW, Mumtaz S, Padder SA, Khan AH, Almohana AI, Almojil SF, Alam SS, Baba TR (2021) Bioenergy: a foundation to environmental sustainability in a changing global climate scenario. J King Saud Univ Sci 34(1):101734. https://doi.org/10.1016/j.jksus.2021.101734

Rentería J, Gallego A, Gamboa D, Cacua K, Herrera B (2022) Effect of amide-functionalized carbon nanotubes as commercial diesel and palm-oil biodiesel additives on the ignition delay: a study on droplet scale. Fuel 338:127202. https://doi.org/10.1016/j.fuel.2022.127202

Saleem M (2022) Possibility of utilizing agriculture biomass as a renewable and sustainable future energy source. Heliyon 8(2):e08905. https://doi.org/10.1016/j.heliyon.2022.e08905

Soudagar MEM, Nik-Ghazali N, Kalam M, Badruddin IA, Banapurmath N, Khan TY, Bashir MN, Akram N, Farade R, Afzal A (2019) The effects of graphene oxide nanoparticle additive stably dispersed in dairy scum oil biodiesel-diesel fuel blend on CI engine: performance, emission and combustion characteristics. Fuel 257:116015. https://doi.org/10.1016/j.fuel.2019.116015

Srinivasan V, Adhikari S, Chattanathan SA, Tu M, Park S (2014) Catalytic pyrolysis of raw and thermally treated cellulose using different acidic zeolites. Bioenergy Res 7(3):867–875. https://doi.org/10.1007/s12155-014-9426-8

Tomar M, Kumar N (2020) Effect of multi-walled carbon nanotubes and alumina nano-additives in a light duty diesel engine fuelled with schleichera oleosa biodiesel blends. Sustain Energy Technol Assess 42:100833. https://doi.org/10.1016/j.seta.2020.100833

Torija M (2003) Effects of fermentation temperature on the strain population of Saccharomyces cerevisiae. Int J Food Microbiol 80(1):47–53. https://doi.org/10.1016/s0168-1605(02)00144-7

Van Eijck J, Faaij APC (2014) Analysis of socio-economic indicators on different bioenergy case studies. In: Socio-economic impacts of bioenergy production. Springer eBooks, pp 267–284. https://doi.org/10.1007/978-3-319-03829-2_16

Vichaphund S, Aht-Ong D, Sricharoenchaikul V, Atong D (2014) Production of aromatic compounds from catalytic fast pyrolysis of Jatropha residues using metal/HZSM-5 prepared by ion-exchange and impregnation methods. Renew Energy 79:28–37. https://doi.org/10.1016/j.renene.2014.10.013

Wang H, Fan W, He Y, Wang J, Kondo JN, Tatsumi T (2013) Selective oxidation of alcohols to aldehydes/ketones over copper oxide-supported gold catalysts. J Catal 299:10–19. https://doi.org/10.1016/j.jcat.2012.11.018

Wu R, Tian X, Xiao Y, Ulstrup J, Christensen HEM, Zhao F, Zhang J (2018) Selective electrocatalysis of biofuel molecular oxidation using palladium nanoparticles generated on Shewanella oneidensis MR-1. J Mater Chem A 6(23):10655–10662. https://doi.org/10.1039/c8ta01318g

Yadav A, Rani P, Yadav DK, Mittal R, Gupta A, Bishnoi NR (2024) Comparative study for enzymatic hydrolysis of sugarcane bagasse using free and nanoparticle immobilized holocellulolytic enzyme cocktail. Waste Manag Bull 2(1):191–202. https://doi.org/10.1016/j.wmb.2024.01.007

Zhang Y, Shen J (2006) Enhancement effect of gold nanoparticles on biohydrogen production from artificial wastewater. Int J Hydrog Energy 32(1):17–23. https://doi.org/10.1016/j.ijhydene.2006.06.004

Zhang Y, Poon K, Masonsong GSP, Ramaswamy Y, Singh G (2023a) Sustainable nanomaterials for biomedical applications. Pharmaceutics 15(3):922. https://doi.org/10.3390/pharmaceutics15030922

Zhang Q, Yang Y, Hou L, Zhu H, Zhang Y, Pu J, Li Y (2023b) Recent advances of carbon-based additives in anaerobic digestion: a review. Renew Sust Energ Rev 183:113536. https://doi.org/10.1016/j.rser.2023.113536

Chapter 9
Evaluating Analysis Effect of Nanotechnology Versus Green Nanotechnology Over Environment and Economy

Rani Padmini Velamakanni, Vagdevi Pally, Jahnavi Alwala, Farheen Naaz, and Ramchander Merugu

Abstract The advent of nanotechnology has enabled widespread applications in fields such as medicine and energy. However, nanotechnological techniques also raise concerns, including toxicity, environmental persistence, high energy demands during synthesis, and increased production costs. These challenges have encouraged and supported interest in green nanotechnology, an approach that integrates the principles of green chemistry and sustainable practices into the design, production, and application of nanomaterials. Green nanotechnology emphasizes the use of environmentally friendly systems such as plants and microorganisms for nanomaterial synthesis. These biological systems are energy-efficient and significantly reduce toxic impacts.

This study highlights the advantages of green nanotechnology over conventional nanotechnology from both environmental and economical perspectives. From an environmental standpoint, green nano technology minimizes hazardous waste generation and enhances biodegradability, thereby reducing long-term ecological damage. Economically, it can lower costs by managing resources efficiently, reducing the number of processing steps, using versatile feedstocks, and the creation of new opportunities for sustainable industrial applications. The study suggests that the transition from conventional nanotechnology to green nanotechnology not only addresses environmental sustainability but also encourages economic viability, paving the way for future innovations. This evaluation highlights the urgent need

R. P. Velamakanni · V. Pally · R. Merugu (✉)
Department of Biochemistry, Mahatma Gandhi University, Nalgonda, Telangana, India

J. Alwala
Department of Biochemistry, Chaitanya Deemed to be University, Himayat Nagar, Hyderabad, Telangana, India

F. Naaz
Center for Gene Regulation in Health and Disease, Department of Biological, Geological, and Environmental Sciences, College of Arts and Sciences, Cleveland State University, Cleveland, OH, USA

N. Srivastava (ed.), *Sustainable Nanostrategies for Biorefineries*, Clean Energy Production Technologies, https://doi.org/10.1007/978-981-95-5875-9_9

for collaborative research, industry–academia collaborations, and supportive policy frameworks to accelerate the adoption of environmentally responsible nanotechnological solutions.

Keywords Green nanotechnology · Nanotechnology · Sustainability · Environment friendly · Economic viability · Bio-based systems.

9.1 Introduction

Nanotechnology involves the design, characterization, production, and application of materials, devices, and systems by precisely manipulating shape and size at the nano scale (Iadiz et al. 2017). Nanotechnology has emerged as a competitive scientific technology with many developed countries striving hard to gain dominance in this particular area. The nano scale is typically defined as the range spanning from 1 to 100 nm. Research and advancements in this field of technology have profoundly influenced nearly every aspect of modern human life (Bhushan 2010). Continuous improvements across diverse areas of research have sustained strong interest among investigators and scientists in adopting this technology. Its applications span multiple sectors, including agriculture, energy, food and packaging, and the petroleum industry, with healthcare and medicine emerging as the most prominent fields of application. In medicine, nanotechnology is considered to be a highly significant and transformative tool. The rapid advancement of nanotechnological techniques, along with their expanding applications and unique properties, has extended their impact across virtually every branch of science, engineering, technology, and industry. The crucial factors that make nanotechnology attractive to researchers include its relatively low production cost, its ability to be manufactured in larger quantities with lower energy requirements, the ability to precisely tailor material properties by controlling particle size, structure, and synthesis conditions, and its generally favorable safety profile for both human health and the environment (Roco 2016). Green nanotechnology is defined as the technology that uses environmentally friendly synthesis approaches, such as plant extracts, microbes, and biowastes including egg shells, fruit and vegetable peels, and agricultural residue to minimize the use of harmful chemicals that pose a risk to the environment (Zhang et al. 2020; Moradi et al. 2021). This approach is inexpensive, simple, safe, and ecologically responsible (Kalpana and Devi 2018; Khodadadi et al. 2021). The primary objective of green nanotechnology is to promote cleaner and more accessible technologies that reduce the release of pollutants and protect the environment. It also aims to convert bioactive resources into ecofriendly nanomaterials. However, finding safe and cost-effective solutions to address to this significant environmental burden associated with conventional methods remains difficult (Murthy et al. 2020; Geetha et al. 2013; Iravani 2014). In recent years, many researchers have explored strategies to benefit from these biological resources by recycling them into environmentally beneficial and acceptable products (Aravind Kumar et al. 2019). Green nanotechnology represents one such strategy, employing recyclable bioresources for nanomaterial production (Hassanisaadi et al. 2021; Bukhari et al. 2021; Singh et al. 2018). This method does not require high pressure, excessive energy input, or toxic substances. It is a biosafe and economical approach for fabricating nanomaterials with applications in catalysis, solar cells, medicines, industrial processes, sensors, water desalination, water purification, and air purification (Aravind et al. 2021; Kumar et al. 2019).

9.2 Nanotechnology Versus Green Nanotechnology

The combination of art and science, involving the manipulation of matter at the nano scale, is called nanotechnology. For thousands of years, humans have understood the world through what was seen, visible, and observable like waterfalls, forests, rocks, etc. But as scientific exploration advanced, new ways of understanding our surroundings also emerged. Theories on atoms, molecules, and their sub-divisions developed and the way we perceived and engaged with our surroundings changed completely (Hullmann and Meyer 2003). Now, the analysis has penetrated even deeper levels to that of the nanometer. A nanometer equals one billionth of a meter (1×10^{-9}) (Enin 2014). Two examples for clarity of the concept: (1) the bacterium has 200 nm of length and the diameter 2 nm for one strand of DNA 0.2) If a standard highway lane represents a string of nanometers, it would require over half a million miles (about 568, 818 miles) to match the length of one yard (0.9144 m). In other words, almost more than 22 times around the globe, or covering more than twice the distance from the earth to the moon. By operating at the nanometer scale, researchers are able to manipulate matter in ways that were previously unimaginable. When we venture into nanotechnology, we generally work with atoms and molecules, enabling precise control at extremely small dimensions. In recent years, scientists have started constructing materials from individual atoms or small clusters of atoms, which is why nanotechnology is sometimes called “atom technology”. They have successfully developed materials that can convert sunlight into electricity or make use of nanoparticles to provide anti-cancer drug delivery. Ongoing research is also focused on creating shatter-resistant glass, batteries that last longer and fabrics that smell fresh even after repeated use Soni 2014). By altering fundamental building blocks, a wide range of surface properties can be engineered into materials. Nanotechnology integrates the principles of physics, chemistry, biology, engineering, and design, making it a powerful interdisciplinary field. Its future impact depends on the extent to which we integrate nanotechnology-enabled products into our lives. This chapter explores this emerging frontier and examines its potential to reshape our future through “The Science of Small”.

Green nanotechnology emphasizes adopting responsible practices from the initial stages of development by designing and utilizing nanomaterials that support sustainability. Nanotechnology involves the imaging, measurement, modeling, and manipulation of materials at the nanoscale, where materials show distinctive physical, chemical, and biological characteristics that are different from those of bulk materials or individual atoms and molecules. Research and development in nanotechnology focus on understanding and designing advanced materials, devices, and systems that exploit novel properties at the nanoscale. This field is expected to lead to cleaner and more sustainable production processes, smaller and faster computing systems, and more effective and stronger ways to treat disease. Nanotechnology holds significant promise for groundbreaking discoveries. The benefits of metal nanoparticles and nanoclay as additives in polymer Nano composites have been reported by several researchers (Sherly et al. 2011). Green nanotechnology is defined as the application of nanotechnology to enhance clean technologies in a way that minimizes the risks to human health and the environment. It combines the use of the products of nanotechnology with environmentally responsible production methods. Green nanotechnology encourages the replacement of conventional products with newly developed nano-based alternatives that are more environmentally friendly (Smith 2011). The word "technology" refers to the practical application of knowledge. The field of green nanotechnology includes continually evolving methods and materials, ranging from energy-generation technologies to non-toxic cleaning products(Fleischer and Grunwald 2008). It is anticipated that green nanotechnology will bring transformative changes to daily life of the same magnitude as those introduced by information technology (Karn 2008). The ultimate goal of green nanotechnology is to develop technologies that support the manufacture of nano-pigments in significantly more eco-friendly methods, using renewable raw materials rather than conventional technologies. To achieve this objective, green nanotechnology aims to fulfill the following goals:

A. Synthesis of nanopigments based on nanoclays and natural and or synthetic dyes (Clarke and Anliker 1980);
B. Development of surface treatment technologies for nanoclays to enhance dye fixation;
C. Surface modification of colored nanoclays for improved lightfastness, weather resistance, and dispersibility;
D. Application of greener nano-pigments in printing inks, paints, and coatings (Bamfield and Hutchings 2010);
E. Utilization of greener nanopigments (Fig. 9.1) in photovoltaic devices;
F. Assessment of the environmental impacts of greener nano-pigments;
G. Commercial exploitation and distribution of greener nanopigments.

9.3 Challenges of Green Nanotechnology

Green nanotechnology (GNT) incorporates the principles of nanotechnology with sustainability in order to reduce impact on the environment and on health. By making use of green chemistry, regenerative resources, and sustainable energy operations, this method guarantees that nanomaterials are produced and utilized in ways that are safe to both humans and the environment (Nasrollahzadeh et al. 2019;

Fig. 9.1 Green nanotechnology for environment and economy

Singh et al. 2022). Vital aspects involve the use of non-hazardous decomposable materials, green solvents, and sustainable synthesis methods that decrease toxic waste and energy consumption (Menegaldo et al. 2023). A life cycle assessment (LCA) approach assures that nanomaterials are developed with minimal toxicity and the capability to recycle in mind, reducing pollution (Ali and Ghareeb 2023) and supporting environmental protection (Tyagi et al. 2016), waste reduction, process optimization, and recycling.

The potential of green nanotechnology spans multiple sectors, addressing global challenges such as sustainable agriculture, clean energy, and water treatment. It plays a key role in advancing the United Nations Sustainable Development Goals (SDGs). In the energy sector, green nanotechnology enables improved energy generation and storage technologies, thereby supporting SDG 7 (Affordable and Clean Energy). Developments in water purification support SDG 6 (Clean Water and Sanitation) (Aithal and Aithal 2021; Dhahri and Omri 2018). Furthermore, green nanotechnology supports climate action (SDG 13), the preservation of marine environments (SDG 14), and the conservation of terrestrial ecosystems (SDG 15). To achieve these objectives, there needs to be collaboration among governments, industries, and civil society, in line with SDG 17 (Partnerships for the Goals) (Lu and Ozcan 2015). In addition, green nanotechnology promotes SDG 14 (the preservation of marine atmospheres), SDG 13 (supporting climate action) and SDG 15 (the conservation of terrestrial ecosystems). Green nanotechnology also assists SDG

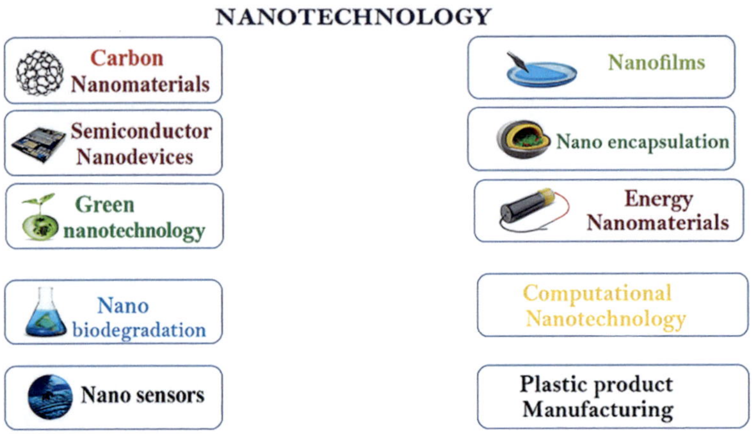

Fig. 9.2 Trends in nanotechnology

12 (Responsible Consumption and Production) by facilitating waste reduction and clean production methods.

As research in green nanotechnology (GNT) advances, it is increasingly recognized as a vital tool for balancing economic growth with environmental protection and human wellbeing. GNT plays a vital role in achieving the Sustainable Development Goals (SDGs) by focusing on pollution reduction, resource efficiency, and promoting a responsible and sustainable balanced future. The objective of sustainable development is to create a balance between social progress, environmental conservation, and economic growth to ensure long-term stability. GNT contributes significantly to these objectives by offering innovative solutions to global challenges all over the globe. GNT is being applied in the design and production of ecofriendly nanomaterials and nano devices that reduce environmental impact, help preserve ecosystems, minimize environmental contamination, and protect natural resources (Tarkeshwar et al. 2023). Moreover, it promotes the efficient use of energy, water, and other resources throughout the lifecycle of nanomaterials and nanodevices (Corsi et al. 2023).

Green nanotehnology helps to promote a more circular economy by minimizing resource use, enhancing energy efficiency, and reducing waste (Omran and Baek 2022). Ncube et al. reviewed the integration of circular economy (CE) and green chemistry (GC), emphasizing approaches that reduce waste, conserve resources, and balance environmental impacts while remaining economically feasible (Fig. 9.2). Despite these benefits, implementing CE and GC faces challenges such as limited investment, insufficient environmental education, and regulatory hurdles (Ncube et al. 2023). Advancing toward a circular economy and green chemistry requires international cooperation, incentives for adoption, educational initiatives, and effective policies to support sustainable production and consumption.

9.4 Sustainability Metrics of the Environment Applied to Green Nanotechnology

The key concepts of green chemistry that can be used in the field of nanotechnology to make it more sustainable:

Avoiding or Reducing the Waste Production Environmentalists have always argued that reducing waste at the source is more effective than dealing with the harmful consequences of improper waste management later. Nanotechnology, with its advanced techniques, contributes to waste reduction during synthesis by using organized, structured, and systematic production methods that yield monodisperse and stable nanoparticles. The use of biological extracts or biological sources can reduce the use of reagents, stabilizing agents, and by-products, making the process more efficient (Silva et al. 2015).

Atom Economy According to the concept of atom theory, sustainability in chemical processes depends not only on high product yield but also on maximizing the incorporation of reactant atoms into the final product (Mulvihill et al. 2011). This principle has been interpreted in various ways and emphasizes considering all the substances involved in a synthetic process, not just the target reaction (Dicks and Hent 2015). Conventional physical synthesis procedures often show lower material and result in highly polydisperse particles with varied shapes and sizes. In contrast, chemical synthesis routes use typically generate more waste and require a greater number of solvents, stabilizers, reducing agents, or surfactants (Palencia et al. 2014a, b). Recent advances such as one-pot synthesis and bioreactor-based approaches have significantly improved the production of homogenous, monodisperse nanoparticles (Ciambelli et al. 2019).

Safer Production Advances in nanotechnology have enabled the development of safer synthesis procedures by replacing dangerous and strong reducing agents with benign alternatives. For example, biological extracts derived from plants or microbioorganisms, including bacteria and algae, can effectively substitute conventional chemical reagents in nanoparticle synthesis. In contrast, chemically mediated processes also can often lead to the release of harmful by-products that can pose a risks to both human health and the environmentbe threatening to the environment (Adelere and Lateef 2016; Ciambelli et al. 2019; Madanayake et al. 2021).

Safer Products With the help of molecular-level design and smart material approaches, it is possible to develop nanoproducts that have minimal adverse effects on humans and on the environment. This approach reduces the likelihood of introducing potentially harmful materials into large-scale manufacturing and commercial use (Dicks and Hent 2015). To support this goal, large data libraries that provide information on nanoparticle properties, structures, catalytic activity, and toxicity are essential. Although nanomaterials exhibit diverse functionalities, their behavior is influenced not only by chemical composition but also by size, shape, surface coatings, and environmental interactions. As a result, assessments of safety must consider these material-specific and context-dependent factors (Harper et al. 2008; Li et al. 2018; Poynton and Robinson 2018; Ciambelli et al. 2019).

Restricting the Use of Solvents This principle emphasizes minimizing the quantities of substances, auxiliaries, and solvents that are needed during production and purification processes. This also promotes the use of solvents that are safe for the environment (Dahl et al. 2007; Coish et al. 2018). The adoption of water-based precursors, ecofriendly solvents that are recyclable or reusable, and ionic liquids for purification techniques such as centrifugation or chromatography, can significantly reduce environmental harm (Khan and Lee 2015; Silva et al. 2015; Upreti et al. 2015; Kumar et al. 2017; Bardos et al. 2018).

Energy Efficiency The principle stresses the need to reduce reliance on non-renewable energy sources, particularly fossil fuels, and to decrease overall energy consumption associated with greenhouse gas emissions. Nanotechnology (Livingston et al. 2020) offers effective solutions, as many nanomaterial synthesis reactions occur under standard conditions, such as room temperature and atmospheric pressure. Also, the synthesis steps can be carried out with the aid of microwaves and ultrasound, thereby reducing the reaction time while improving efficiency. Green synthesis methods demonstrate promising outcomes when carried out at room temperature and atmospheric pressure (Upreti et al. 2015).

Renewable Sources Nanobiotechnology makes use of renewable biological resources through green synthesis approaches that rely on biomolecules and sustainable materials. This reduces the need for conventional chemical systems while allowing the production of nanoparticles with varied catalytic, biological, and optical properties (Dicks and Hent 2015; Khan and Lee 2015; Upreti et al. 2015; Adelere and Lateef 2016; Madanayake et al. 2021).

Minimizing the Reaction Steps Protocols involving multiple reaction steps and derivatization typically require large amounts of chemical agents and reduce the final elemental efficiency of the process. Advances in nanotechnology, such as using the one-pot approach, reduce the over-use of reagents and make the purification process easier (Dicks and Hent 2015; Silva et al. 2015).

Catalysis Catalysis involves substances with high surface areas that enable efficient and selective chemical reactions. Nanoparticles, due to enhanced surface area and reactivity, are often excellent catalytic agents. However, it is crucial that these catalysts do not remain in the final products. While laboratory-scale reactions typically require only small amounts of catalysts, large-scale production may necessitate greater quantities, increasing the risk of catalyst contamination in the products. This contamination can pose environmental hazards or toxicity concerns (Dicks and Hent 2015; Ciambelli et al. 2019).

Procedure for Final Disposal This concept emphasizes the careful management of materials after the final stage, focusing on recycling, reuse, and recovery. Using recyclable or recoverable materials helps reduce environmental risks and minimizes the need for detoxification. Nanotechnology has enabled the development of tools for efficient recycling and recovery of raw materials. Additionally, computer-based programs make it possible to analyze and optimize these processes, supporting the establishment of a sustainable life cycle for nanomaterials (Dahl et al. 2007; Dicks and Hent 2015; Sanjay 2019).

Real-Time Monitoring This concept highlights the development of advanced analytical tools that track reactions and processes to prevent potential contamination. Nanoparticles have been utilized to create highly sensitive monitoring devices, leveraging their surface plasmon resonance and resistance to unwanted interactions. The low concentration and high sensitivity of these materials lead to rapid changes, highlighting the limitations of monitoring under unregulated conditions (Bhandari 2018; Ciambelli et al. 2019).

Preventive Measures This principle emphasizes accident prevention through the use of inherently safer substances (Dicks and Hent 2015; Khan 2020). Significaant advances have been achieved in transmaterialization, which involves transforming a material into a functional product. Using this traditional method, nanomaterials can be synthesized with properties similar to those of the solvents and reagents used; however, this method cannot be applied to all synthesis procedures as some methods involve the use of metallic precursors, untenable methods for extractions, and inherent toxicity (Ciambelli et al. 2019; Livingston et al. 2020). The release of nanomaterial into the environment can be averted by preparing nanocomposites, which also aid in improving the mechanical and functional properties of the prepared particles (Silva et al. 2015; Córdoba and Palencia 2017; Villalobos et al. 2019a, b).

Understanding the sustainability of chemical processes is essential in green chemistry. Currently, there is no single method to measure green chemistry in a way that clearly identifies the best processes for nanomaterial synthesis (Muñoz 2012; Constable and Jiménez-González 2018). This laborious method to define a process as "green" makes obligatory the use of metrics on the basis of the scientific theories that include material, energy, and environmental impact. These metrics help academic, industrial, and government organizations set guidelines and monitor the release of particles into the environment (Mulvihill et al. 2011; Gilbertson et al. 2015). Various metrics have been created and modified to assess how well a synthesis aligns with sustainability principles and green chemistry. These metrics primarily focus on mass and energy efficiency, molar productivity, and environmental and ecotoxicological impacts (Andraos 2008; Constable and Jiménez-González 2018; Zhang et al. 2018). Currently, there is an extensive range of tools available to analyze and study specific aspects of green chemistry across various procedures. Among the most common are Atom Economy, E-Factor, Process Mass Intensity, Effective Mass Yield, Material Recovery, Energy Efficiency, and Life Cycle Analysis—the latter provides a context-sensitive, multivariable evaluation (Manley et al. 2008; Albini and Protti 2016; Reid and Reed 2016; Constable and Jiménez-González 2018).

Despite this diversity, standardizing the evaluation of synthetic processes remains challenging. These metrics have found more consistent applications in organic and

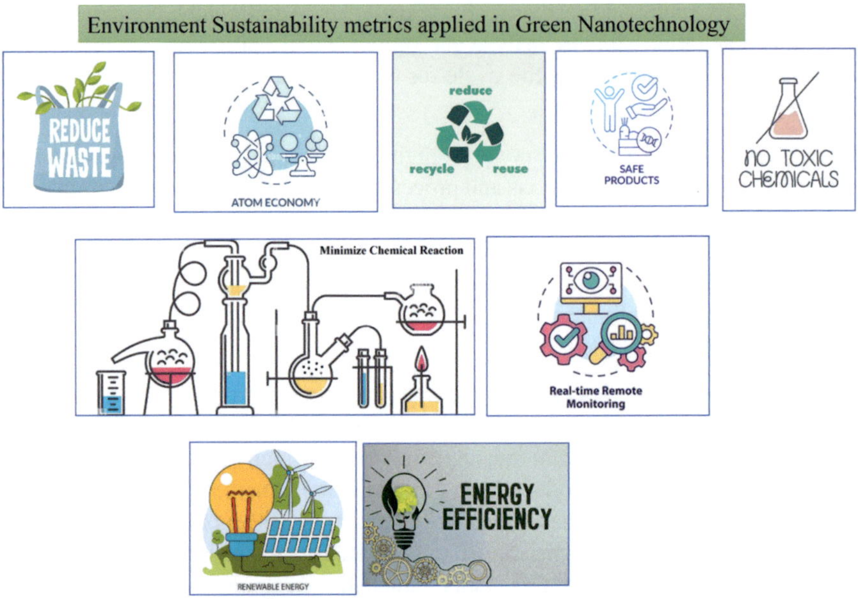

Fig. 9.3 Sustainability metrics in green nanotechnology

analytical chemistry, where procedures typically involve numerous auxiliary reagents and steps aimed at obtaining pure or homogeneous substances. In contrast, applying these same metrics to nanomaterial synthesis is more problematic. This is largely due to factors such as the intrinsic polydispersity and heterogeneity of nanomaterials, the difficulty in establishing precise stoichiometric relationships, the unknown concentrations of nanomaterials in real-world settings, and limited knowledge about their environmental interactions (Reid and Reed 2016).

Moreover, a frequent lack of essential data in published studies limits the feasibility of drawing conclusions through prospective assessments or meta-analyses based on the twelve green chemistry principles. As a result, many research teams do not regularly incorporate green metrics into their evaluations. Instead, analyses often emphasize product characteristics, synthetic efficiency, and in many cases, equate the term "green" with bio-based or nanobiotechnological methods.

This situation highlights a substantial gap between theoretical discussions of sustainability and the practical implementation of green chemistry in scientific and technological research. Additional obstacles include the lack of standardized protocols for nanomaterial synthesis, characterization, toxicity testing, and environmental impact assessment (Fig. 9.3) within the field of nanotechnology (Gilbertson et al. 2015; DeVierno Kreuder et al. 2017). Consequently, the scientific community has been working to adapt established green chemistry metrics—such as Atom Economy, E-Factor, Energy Intensity, Process Mass Intensity, and Life Cycle Analysis—for application in nanomaterial manufacturing (Gilbertson et al. 2015; Reid and Reed 2016; DeVierno Kreuder et al. 2017).

9.5 Approach Towards Social and Environmental Impact

There are a lot of opportunities to use this new technology. Today, nano technology is mentioned in the world as key and influential in science, technology, and industry (Alagumalai et al. 2021). A wide range of technologies and expertise from fields including physics, chemistry, biology, and engineering are used in nanotechnology. The production of nanomaterials, nanotubes, nano composites, nano filters and nanoparticles are examples of the use of nanotechnology to build systems and significant applications in environmental issues (Fang and Guo 2018). The production of nanocomposites using nanotechnology has led to the production of highly durable and lightweight raw materials that are able to replace heavy metal parts and significantly reduce the weight of equipment and auto parts and consequently significantly reduce energy consumption and ultimately reduce air pollution (Dong et al. 2021). Nanotechnology is in the field of lighting, which in turn will reduce air pollution (Rahimi et al. 2021). Biofuel cells that use nanotechnology in their production have the ability to convert biochemical energy directly into electrical energy (Moreno et al. 2021). Utilizing waste like carbon dioxide and human wastewater is one of the distinctive and appealing features of these cells. On the other hand, some believe that the use of nanotechnology itself could lead to new environmental problems such as new toxic substances and related biological hazards (Hosseini et al. 2018). Nanotechnology, the science of manipulating materials at the nanoscale (1–100 nm), has emerged as a transformative tool across multiple disciplines. Its applications in environmental science have garnered significant interest, offering innovative solutions to pollution, energy efficiency, waste management, and sustainability. However, its deployment also raises ecological and ethical concerns that require thorough evaluation.

Environmental Impacts of Nanotechnology

More accessible and affordable pollutant detection technology leads to increased environmental decision-making, ecosystem monitoring, and process management. Fast and accurate sensors that are able to detect pollutants at the molecular level increase human ability to support sustainable human health and the environment. A sensor is basically a kind of energy converter that can identify specific physical, chemical, mechanical, and other characteristics or events in its environment and show them as an output signal, typically in the form of an electrical or optical signal.

As a result, numerous sensors have been created across a range of industries and have numerous uses. Nano sensors are currently among the most widely used sensors.

Basically, nano sensors are chemical, physical, or biological sensors that can detect changes at the nanoscale qualitatively or quantitatively with extremely high sensitivity and precision. High sensitivity, high detection power, and the capacity to measure multiple species at once are the key features that have contributed to the high level of confidence in the data gathered by sensors and nano sensors.

Positive Impacts on Environmentally Related Nanomaterials

The world's biggest issues are energy supply and environmental protection since fossil fuels are depleting, and their use pollutes the environment. Nanotechnology holds promise as a means of controlling and treating pollution while producing clean, sustainable energy. Nanotechnology offers new ways to optimize industrial processes, cut down on greenhouse gas emissions and dangerous chemicals, decrease the use of plastics and replace them with biodegradable materials. The photocatalytic breakdown of organic contaminants in air and water is facilitated by nanomaterials such as carbon nanotubes, zinc oxide, and titanium dioxide. Nano-scale zero-valent iron is efficient in cleansing the heavy metals and chlorinated compounds in groundwater. Magnetic nanoparticles serve an important role in the separation and removal of contaminants from water.

Silver and copper nanoparticles mainly act on the antimicrobial agents, enabling effective water disinfection. Nano-membranes also enhance desalination and filtration efficiency due to their occupancy and permeability across the membrane in water purification. Nanotechnology also contributes green energy through improvised solar cells on quantum dots, and hydrogen storage using carbon nanostructures enhances the battery backup and capacitor performance. These conditions aid in reducing greenhouse gas emissions to clear the industrial effluents. Sustainability in agriculture also improves crop yield, reduces the amount of chemicals leaching, and minimizes environmental damage by using nano-fertilizers and nano-pesticides. Expandable monitoring of soil fertility and environmental conditions was enabled by nano-sensors. Recycling plastics, metals, and electronic waste is efficiently enabled by nanomaterials. Waste is reduced by using nano-coatings for hazardous agents and polluted water.

Environmental Risks and Concerns

Toxic substances that are discharged in the ecosystems of aquatic and terrestrial life may not be regulated environmentally. Disruption of microbes in soil and water may be exemplified by silver nanoparticles. Long-term effects are accumulated in the microbes and persist. Nanoparticles tend to bioaccumulate within organisms and persist in the environment, leading to potential long-term effects. Carbon-based nanomaterials can disrupt food chains and resist natural degradation processes. Without universal standards, nano-related risks may not be regulated. Current risk assessment protocols often fall short, limiting the effective implementation of safe disposal practices. Additionally, nanotechnology production can be resource-sensitive, requiring high energy inputs and the use of potentially toxic materials.

Role of Nanomaterial in the Environment

Nanotechnology and its applications on the environment to monitor the pollutants related to various types of pollution and their extent of spreading results in the determination of specified aspects. Nano-sensors may monitor the contamination in the targeted areas economically and on the large-scale manufacturing aspects. Nano-sensors transmit the information by consuming less energy and consumes low voltage with direct power sources (Forruque et al. 2022). In various countries, huge methods have been used to reduce the pollutants present in the environment. The usage of nanotechnology reduces many pollutants in cities and industries for the treatment of industrial and urban effluents to prevent water pollution and minimize soil pollutants (Upadhyayula et al. 2009).

9.6 Future Research Directions

Advanced Characterization Techniques:

In Situ *Characterization*: Developing in situ characterization techniques will allow researchers to study the dynamic behavior of NPs during reactions. This understanding can lead to the design of more efficient and robust NPs for various applications (Xu and Wang 2016).

Nanoscale Imaging: Techniques such as high-resolution transmission electron microscopy (HRTEM) and atomic force microscopy (AFM) will provide detailed insights into the structural and compositional properties of NPs, aiding in their precise design (Hansma and Hoh 1994).

Computational Modeling and Simulation:

Predictive Modeling: Advanced computational models can predict the properties and behavior of NPs, guiding the design of new nanoparticles with desired characteristics. Machine-learning algorithms can further enhance these models by analyzing large datasets to identify trends and optimize synthesis processes (Butler et al. 2018).

Multiscale Simulations: Combining molecular dynamics simulations with quantum mechanical calculations can provide a comprehensive understanding of the interactions between metals at the atomic level, informing the synthesis of BNPs with tailored properties (Lin et al. 2019).

Green and Sustainable Synthesis:

Eco-friendly Methods: Research on environmentally friendly synthesis techniques that use bio-based systems, such plants, and microbes, including bacteria, algae, and fungi, can help lessen the impact of producing nanoparticles on the environment. These methods should be optimized for scalability and reproducibility to meet industrial demands (Ahmed et al. 2016).

Resource Efficiency: Developing synthesis techniques that minimize the use of precious metals and maximize resource efficiency will reduce costs and the environmental footprint, making NPs more accessible for various applications (Iravani and Varma 2020).

Interdisciplinary Collaboration:

Cross-Disciplinary Research: Collaboration between chemists, physicists, biologists, materials scientists, and engineers will drive innovation in NP research. Interdisciplinary approaches can lead to new insights and applications that leverage the unique properties of NPs (Whitesides 2003)).

Academic-Industrial Alliance: Partnerships between educationa institutions, research organizations, and industry will ease the transition of NP research work into commercialized products. These collaborations can accelerate the development of practical applications and address real-world challenges (Geuna and Muscio 2009)

Nanoparticles offer a plethora of advantages that extend their applicability across numerous fields, including catalysis, medicine, environmental science, energy, and electronics. The enhanced catalytic activity, improved stability, unique optical and magnetic properties, and versatility in functionalization make NPs an invaluable asset in advancing technology and addressing global challenges.

The future of NPs lies in continued innovation and interdisciplinary collaboration. The successful use of nanoparticles in practical applications depends on resolving current challenges involving toxicity, ecological concerns, scalability, cost, and regulatory compliance. Advances in synthesis techniques, characterization methods, computational modeling, and sustainable practices will further enhance the capabilities and applications of BNPs.

By embracing the unique properties of bimetallic nanoparticles and investing in research and development, we can unlock their full potential to create sustainable solutions, enhance technological capabilities, and improve the quality of life. The journey of NPs is just beginning, and their transformative impact on various industries promises a brighter and more sustainable future.

9.7 Conclusion

Nanotechnology involves the design and manipulation of materials at the molecular scale and has applications across many branches of science and technology. It acts as a double-edged tool for the environment, offering solutions for pollution control, clean energy, and resource conversation, while also posing ecological risks due to limited understanding of its biological interactions and regulatory impacts. A sustainable and ethical approach promotes advances in nanotechnology and emphasizes green nanotechnology, which focuses on the development of biodegradable and eco-friendly nanomaterials. Green nanotechnology is essential to the advancement of sustainability and the environment and addresses major challenges in chemistry, biology, and agriculture. Guided by the principle of green chemistry, it enables the design of sustainable products and processes with minimal environment impact. There are government policies to ensure environmental safety and eco-friendliness of nanomaterials, particularly in applications such as hospital waste management.

These policies address regulatory challenges and promote public awareness of innovative nanotechnology solutions that contribute to global economic development. Overall, ongoing innovations in nanotechnology position it as a powerful driver of a greener and cleaner world.

References

Adelere IA, Lateef A (2016) A novel approach to the green synthesis of metallic nanoparticles: the use of agro-wastes, enzymes, and pigments. Nanotechnol Rev 5:567–587. https://doi.org/10.1515/ntrev-2016-0024

Ahmed S, Ahmad M, Swami BL, Ikram S (2016) A review on plant extract mediated synthesis of silver nanoparticles for antimicrobial applications: a green expertise. J Adv Res 7(1):17–28. https://doi.org/10.1016/j.jare.2015.02.007

Aithal S, Aithal PS (2021) Green nanotechnology innovations to realize UN sustainable development goals 2030. Int J Appl Eng Manag Lett. https://doi.org/10.47992/ijaeml.2581.7000.0106

Alagumalai A, Mahian O, Hollmann F, Zhang W (2021) Environmentally benign solid catalysts for sustainable biodiesel production: a critical review. Sci Total Environ 768:144856. https://doi.org/10.1016/j.scitotenv.2020.144856

Albini A, Protti S (2016) Paradigms in green chemistry and technology. Springer International Publishing, Cham. https://doi.org/10.1007/978-3-319-25895-9

Ali Q, Ghareeb O (2023) Achieving sustainable development goals by wastewater management: review. Eurasian J Eng Technol 19:99–108

Andraos J (2008) Application of green metrics analysis to chemical reactions and synthesis plans. In: Lapkin A, Constable DJC (eds) Green chemistry metrics: measuring and monitoring sustainable processes. Wiley, Chichester, pp 69–199. https://doi.org/10.1002/9781444305432.ch4

Aravind Kumar J, Joshua Amarnath D, Anuradha Jabasingh S, Senthil Kumar P, Vijai Anand K, Narendrakumar G, Karthick Raja Namasivayam S, Krithiga T, Sunny S, Purna Pushkala S, Yuvarajan D (2019) One-pot green synthesis of nano magnesium oxide-carbon composite: preparation, characterization and application towards anthracene adsorption. J Clean Prod 237:117691. https://doi.org/10.1016/j.jclepro.2019.117691

Aravind KJ, Krithiga T, Vijai AK, Sathish S, Karthick Raja Namasivayam S, Renita A, Hosseini-Bandegharaei A, Praveenkumar T, Rajasimman M, Bhat N, Dutta S (2021) Kinetics and regression analysis of phenanthrene adsorption on the nanocomposite of CaO and activated carbon: characterization, regeneration, and mechanistic approach. J Mol Liq 334:116080. https://doi.org/10.1016/j.molliq.2021.116080

Bamfield P, Hutchings MG (2010) Chromic phenomena: technological applications of colour chemistry. Royal Society of Chemistry

Bardos P, Merely C, Kvapil P, Koschitzky HP (2018) Status of nanoremediation and its potential for future deployment: risk-benefit and benchmarking appraisals. Remediation: J Environ Cleanup Costs Technol Techniq 28:43–56. https://doi.org/10.1002/rem.21559

Bhandari G (2018) Environmental nanotechnology: applications of nanoparticles for bioremediation. In: Prasad R, Aranda E (eds) Approaches in bioremediation: the new era of environmental microbiology and nanobiotechnology. Springer International Publishing, Cham, pp 301–315. https://doi.org/10.1007/978-3-030-02369-0_13

Bhushan B (ed) (2010) Springer handbook of nanotechnology. Springer. [Covers global investment, impact on society]

Bukhari SI, Hamed MM, Al-Agamy MH, Gazwi HSS, Radwan HH, Youssif AM (2021) Biosynthesis of copper oxide nanoparticles using Streptomyces MHM38 and its biological applications. J Nanomater 16:6693302. https://doi.org/10.1155/2021/6693302

Butler KT, Davies DW, Cartwright H, Isayev O, Walsh A (2018) Machine learning for molecular and materials science. Nature 559(7715):547–555. https://doi.org/10.1038/s41586-018-0337-2
Ciambelli P, La Guardia G, Vitale L (2019) Nanotechnology for green materials and processes. In: Basile A, Centi G, De Falco M, Iaguaniello G (eds) Catalysis, green chemistry and sustainable energy. Elsevier, Oxford, pp 97–116. https://doi.org/10.1016/B978-0-444-64337-7.00007-0
Clarke EA, Anliker R (1980) Organic dyes and pigments. In: Anthropogenic compounds. Springer, Berlin Heidelberg, pp 181–215
Coish P, McGovern E, Zimmerman JB, Anastas PT (2018) The value-adding connections between the management of ecoinnovation and the principles of green chemistry and green engineering. In: Török B, Dransfield T (eds) Green chemistry. Elsevier, pp 981–998. https://doi.org/10.1016/B978-0-12-809270-5.00033-9
Constable DJC, Jiménez-González C (2018) Tying it all together to drive sustainability in the chemistry enterprise. In: Constable DJC, Jimenéz-González C (eds) Handbook of green chemistry. Green metrics 11. Wiley BCH, Weinheim, pp 281–290. https://doi.org/10.1002/9783527628698.hgc132
Córdoba A, Palencia M (2017) Development of nanostructured crosslinkers with antimicrobial properties for free-radical polymerization. J Sci Technol Appl 2:54–64. https://doi.org/10.34294/j.jsta.17.2.14
Corsi I, Venditti I, Trotta F, Punta C (2023) Environmental safety of nanotechnologies: the eco-design of manufactured nanomaterials for environmental remediation. Sci Total Environ. https://doi.org/10.1016/j.scitotenv.2022.161181
Dahl JA, Maddux BLS, Hutchinson JE (2007) Toward greener nanosynthesis. Chem Rev 107: 2228–2269. https://doi.org/10.1021/cr050943k
DeVierno Kreuder A, House-Knight T, Whitford J, Ponnusamy E, Miller P, Jesse N, Rodenborn R, Sayag S, Gebel M, Aped I, Sharfstein I, Manaster E, Ergaz I, Harris A, Nelowet Grice L (2017) A method for assessing greener alternatives between chemical products following the 12 principles of green chemistry. ACS Sustain Chem Eng 5:2927–2935. https://doi.org/10.1021/acssuschemeng.6b02399
Dhahri S, Omri A (2018) Entrepreneurship contribution to the three pillars of sustainable development: what does the evidence really say? World Dev 106:64–77. https://doi.org/10.1016/j.worlddev.2018.01.008
Dicks AP, Hent A (2015) Green chemistry metrics. Springer International Publishing, Cham. https://doi.org/10.1007/978-3-319-10500-0
Dong J, Xu W, Liu S, Du L, Chen Q, Yang T, Gong Y, Li M, Tan X, Liu. Y (2021) Recent advances in applications of nonradical oxidation in water treatment: mechanisms, catalysts and environmental effects. J Clean Prod 321. https://doi.org/10.1016/j.jclepro.2021.128781
Enin H A A (2014) Nanotechnology – a review article. Int J Pharm Sci Rev Res 29(1):247–257
Fang Y, Guo Y (2018) Copper-based non-precious metal heterogeneous catalysts for environmental remediation, Cuihua Xuebao/Chinese. J Catal 39:566–582. https://doi.org/10.1016/S1872-2067(17)62996-6
Fleischer T, Grunwald A (2008) Making nanotechnology developments sustainable. A role for technology assessment. J Cleaner Prod (2011, September) 16(8):889–898
Forruque S, Mofijur M, Rafa N, Tasnim A, Chowdhury S, Nahrin M, Islam ABMS, Chyuan H (2022) Green approaches in synthesising nanomaterials for environmental nanobioremediation: technological advancements, applications, benefits and challenges. Environ Res 204:111967. https://doi.org/10.1016/j.envres.2021.111967
Geetha AR, George E, Srinivasan A, Shaik J (2013) Optimization of green synthesis of silver nanoparticles from leaf extracts of *Pimenta dioica* (Allspice). Sci World J 5:362890. https://doi.org/10.1155/2013/362890
Geuna A, Muscio A (2009) The governance of university knowledge transfer: a critical review of the literature. Minerva 47(1):93–114. https://doi.org/10.1007/s11024-009-9118-2
Gilbertson LM, Zimmerman JB, Plata DL, Hutchison JE, Anastas PT (2015) Designing nanomaterials to maximize performance and minimize undesirable implications guided by the

principles of green chemistry. Chem Soc Rev 44:5758–5777. https://doi.org/10.1039/c4cs00445k
Hansma PK, Hoh JH (1994) Biomolecular imaging with the atomic force microscope. Annu Rev Biophys Biomol Struct 23(1):115–139. https://doi.org/10.1146/annurev.bb.23.060194.000555
Harper SL, Dahl JA, Maddux BLS, Tanguay RL, Hutchison JE (2008) Proactively designing nanomaterials to enhance performance and minimize hazard. Int J Nanotechnol 5:124. https://doi.org/10.1504/ijnt.2008.016552
Hassanisaadi M, Bonjar GHS, Rahdar A, Pandey S, Hosseinipour A, Abdolshahi R (2021) Environmentally safe biosynthesis of gold nanoparticles using plant water extracts. Nano 11(8). https://doi.org/10.3390/nano11082033
Hosseini M, Abdelrazek AH, Sadri R, Mallah AR, Kazi SN, Chew BT, Rozali S, Yusoff N (2018) Numerical study of turbulent heat transfer of nanofluids containing eco- friendly treated carbon nanotubes through a concentric annular heat exchanger. Int J Heat Mass Tran 127:403–412. https://doi.org/10.1016/j.ijheatmasstransfer.2018.08.040
Hullmann A, Meyer M (2003) Publications and patents in nanotechnology. Scientometrics 58(3): 507–527
Iadiz MAR, Bamedi M, Fakour SR (2017) Periodontal diseases and recently applied nanotechnology: a review article. Health 9:345–351
Iravani S (2014) Bacteria in nanoparticle synthesis: current status and future prospects. Int Scholar Res Notic 18:359316. https://doi.org/10.1155/2014/359316
Iravani S, Varma RS (2020) Greener synthesis of bimetallic nanoparticles and their applications. Green Chem 22(8):2643–2661. https://doi.org/10.1039/D0GC00829A
Kalpana V, Devi RV (2018) A review on green synthesis, biomedical applications, and toxicity studies of ZnO NPs. Bioinorg Chem Appl 12:3569758. https://doi.org/10.1155/2018/3569758
Karn B (2008) The road to green nanotechnology. J Ind Ecol 12(3):263–266
Khan SH (2020) Green nanotechnology for the environment and sustainable development. In: Naushad M, Lichtfouse E (eds) Green materials for wastewater treatment. Springer International Publishing, Cham, pp 13–46. https://doi.org/10.1007/978-3-030-17724-9_2
Khan SA, Lee C (2015) Green biological synthesis of nanoparticles and their biomedical applications. In: Basiuk VA, Basiuk EV (eds) Green processes for nanotechnology. Springer International Publishing, Cham, pp 207–235. https://doi.org/10.1007/978-3-030-4
Khodadadi S, Mahdinezhad N, Fazeli-Nasab B, Heidari MJ, Fakheri B, Miri A (2021) Investigating the possibility of green synthesis of silver nanoparticles using *Vaccinium arctostaphlyos* extract and evaluating its antibacterial properties. Biomed Res Int 13:5572252. https://doi.org/10.1155/2021/5572252
Kumar P, Burman U, Kaul RK (2017) Ecological risks of nanoparticles. In: Tripathi DK, Ahmad P, Sharma S, Chauhan DK, Dubey NK (eds) Nanomaterials in plants, algae and microorganisms. Academic press, San Diego, pp 429–452. https://doi.org/10.1016/B978-0-12-811487-2.00019-0
Kumar JA, Amarnath DJ, Sathish S, Jabasingh SA, Saravanan A, Hemavathy R, Anand KV, Yaashikaa P (2019) Enhanced PAHs removal using pyrolysis-assisted potassium hydroxide induced palm shell activated carbon: batch and column investigation. J Mol Liq 279:77–87. https://doi.org/10.1016/j.molliq.2019.01.121
Li Y, Wang J, Zhao F, Bai B, Nie G, Nel AE, Zhao Y (2018) Nanomaterial libraries and model organisms for rapid high-content analysis of nanosafety. Natl Sci Rev 5:365–388. https://doi.org/10.1093/nsr/nwx120
Lin Z et al (2019) Multiscale modeling of nanoparticles: from atomistic to coarse-grained simulations. Chem Rev 119(5):2823–2878. https://doi.org/10.1021/acs.chemrev.8b00243
Livingston A, Trout BL, Horvath IT, Johnson MD, Vaccaro L, Coronas J, Babbitt CW, Zhang X, Pradeep T, Drioli E, Hayler JD, Tam KC, Kappe O, Fane AG, Szekely G (2020) Challenges and directions for green chemical engineering—role of nanoscale materials. In: Livingston A, Szekely G (eds) Sustainable nanoscale engineering: from materials design to chemical

processing. Elsevier, Cambridge, pp 1–18. https://doi.org/10.1016/B978-0-12-814681-1.00001-1
Lu Y, Ozcan S (2015) Green nanomaterials: on track for a sustainable future. Nano Today 10(4): 417–420. https://doi.org/10.1016/j.nantod.2015.04.010
Madanayake NH, Hossain A, Adassooriya NM (2021) Qual Safe Crops Foods 13:20–36. https://doi.org/10.15586/qas.v13i1.838
Manley JB, Anastas PT, Cue BW (2008) Frontiers in green chemistry: meeting the grand challenges for sustainability in R&D and manufacturing. J Clean Prod 16:743–750. https://doi.org/10.1016/j.jclepro.2007.02.025
Menegaldo M, Livieri A, Isigonis P, Pizzol L, Tyrolt A, Zabeo A, Semenzin E, Marcomini A (2023) Environmental and economic sustainabilityin cultural heritage preventive conservation: LCA and LCC of innovative nanotechnology-based products. Clean Environ Syst 9:100124. https://doi.org/10.1016/j.cesys.2023.100124
Moradi F, Sedaghat S, Moradi O, Arab SS (2021) Review on green nano-biosynthesis of silver nanoparticles and their biological activities: with an emphasis on medicinal plants. Inorganic Nano-Metal Chem 51(1):133–142. https://doi.org/10.1080/24701556.2020.1769662
Moreno-v MJ, Rodríguez-f F, AG L, CL, Del-toro-S (2021) Sustainable-green synthesis of silver nanoparticles using Saf Fl Ower (Carthamus tinctorius L.) waste extract and its antibacterial activity. Heliyon 7. https://doi.org/10.1016/j.heliyon.2021.e06923
Mulvihill MJ, Beach ES, Zimmerman JB, Anastas PT (2011) Green chemistry and green engineering: a framework for sustainable technology development. Ann Rev 36:271–293. https://doi.org/10.1146/annurev-environ-032009-095500
Muñoz I (2012) LCA in green chemistry: a new subject area and call for papers. Int J LCA 17:517–519. https://doi.org/10.1007/s11367-012-0410-2
Murthy HCA, Desalegn T, Kassa M, Abebe B, Assefa T (2020) Synthesis of green copper nanoparticles using medicinal plant *Hagenia abyssinica* (Brace) JF. Gmel. leaf extract: antimicrobial properties. J Nanomater 12:3924081. https://doi.org/10.1155/2020/3924081
Nasrollahzadeh M, Sajjadi M, Sajadi SM, Issaabadi Z (2019) Green nanotechnology. Interface. Sci Technol 28:145–98. https://doi.org/10.1016/B978-0-12-813586-0.00005-5
Ncube A, Mtetwa S, Bukhari M, Fiorentino G (2023) Energies RP-, and 2023 undefined. Circular economy and green chemistry: the need forradical innovative approaches in the design for new products. mdpicom. https://doi.org/10.3390/en16041752
Omran BA, Baek KH (2022) Valorization of agro-industrial biowaste to green nanomaterials for wastewater treatment: approaching green chemistry and circular economy principles. J Environ Manag. https://doi.org/10.1016/j.jenvman.2022.114806
Palencia M, Córdoba A, Rivas BL (2014a) Concentration–polarization effect of poly(sodium styrene sulfonate) on size distribution of colloidal silver nanoparticles during diafiltration experiments. Colloid Polym Sci 292:619–626. https://doi.org/10.1007/s00396-013-3096-5
Palencia M, Rivas BL, Valle H (2014b) Size separation of silver nanoparticles by dead end ultrafiltration: description of fouling mechanism by pore blocking model. J Membr Sci 455:7–14. https://doi.org/10.1016/j.memsci.2013.12.051
Poynton H, Robinson WE (2018) Contaminants of emerging concern, with an emphasis on nanomaterials and pharmaceuticals. In: Török B, Dransfield T (eds) Green chemistry. Elsevier, pp 291–315. https://doi.org/10.1016/B978-0-12-809270-5.00012-1
Rahimi M, Mehdinavaz R, Heydarzadeh M, Hossein A, Ettelaei M (2021) Surface & Coatings Technology Improving biocompatibility and corrosion resistance of anodized AZ31 Mg alloy by electrospun chitosan/mineralized bone allograft (MBA) nanocoatings. Surf Coat Technol 405:126627. https://doi.org/10.1016/j.surfcoat.2020.126627
Reid BT, Reed SM (2016) Improved methods for evaluating the environmental impact of nanoparticle synthesis. Green Chem 18:4263–4269. https://doi.org/10.1039/c6gc00383d
Roco MC (2016) Building foundational knowledge and infrastructure for nanotechnology: 2000-2030. Chapter 4. In: Cheng et al (eds) Nanotechnology: delivering the promise, vol 1. ACS, Washington, DC, pp 39–52

Sanjay SS (2019) Safe nano is green nano. In: Shukla AK, Iravani S (eds) Green synthesis, characterization and applications of nanoparticles. Elsevier, London, pp 27–36. https://doi.org/10.1016/B978-0-08-102579-6.00002-2

Sherly ED, Madgula K, Kakkar R (2011) Green nanotechnology for cleaner environment present and future research needs. Curr World Environ 6(1):177–181

Silva LP, Garcez I, Bonatto CC (2015) Green synthesis of metal nanoparticles by plants. In: Basiuk VA, Basiuk EV (eds) Green processes for nanotechnology. Springer International Publishing, Cham, pp 259–275. https://doi.org/10.1007/978-3-319-15461-9_9

Singh J, Dutta T, Kim K-H, Rawat M, Samddar P, Kumar P (2018) "Green" synthesis of metals and their oxide nanoparticles: applications for environmental remediation. J Nanobiotechnol 16(1): 84–24. https://doi.org/10.1186/s12951-018-0408-4

Singh BP, Sharma K, Tyagi S, Gautam D, Chaudhary M, Kumar A, Vikal S, Gautam YK (2022) Green and sustainable technology for clean energy production: applications. Handb Green Sustain Nanotechnol. https://doi.org/10.1007/978-3-030-69023-6_64-1

Smith GB (2011) Green nanotechnology. In: SPIE NanoScience+ engineering. International Society for Optics and Photonics, pp 810402–810402

Soni S (ed) (2014) Handbook of research on diverse applications of nanotechnology in biomedicine, chemistry, and engineering. IGI Global

Tarkeshwar PMA, Kapinder BK, Kaur J (2023) Microbial nanotechnology: a potential tool for a sustainable environment. Environ Appl Microb Nanotechnol. https://doi.org/10.1016/b978-0-323-91744-5.00010-2

Tyagi PK, Mishra M, Khan N, Tyagi S, Sirohi S (2016) Toxicological study of silver nanoparticles on gut microbial community probiotic. Environ Nanotechnol Monit Manag 5:36–43. https://doi.org/10.1016/j.enmm.2016.01.003

Upadhyayula VKK, Deng S, Mitchell MC, Smith GB (2009) Application of carbon nanotube technology for removal of contaminants in drinking water: a review. Sci Total Environ 408:1–13. https://doi.org/10.1016/j.scitotenv.2009.09.027

Upreti G, Dhingra R, Naidu S, Atuahene I, Sawhney R (2015) Life cycle assessment of nanomaterials. In: Basiuk VA, Basiuk EV (eds) Green processes for nanotechnology. Springer International Publishing, Cham, pp 393–408. https://doi.org/10.1007/978-3-319-15461-9_14

Villalobos RJ, Medina C, Canales C, Melendrez M, Flores P (2019a) Development of nanocomposite materials with thermostable matrix from nanoreinforcement of titanium dioxide. J Sci Technol Appl 6:65–78. https://doi.org/10.34294/j.jsta.19.6.42

Villalobos RJ, Rojas D, Díaz A, Ramírez J, Meléndrez M, Flores P (2019b) Synthesis and characterization of 1D metal oxide nano-structures: titanium dioxide nanotubes (TNTs) and zinc oxide nanobars (ZNBs). J Sci Technol Appl 6:79–95. https://doi.org/10.34294/j.jsta.19.6.43

Whitesides GM (2003) The 'right' size in nanobiotechnology. Nat Biotechnol 21(10):1161–1165. https://doi.org/10.1038/nbt871

Xu C, Wang H (2016) In situ techniques for characterization of nanomaterials. Chem Soc Rev 45(20):6124–6144. https://doi.org/10.1039/C6CS00142A

Zhang L, Wang Q, Hessel V (2018) Green chemistry metrics and life cycle assessment for microflow continuous processing. In: Constable DJC, Jimenéz-González C (eds) Handbook of green chemistry. Green metrics 11. Wiley BCH, Weinheim, pp 157–205. https://doi.org/10.1002/9783527628698.hgc129

Zhang D, Ma X-L, Gu Y, Huang H, Zhang G-W (2020) Green synthesis of metallic nanoparticles and their potential applications to treat cancer. Front Chem 8. https://doi.org/10.3389/fchem.2020.00799

Chapter 10
Economical and Societal Debut on Sustainable Environment and Industry Using Bioenergy and Nanotechnology

Waseem Ahmed Khattak, Muhammad Anas, Erdogan E. Hakki, Javed Iqbal, Banzeer Ahsan Abbasi, and Akhtar Munir

Abstract This chapter explores the intersection of bioenergy and nanotechnology in advancing sustainable industrial development and environmental resilience. It examines economic, societal, and technological dimensions of integrating bio-based and nano-enabled solutions to address global challenges such as climate change, energy security, and resource scarcity. The discussion begins by framing sustainability in the context of modern economies, highlighting the interdisciplinary role of renewable energy and material innovation. It then delves into the cost-benefit analysis of bioenergy systems, the efficiency gains enabled by nanotechnology, and the socioeconomic implications of their adoption. Public perception, policy frameworks, and community engagement are emphasized as critical factors influencing technology acceptance. Case studies demonstrate the successful implementation of nano-enhanced bioenergy solutions in both developed and developing regions. The chapter also presents hybrid energy-waste systems, smart manufacturing models, and global strategies adopted by leading countries. Challenges such as regulatory gaps, safety concerns, and technological scaling are addressed alongside future directions for research, innovation, and policy. Overall, the chapter argues that the convergence of bioenergy and nanotechnology provides a transformative

W. A. Khattak
Department of Plant Sciences, Faculty of Biological Sciences, Quaid-i-Azam University, Islamabad, Pakistan
e-mail: waseemkhattack@bs.qau.edu.pk

M. Anas · E. E. Hakki
Department of Soil Science and Plant Nutrition, Selçuk University, Konya, Turkey
e-mail: muhammed.anas@selcuk.edu.tr; eehakki@selcuk.edu.tr

J. Iqbal (✉)
Department of Botany, Bacha Khan University, Charsadda, Khyber Pakhtunkhwa, Pakistan

B. A. Abbasi
Department of Botany, Rawalpindi Women University, Rawalpindi, Pakistan

A. Munir
Department of Chemistry, Quaid-i-Azam University, Islamabad, Pakistan
e-mail: akhtarmunir@qau.edu.pk

N. Srivastava (ed.), *Sustainable Nanostrategies for Biorefineries*, Clean Energy Production Technologies, https://doi.org/10.1007/978-981-95-5875-9_10

platform for achieving sustainability goals—offering integrated, efficient, and inclusive pathways for industry and society. By fostering cross-sector collaboration, responsible innovation, and global cooperation, these technologies can support a just transition toward a circular and carbon-neutral economy.

Keywords Bioenergy · Nanotechnology · Sustainable industry · Hybrid energy systems · Circular economy

10.1 Introduction

The twenty-first century has ushered in an era marked by unprecedented technological advancement, rapid urbanization, and intense industrial growth (Scott 2012). However, this progress has come at a significant environmental and societal cost, prompting global economies to reevaluate their developmental models. The movement from conventional resource heavy systems toward sustainable economic infrastructure is no longer a choice, but it's a worldwide priority. Sustainability in contemporary economies is a multifaceted term that encompasses environmental stewardship, economic vitality, and social equity (Hariram et al. 2023). It underlines that current needs should be met while not reducing the capability of future generations to meet their own needs. This task needs to achieve a sustainable trade-off between development and conservation, which will be facilitated by novel, inclusive, and sensible policies and measures. The challenge is to tackle climate change, erosion of biodiversity, pollution, depletion of resources, social inequality, and to do so and yet remain competitive and vibrant. Part of such transformation is using green technologies specifically those that would result in clean energy, clean use of resources, and minimal ecological footprints. Bioenergy and nanotechnology have entered as the leading technologies, which could potentially be the route to novel industrial processes, energy regimes, and environmental practices, long term (Vasantha et al. 2021). They are also a unique interdisciplinary proposition that unites science, technology, and policy in a synergistic way to provide cross-cutting solutions that work across economic and environmental challenges.

10.1.1 Overview of Sustainability in Modern Economies

Sustainability in modern economies is rooted in a systemic approach that transcends mere environmental concern. It encompasses redefined growth metrics that include ecological health, social well-being, and long-term viability (Hariram et al. 2023). Traditional metrics like GDP fail to account for environmental degradation and resource depletion. It includes redefined measurements of growth that reflect environmental well-being, societal prosperity, and long-term sustainability. The classic indicators, such as GDP, do not include the negative effects of environmental degradation or resource depletion (Gorus and Aslan 2019). Instead, sustainable economic models favor the circular economy approach, which aims to reduce waste and keep products, materials, and resources in use as long as possible. Nations and companies have set their objective according to global frameworks, such as the UN Sustainable Development Goals, specifically SDG 7 (Affordable and Clean Energy), SDG 9 (Industry, Innovation, and Infrastructure), SDG 12 (Responsible Consumption and Production), and SDG 13 (Climate Action). To move toward sustainability, one must begin to think and plan in decades for the long processes of energy change, which also require integrating clean energy into national grids, closing or converting high-carbon plants, improving data on clean energy, building sophisticated supply chains for low-carbon technologies, and patiently cultivating new technologies and techniques, industries, entrepreneurs to bring it all to market. There are also changing market dynamics, with businesses that are environmentally conscious becoming favorites among both investors and consumers. (ESG) Environmental, Social and Governance criteria have become entrenched in investment decisions, while green bonds have become indispensable to back sustainable development projects. In addition, today's economies are adopting digital tools, e.g., artificial intelligence and big data analytics, to promote efficiency and responsibility in sustainability (Rane et al. 2024). However, sustainability is not a function of those economic tools or policy changes. It requires new technology to drive system-level change. This is where bioenergy and nanotechnology and suchlike come in as interdisciplinary kinds of solutions across sectors. Bioenergy produced from organic matter such as agricultural residues, forestry wastes, and even municipal waste is a renewable energy that complements the concept of a carbon-neutral planet (Barot 2022). It contributes to energy system decentralization, promotes rural development, and displaces fossil fuel use. At the same time, nanotechnology, which involves manipulating matter at molecular and atomic scales, promises an unprecedented accuracy of materials science, energy storage, water purification, and pollution removal (Das et al. 2025; Prasad et al. 2025). These technologies, when combined, have the potential to mitigate essential sustainability problems related to energy, agriculture, industrial production, and environment.

10.1.2 Interdisciplinary Role of Bioenergy and Nanotechnology

Bioenergy and its place in the modern economy is of an intradisciplinary nature (Zabaniotou 2018). Unlike conventional renewables such as solar and wind that are intermittent and location specific, bioenergy is storable and can be applied across sectors in rural and urban areas. Biomass conversion technologies are scalable, and they can be used to produce heat, electricity, and biofuels from a wide range of biomass feedstock (Adams et al. 2018). Such biofuels encompass biodiesel, bioethanol, and biogas and have been utilized for transportation, industrial, and power-generating purposes. District heating systems in Sweden largely rely on bioenergy, and Brazil has the world's most efficient ethanol transport system. The economic impact is substantial: bioenergy creation provides local jobs, boosts energy independence, and provides a route for waste valorization. In a country like India, the National Bio-Energy Mission has been launched converting agri-waste to energy which is a single solution to both the local pollution and energy access problems (Verma et al. 2024). Nanotechnology, by contrast, brings the ability to revolutionize industry and the environment by improving systems in ways that could not have been previously thought possible. Nanomaterials have distinctive mechanical, electrical, thermal, and catalytic properties that may characterize applications not possible by materials other than nanomaterials. In the field of energy, nanotechnology helps enable higher capacity batteries, more efficient solar panels, as well as more durable fuel cells to name a few. Single-walled carbon nanotube-based supercapacitors enable extremely fast charge-discharge rates, and quantum dot solar cells have already surpassed the Shockley-Queisser efficiency limit (Davis et al. 2024). In environmental cleanup, nano-filtration system open opportunity to clean up the drinking water from heavy metal and pathogens with key variety. Nanocatalysts are crucial to the degradation of pollutants, and nano-sensors are enabled to monitor the environment at real time. In addition, nanotechnology for agriculture such as nano-fertilizers and smart delivery systems improves yields by reducing resource consumption and environmental footprints. These advancements reduce overall energy and resource consumption, cut back on waste, and allow for cleaner production methods (Manikandan et al. 2024).

The real opportunity is in the coupling of bioenergy and nanotechnology to produce integrated systems that are more than simply the sum of the constituent components (Thanigaivel et al. 2022). Bioenergy processes can also benefit from nanotechnology that leads to more efficient catalyst, faster reactions, and lower energy inputs. For instance, the nano-catalysts used in transesterification are capable of greatly increasing the yield of biodiesel and reducing the cost of production (Zhang et al. 2022). Likewise, nano-engineered membranes can enhance the performance of biogas purification and water recovery processes pertinent to bioenergy plants. The reactor walls can be coated with nanomaterials to minimize biofouling and enhance thermal stability. These progresses make the bioenergy development more promising and also are economically attractive for the sustainable energy systems. Additionally, sensors enabled by nanotechnology integrated in bioenergy plants can allow real-time monitoring and control, resulting in predictive maintenance and minimized operational downtime. From a market perspective, the combination of the bioenergy and nanotechnology provides synergistic opportunities for new markets and business platforms. Hybrid technologies, such as enhanced biomass conversion with nanomaterials, are becoming popular among start-ups and academic institutions for investment. This has resulted in the green industrial ecosystems, in which sustainable materials, energy generation, and waste recycling cohabit in synergy. Governments are trying to promote these ecosystems by way of policies, subsidies, and cleantech/circular economy innovation hubs. There are also possible job creation opportunities since there's a demand for educated people in these two areas of chemistry, biology, engineering, materials science, and environmental policy (Matus et al. 2012). Furthermore, the industrial adoption of these electronics makes the supply chain greener, decreasing the carbon footprint and thus increasing the brand value with a growing environmentally friendly customer base. But this interaction between fields is not without its problems: Scale-up of nanotechnology, safety, toxicology, and the need for regulation. Because of their small size and reactivity, nanoparticles could have some unknown impact on human health and ecosystems. Hence, risk assessment frameworks that are sturdy and responsible innovation protocols must be in place. For bioenergy, the availability of feedstocks, land service conflicts, and conversion efficiency are the bottleneck.

Moreover, connecting the bioenergy system to the field of nanotechnology requires cross-sector cooperation, standard protocols, and interdisciplinary education (Mangal et al. 2025). Universities and training institutes need to update their curriculums with topics such as bio-nano interfaces, life-cycle analysis, sustainability metrics, and system thinking. In the next step, international partnership will be a key factor in the further interdisciplinary application of bioenergy and nanotechnology. Both innovation and diffusion could be sped up through international networks of research, instruments to promote funding based on efficiency, and platforms to support knowledge sharing. There are already institutions like the International Renewable Energy Agency (IRENA), the International Energy Agency (IEA), and the United Nations Environment Programme (UNEP) (Koutoudjian et al. 2021) that are promoting discussions and partnerships among different countries and sectors. In addition, open-access repositories, patents databases, and collaborative testbeds will enable iterative development and deployment of these technologies in different geographical and socioeconomic settings.

10.2 Economic Impacts of Sustainable Technologies

The global transition to sustainable technologies is not only a response to environmental imperatives but also a dynamic economic transformation that promises long-term prosperity and resilience (Prabhakar 2025). The emergence of clean and green technologies, such as bioenergy and nanotechnology, is transforming economic governance, labor markets, and patterns of industrial production. They are closely associated with cost levels, production efficiency, job creation, and market competitiveness. Though detractors often point to the high initial investment required to deploy enhanced technologies, a net economic gain can be generated over an infinitely long-time horizon primarily due to savings in energy consumption, operational efficiency, forgoing environmental costs, and the creation of new markets. In a climate where sustainability is increasingly also an economic driver, the strategic use of bioenergy and nanotechnology is a cost-effective route to low-carbon growth, innovation, and future-proofing of industries in a world with mounting resource constraints (Erhueh et al. 2024).

10.2.1 Cost-Benefit Analysis of Bioenergy Implementation

The economic value of bioenergy is also in the valorization of wastes and by-products. Agri-waste, widely burnt, thrown away, can be turned into high-value energy products. For instance, in Brazil, sugarcane bagasse has been used for cogeneration and the energy self-sufficiency of sugar mills has been intensified with the surplus energy that fed into the grid (Fioranelli and Bizzo 2023). Likewise, the increasing use of biogas from food and farm waste in Europe fosters circular economy practices, combats methane emissions, and provides an additional income stream for farmers. An India case study where 1 ton of paddy straw transformed into bio-CNG can abate up to 400 l of diesel unearths economic value realization by the producer and consumer. Furthermore, government backing offered via feed-in tariffs, carbon credits, and incentives for green energy makes bioenergy ventures an even more attractive investment (Nuhu et al. 2023). The overall societal gain considering better air quality, jobs in the countryside, and reduced dependence on the fossil is much higher than the capital cost (Fig. 10.1).

10.2.2 Nanotechnology in Industrial Efficiency and Waste Reduction

nanotechnology has been established as a strong enabler of industrial efficiency and waste reduction offering potential savings and gains in productivity. Nanotechnology, in essence, involves controlling matter on a scale between about 1 and 100 nm and the exploitation of material properties that may not exist at macro scales (Anas et al. 2024). It's this precision that makes them and us go faster with less input and is an economic game changer for manufacturing, energy, construction, and environmental sectors. The most obvious economic one in energy efficiency. Such nano-structured materials are applied in thermal insulation, photovoltaic cells, or LED lighting, with potential energy savings being more than that of any conventional low-energy lighting. Nano-enhanced window coatings and construction materials, for example, can reflect infrared radiation, meaning less need for air-conditioning and heating and lower electricity bills and carbon emissions.

In automotive production, the use of nanomaterials, such as graphene, carbon nanotubes, and nanosilica enhance product strength, robustness, and wear-resistance, which in turn extend the life of the product, causing consumers to spend less to maintain them. Nanotechnology has already been adopted by the automotive and aerospace industries to produce components that are lightweight and strong, resulting in increased fuel economy and reduced operation costs (Wong and Paddon 2017). Recently, transistors and memory devices at the nanoscale have been employed to reduce the size of electronic devices and enhance their function without the need of additional energy consumption in agreement with the guiding principles of sustainable design. Furthermore, nanotechnology help to minimize

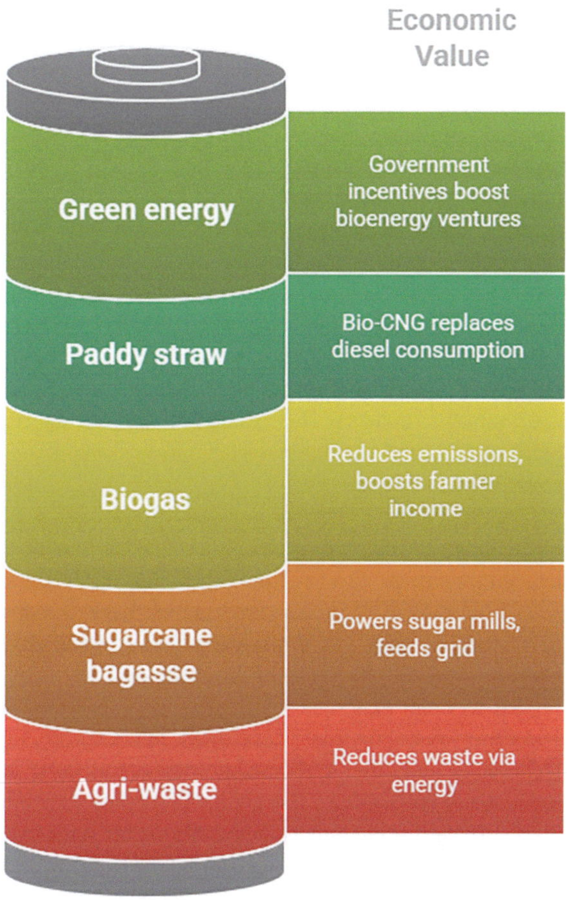

Fig. 10.1 Economic value chain of agri-waste to green energy: This illustration highlights how agricultural residues such as paddy straw, biogas, sugarcane bagasse, and other agri-waste contribute to green energy production, reduce emissions, power rural industries, and enhance farmer income through government incentives and bio-CNG alternatives

waste through enhancing catalytic effectiveness in processes of chemicals (Anas et al. 2025). Nanocatalysts in refining and manufacturing processes help to consume less raw materials and create less by-products, thereby reducing costs associated with waste treatment and waste disposal. Another important application is that of water treatment, for which nanofiltration and nano-adsorbents provide inexpensive methods for the removal of heavy metals, pathogens, and organic pollutants (Kousar et al. 2025). These systems provide expressiveness in the treatment of industrial effluents in which meeting regulations or having responsibility to the environment is a costly task. A comparative study by the World Bank found that nano-filtration, in textile and dye works, can reduce wastewater treatment cost by 30%–50%. By having nano-sensors embedded in production facilities, systems can benefit from real-time monitoring, and from predictive maintenance, hence limiting standstill and overall costs of operation. So, nanotechnology not only makes for a more efficient process, but also brings with it reduced externalization of environmental expenses, which do not get accounted for in traditional economic analyses.

10.2.3 Employment and Market Trends in Green Technologies

Another important aspect of the economic impact of green technologies is represented by the employment and market trends in this field. As economies transition from those powered by fossil fuels to those that rely on sustainable sources, green technology has become a key driver of both job creation and economic growth. It is estimated by the International Labour Organization (ILO), for instance, that the green economy will create up to 24 million new jobs worldwide by 2030, replacing jobs lost in carbon. Bioenergy, by itself, provides over 3.6 million jobs globally in the feedstock supply chain, construction, and operation and maintenance of plants. In countries such as the United-States and Germany, the bioenergy distribution is not only part of the national energy supply, but is also an important employer in rural regions, which improves social and economic resilience (Jenssen et al. 2014).

Nanotechnology (which includes more research than some of the other options listed above) is also seeing strong job growth in nanomedicine, nanoelectronics, and nanoengineering. The world market for nanotechnology is expected to exceed $125 billion by 2025 on account of rising R&D initiatives and the commercialization of nanotechnology-based products, along with growing demand across various end-use applications from different sectors. As a consequence, specialist positions related to nano-materials synthesis, nano-characterization, regulatory compliance, and nano-toxicology have developed. Furthermore, the intersections between nanotechnology and other green technologies (e.g., solar, hydrogen, and battery storage) are creating innovative value chains that necessitate interdisciplinary capabilities and innovative environments (Kiani 2025). And startups, university spin-offs, and tech incubators are the driving force behind the transformation, funded by venture capital, government grants, and public-private partnerships.

Critically, the emergence of new, sustainable technologies is transforming the competitive conditions in sectors. Society itself, in the end, is warming up to the idea of eco-friendly products as consumers and companies are both acquiring a sustainable business model. Green certification, carbon labeling, and lifecycle assessments are driving consumer decisions, and ESG reporting has become a universal mandate with investors. This trend is creating both competition and innovation, by encouraging companies to use the most up-to-date technologies and methods to increase efficiency and decrease their environmental impact. In such a competitive environment, bioenergy and nanotechnology are strategic elements that allow companies meet regulations, add value to brand, and gain new segments of market. Furthermore, governments are promoting this shift by implementing renewable energy targets (RET), tax incentives for the adoption of clean technology, and carbon market measures. However, to unlock the economic potential of sustainable technologies, obstacles need to be overcome. These include the high R&D costs and intellectual property barriers, lack of SME access to financing, and regulatory risk. For example, nanotechnology tends to encounter public safety concerns in commercialization that must be addressed in terms of regulatory standard and testing protocol. Similarly, bioenergy deployment could face issues like feedstock availability, land use competition, or competition with food production. Policy encompassing innovation support, environmental regulation, financial incentives, and stakeholder engagement is needed to address these challenges in an integrated manner. Education and skills training must be part of workforce preparation for a green future, as well. VET, industry-university collaboration, and lifelong learning allow us to secure a just and inclusive transition to a sustainable economy.

10.3 Societal Implications and Adoption

The societal considerations for the use of bio-based and nano-supported solutions are broader than only technical possibilities and environmental potential. But these are not simply scientific achievements: they are the potential for a new way that humans can interact with technology, with their environment, and with each other. Therefore, public opinion, social acceptance, policy frameworks, education, and effective community involvement are key to challenging or fulfill the promise of clean technologies. Despite their potential contributions toward climate change mitigation, resource efficiency, and economic growth, technologies like bioenergy and nanotechnologies are facing social barriers related to lack of awareness, security considerations, ethical access, and equity. The Covid-19 pandemic underscored the need for a holistic approach—one that is advanced, inclusive, transparent, and grounded in participatory processes—to build societal trust, ensure successful adoption of new technologies, and enable these innovations to genuinely serve the public interest.

10.3.1 Public Perception of Bio-based and Nano-driven Solutions

The public credibility is a critical factor that mainly affects the integration of bio-based and nano-enabled technologies into the society. Public perceptions of these breakthroughs significantly impact consumer behavior, regulatory policy, and investment economics. Bioenergy (e.g., biochemicals, biofuels from agricultural residues, forest biomass, algae, and organic refuse) is recognized primarily with positive connotations because of its renewable character as well as linkage with rural development and waste valorization. Bioenergy technologies are accepted as part of the solution in many regions, particularly for addressing energy poverty. or crop-burning community (and thus suffering from feeds into the energy services sectors problem where supply of energy has a direct effect on human development to attain environments de-tormented by wood, biomass, and cow dung combustion). Biogas projects in India and sub-Saharan Africa have increased energy availability and reduced the amount of indoor air pollution (Gebreegziabher et al. 2014). But bioenergy is not free from doubt in some cases either, notably if it competes with food or involves massive land use changes. Public concern is increased when feedstock production is accompanied by deforestation or the loss of biodiversity, following discussions of palm oil biodiesel in Southeast Asia. These concerns illustrate the importance of products with sustainability certifications and transparent supply chains, but also the importance of place-based communication efforts.

Nanotechnology, although just as revolutionary in potential, has encountered a more circumspect and divided welcome. Though the public appreciates its promise in medicine, clean energy, and water purification, issues of health risks, ethical dilemmas, and environmental exposure are still top of mind. The prefix "nano" can invoke ambiguity as there is little public knowledge of the science at the nanoscale, and what constitutes invisible. This apprehension has been fueled by high-profile media reportage of nanoparticle toxicity in available consumer products including sunscreens, cosmetics, and food packaging sources. In addition, ethical public deliberations on data generation, human-enhancement, and synthetic biology also contribute to ambivalence among the public (Fisher 2021). In a Eurobarometer poll, many citizens indicated they would not be willing to accept nanotechnology unless it was well regulated and offered obvious benefits. Those attitudes aren't fixed either; they change with exposure and education and everyday life. Consequently, promoting the public trust is about more than just risk communication—it is about conversation, participatory government, and visible social value.

10.3.2 Role of Policy, Education, and Community Engagement

Policy, education, and community engagement are critical levers for shaping public understanding and ensuring inclusive adoption of sustainable technologies. At the policy level, governments play a pivotal role in setting the tone for public discourse through legislation, funding, and institutional support. At the level of policy, governments have a critical role in reframing the context for public discussion with legislation, funding, and institutional commitment. It is important to establish regulatory clarity and continuity to ensure confidence among developers and users. For bioenergy, renewable energy targets, feed-in tariffs, carbon price, and licensing (biofuels mandates) schemes have driven the uptake of the technology by establishing market confidence and economic opportunity (Baumber 2017). But in the absence of well-defined sustainability standards and monitoring mechanisms, such policies may have unintended effects, such as monoculture plantations and displacement of rural people. Hence social safety nets and participatory policy making are crucial if benefits are to be shared and communities are to be strengthened rather than disempowered. The governance of nanotechnology is even more complicated because it straddles sectors and changes quickly by its nature. While agriculture and bioenergy belong to familiar paradigms, nanotechnology crosses health, the environment, materials, and electronics and thus requires regulatory harmonization. Regulatory guidelines for nanoscale have been issued by countries including the United States (US), the European Union (EU), and Japan, although there are still gaps in risk assessment, labeling, and lifecycle analysis (Malik and Patil 2020). Public policy must balance the incentives for innovation against precautionary concerns, and all voices in this dialog scientists, industry, civil society, the public need to be heard. The notion of "responsible innovation" developed as an overarching framework focusing on anticipation of impacts, reflection of researchers, inclusively deliberated options, and responsiveness to societal values.

Education, scientific communication, and literacy are essential for transforming public perception into informed engagement. An educated public is more likely to favor technological uptake, deal constructively with risk, and engage in decision-making. However, educational programs should not focus on mere technical issues but contextualize technologies within societal objectives, such as climate change mitigation, health equity, and sustainable development. The central issue is that HIV/AIDS represents not only a complex medical condition but also a multifaceted social challenge, which has hindered the effectiveness of past educational and intervention efforts. Bioenergy and nanotechnology should be part of the STEM programs of schools/universities/vocational institutions with emphasis on ethical, economic, and environmental aspects. Moreover, "informal" education that takes places via museums, media, public dialogs, and local workshops can help demystify complex technologies and generate trust (Degner et al. 2022). Models such as "citizen science" and public research labs have shown great value in establishing connections between researchers and the public. Equally important is the requirement to engage communities so that technology deployment is compatible with local needs, values, and capabilities. Community involvement can also be of different types: stakeholder consultations, participatory planning, co-designing the solutions, and collaborative monitoring (Aguilar 2022). Community ownership models—such as cooperatives or decentralized management—have proven more successful in bioenergy projects, for instance, by stimulating local investment, sharing knowledge, and fostering accountability. Lessons from rural bioenergy projects in Kenya and Nepal demonstrate that involvement of communities from the beginning, in terms of design, installation, and governance, greatly helps in increasing the rate of adoption and ensuring sustainability of the system. Engaging stakeholders in the debate is more difficult in the case of nanotechnology because of its abstract character, but participatory foresight initiatives, public deliberation, and ethics panels represent interesting directions to establish reciprocal understanding and societal control.

Digital infrastructure and social media also provide new ways to engage the community in meaningful ways, as channels for real-time feedback, transparency, and the ability for people to solve problems together. But they also have to be used responsibly to reduce misinformation and polarization. Governments, academia, and industry should work together to ensure that public engagement is not tokenistic but substantial, helping to make communities inform technological paths and redress power imbalances in innovation systems (Eze et al. 2024). Equity and inclusion also deserve attention, with an eye toward making sure all voices and interests, particularly those of marginalized women, indigenous peoples, and people in the Global South are heard and well represented in the adoption process. Intersectional analyses that account for social, cultural, and economic diversity can result in more equitable and sustainable end states.

10.4 Bioenergy in Sustainable Industry

Bioenergy is a key element in sustainable industrial development providing renewable, decentralized, and flexible energy resources that link with the goals of circular economy and climate mitigation. Bioenergy, which comprises biofuels, is obtained from natural materials such as agriculture and forest residues, municipal solid waste, energy crops, and even algae, offering a cleaner alternative to fossil energy and simultaneously converting organic waste streams into a commodity. For the current industry, it is a potentially very interesting solution for the generation of thermal energy, power, and transport fuels to contribute to decarbonizing and minimizing the environmental impact of industry. Its potential to complement existing industrial systems, satisfy baseload electricity demand, and add value to the waste management landscape position Advanced Catalytic Thermo-Gasification System (ACTGAS) is a technology designed to convert waste materials into clean syngas through high-efficiency catalytic processes as an ideal technology for the transition of industrial processes to a sustainable future (Fig. 10.3). Widespread deployment of bioenergy requires a myriad of factors such as technology development and deployment, resource management, life-cycle analysis, and policy formulation (Fig. 10.2) (Chopra et al. 2024).

10.4.1 Biomass Resources and Conversion Technologies

Biomass resources form the foundation of the bioenergy supply chain, and their diversity is both a strength and a logistical challenge. There are different kinds of biomass: lignocellulosic sources (such as straw, wood, and animal manure), energy crops (like switchgrass, miscanthus, Alceno, and jatropha), and organic wastes (manure, sewage, and food waste), and aquatic biomass such as algae (Anas et al. 2025). The calorific value, moisture content, chemical composition, and availability of various feedstocks make them suitable in different conversion processes. In countries like India, China, and Brazil, which have an agricultural economy, large quantities of crop residues, such as rice husk, corn stover, and sugarcane bagasse, are produced which can act as a consistent and untapped source of energy. Similarly, Canada and Scandinavia use forestry residues for wood pelletization and biomass co-generation (García et al. 2019). Some people are interested in algae though it is at

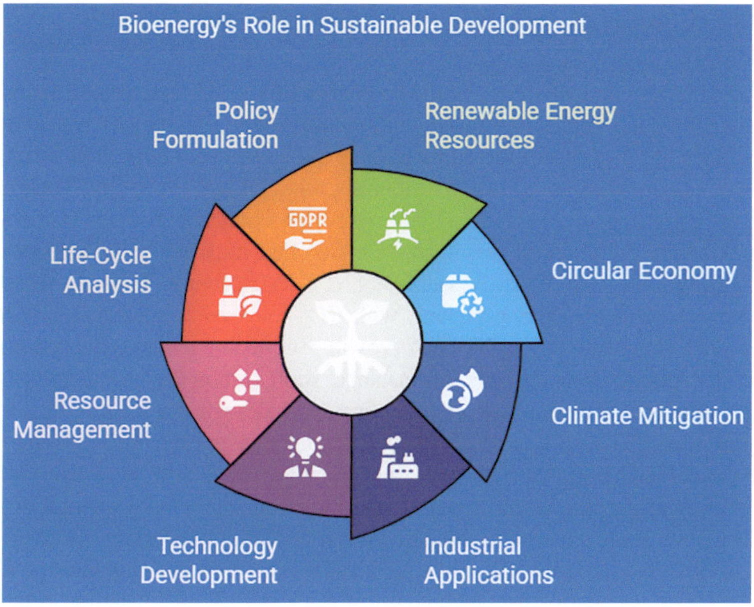

Fig. 10.2 Key roles of bioenergy in sustainable development, including policy support, circular economy, climate action, and industrial innovation

a research and development stage because of its high lipid content, high growth rate, and little need for land. The sustainability of biomass supply is crucially linked to land management practices, preservation of biodiversity, and the absence of food production competition; hence, feedstock choice is key strategic decision.

Diversified conversion technologies are used to transform biomass for the purpose of utilizing energy, and the technologies can be divided into thermochemical, biochemical, and physicochemical methods. Among these thermochemical processes are combustion, gasification, and pyrolysis. Combustion is the oldest and most well-known process in which biomass is burned directly to form steam that turns turbines to produce heat or electricity. In gasification, biomass is incompletely oxidized at higher temperatures, producing syngas (CO, H_2, and CH_4), which can be used as the combined heat and power (CHP) or fuels feedstock. Pyrolysis, which is the thermal treatment of biomass in oxygen-free conditions, results in the formation of either bio-oil, biochar, or syngas and has the potential to serve as an option for energy and Carbon sequestration (Al-Rumaihi et al. 2022). Biochemical conversion processes (like anaerobic digestion and fermentation) are among the most appropriate for humid organic waste and rely on microbial action in order to generate biogas (principally CH_4 and CO_2) or bioethanol. Transesterification is a common physicochemical process to produce biodiesel from vegetable oils or animal fats. The selection of a technology is based on the feedstock and the end-use application as well as the economic and environmental infrastructure. In addition, hybrid systems that combine several conversion paths appear, with the aim to increase the efficiency and optimize the usage of resources.

10.4.2 Bioenergy Lifecycle and Environmental Impact

A comprehensive understanding of the bioenergy lifecycle is essential to evaluate its true environmental impact and sustainability credentials. Bioenergy has a number of stages in its life cycle from the cultivation or collection of the biomass through its transport, conversion/processing, distribution, and end use (Tong and Tong 2019). There are input resources (land, water, energy) and output emissions (Greenhouse Gases (GHGs), particulates, waste) at each step that must be evaluated systematically through an Life Cycle Assessment (LCA) structure. It is assumed that bioenergy systems are carbon-neutral, with CO_2 released in combustion balanced toward the CO_2 uptake during biomass growth. But this neutrality assumption holds only under special circumstances, for instance, short carbon cycles, sustainable land management, and little land use change. For example, when allocating residues or waste biomass that would have otherwise decomposed and released methane, there are obvious GHG mitigation benefits, whereas clearing forests on a large scale to grow energy crops may result in net carbon emissions and environmental degradation. The environmental effects on air and water quality, biodiversity, and soil health are also stretches. Bioenergy conversion by combustion can produce pollutants (NOx, SOx, particulate) that demand sophisticated flue gas cleaning systems to meet increasingly strict environmental regulations (Larki et al. 2023). Another important consideration to include is water use, particularly when energy crops and biorefineries are concerned. Combining wastewater treatment and water reuse with bioenergy operations can provide an approach to address these challenges. Biochar, which is a byproduct of pyrolysis, has co-benefits as it enhances soil fertility and also sequesters carbon, supporting regenerative agriculture (Yrjälä et al. 2022). To ensure environmental performance optimization, the holistic approach of feedstock selection, logistics, advanced processing, and circular system design is required.

10.4.3 Industrial Applications and Scaling Challenges

Industrial applications of bioenergy are wide-ranging, encompassing sectors such as food processing, pulp and paper, cement, chemical manufacturing, and transportation (Behera and Varma 2019). Many of these industries require high-temperature heat, which biomass can provide more sustainably than fossil fuels. For example, biomass-fired boilers are increasingly replacing coal-fired systems in food and beverage manufacturing, where steam is used for sterilization, drying, and pasteurization. In the pulp and paper industry, black liquor, a lignin-rich waste stream, is used for in-house cogeneration, reducing energy costs and waste disposal needs. In cement production, biomass co-firing reduces both emissions and reliance on coal, particularly in developing countries where agro waste is abundant. The transport sector is another key area, with biofuels such as biodiesel, ethanol, and Compressed

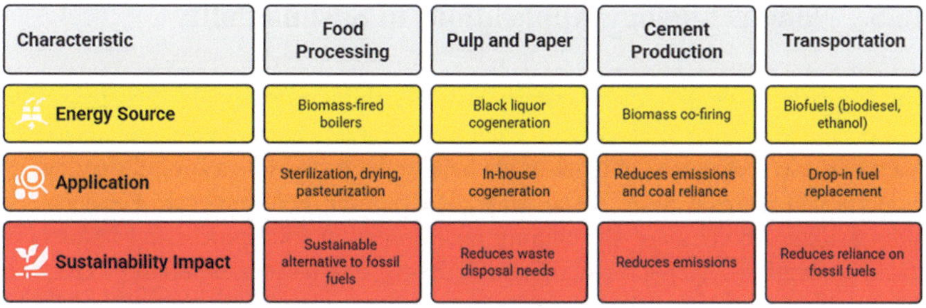

Fig. 10.3 Sector-wise applications of bioenergy, highlighting its role as a sustainable energy source across food processing, pulp and paper, cement production, and transportation industries

Natural Gas from biomass (bio-CNG) (Mangal et al. 2025) offering drop-in replacements or supplements to conventional fuels (Munagala et al. 2022). Advanced biofuels derived from lignocellulosic materials and algae are being developed to address concerns related to food vs. fuel and to improve energy yields (Fig. 10.3).

Despite these promising applications, scaling up bioenergy faces several challenges—technical, economic, logistical, and regulatory. Feedstock supply chains are often fragmented, seasonal, and region-specific, requiring investments in storage, preprocessing, and transport infrastructure. The heterogeneous nature of biomass complicates standardization and quality control, affecting combustion efficiency and emissions. Technologically, improving conversion efficiencies and developing modular, decentralized systems remain priorities. Economically, bioenergy must compete with low-cost fossil fuels and increasingly competitive solar and wind power, particularly in electricity markets. Policy support in the form of subsidies, carbon pricing, and renewable portfolio standards can help bridge this gap. Another challenge lies in public perception and acceptance, especially when bioenergy projects involve land acquisition or alter traditional livelihoods. Furthermore, the institutional landscape often lacks coordination between energy, agriculture, environment, and industrial sectors, impeding integrated bioenergy planning. Addressing these challenges requires a systems-based approach involving multi-stakeholder collaboration, public-private partnerships, and investment in innovation and capacity building. Countries that have successfully scaled up bioenergy such as Sweden, Brazil, and Finland demonstrate the importance of long-term policy commitment, clear regulatory frameworks, and active engagement with local communities and industry players (Junginger et al. 2019; Silveira and Johnson 2016). International cooperation through knowledge sharing, technology transfer, and harmonized sustainability standards can accelerate the global uptake of sustainable bioenergy solutions.

10.5 Nanotechnology Applications in Sustainability

The industrial utilization of bioenergy is abundant due to its availability, from food processing, pulp and paper, cement, chemical production, transportation, and others. A lot of these sectors need higher-grade heat in the form of heat for industry, which biomass can provide more sustainably than burning coal and gas. For instance, biomass-based boilers are substituting coal boilers in food processing industry as steam generation systems for sterilization, drying, and pasteurization of products. In the pulp and paper practices, black liquor—a lignin-rich waste—is utilized for an in-house cogeneration, thus lowering energy costs over and waste disposal requirements. The substitution of agro waste with coal in cement production decreases the emissions, being of great importance for the developing countries with an easy access to the agro waste. Another major sector is transport where biofuels including biodiesel, ethanol, and bio-CNG provide drop-in or supplementary solution to traditional fuels. The development of advanced biofuels produced from lignocellulosic feedstocks or algae has been proposed to overcome issues of food vs. fuel concerns and to increase the energy content of the biofuel.

While exciting work is ongoing in all these areas, the scaling-up of bioenergy still faces many hurdles—technical, economic, logistical, and regulatory. Supply chains from feedstock to end use are typically broken, disjointed, seasonal and geographical requiring storage, pre-processing, and transport infrastructure. The diversification of biomass makes standardization and quality control difficult for biomass, which impacts the combustion efficiency and emissions. From a technical point of view, optimization of the conversion efficiencies and modularity and decentralization of the systems are still necessary. Economically, bioenergy has to compete with cheap fossil fuels and solar and wind power, which are becoming more and more competitive, especially in markets for electricity. Policymaker support, including subsidies, carbon pricing, and renewable portfolio standards, can help make up the difference. An additional challenge is public perception and acceptance when bioenergy projects require land acquisition or change conventional livelihoods. In addition, the institutional setting is usually characterized by the non-coordination of the energies, agriculture, environment, and industries, which is detrimental to integrated bioenergy planning. Overcoming such challenges requires a comprehensive and coordinated strategy, involving cross-sectoral cooperation, public-private partnerships, and innovative initiatives to build capacity and address these challenges effectively. Countries that have successfully promoted bioenergy, such as Sweden, Brazil, and Finland, understand the importance of long-term commitment to ensuring that regulatory and policy frameworks, as well as rules for engaging with communities and other industries, are clear, consistent, and applied uniformly. Knowledge exchange, technology transfer, and standardized sustainable criteria through international collaboration also facilitate the global deployment of sustainable bioenergy solutions.

10.5.1 Nanomaterials for Pollution Control and Resource Management

Nanomaterials have already had a significant impact in pollution abatement and resource utilization. Nanomaterials, with a high surface-area-to-volume ratio and controllable surface chemistry, can adsorb, capture, and degrade pollutants in a molecular scale, providing a new way for highly efficient detection, capture, degradation, endocrine disrupting, wastewater treatment, etc. (Srivastava and Mishra 2022). In air pollution mitigation, nanoparticles, including titanium dioxide (TiO_2) and zinc oxide (ZnO), are employed as photocatalysts for the decomposition of volatile organic compounds (VOCs), nitrogen oxides (NO_x), and soot (David and Niculescu 2021). Upon light illumination, these materials produce reactive oxygen species that oxidize pollutants present in the air to less toxic materials. Nanostructured filters and coatings, additionally, are used in industrial chimneys and vehicle fumes to minimize their emission and meet environmental regulations. In wastewater, treatment of heavy metals, dyes, and organic pollutants, ultra-efficient sorption is achieved with carbon nanotubes, nano-iron particles, and graphene oxide. These nanomaterials offer greater capacity and selectivity than conventional filtration media; therefore, they can treat highly contaminated industrial effluents as well as recover and reuse water (Ahmed et al. 2022).

In the area of solid waste management and resource recycling, the selective adsorption and catalytic reaction of nanomaterials make it possible to recover precious metals from electronic waste and industrial residues. To put it another way, nano-engineered desalination and water treatment membranes can remove salts, pathogens, and micropollutants at lower operation costs compared to traditional reverse osmosis systems (Anas et al. 2024). Such membranes are more and more implemented in water-stressed areas and off-grid applications, which is very beneficial to both the environment and public health. Nanotechnologies also provide health advantages in construction: they allow for high-performance concrete with lower cement content, with a lower carbon footprint, which in turn can address one of the most emission intensive sectors of industry. Together, these applications demonstrate the agility of nanomaterials to control pollution and maintain resource efficiency across different industries (Solomon et al. 2024).

10.5.2 Energy Harvesting and Storage with Nano-enabled Devices

Energy harvesting and storage with nano-enabled devices is another important area, and here nanotechnology is also stimulating advancements in renewable energy systems and grid resiliency. In photovoltaics, nanostructured materials, such as quantum dots, perovskite, and plasmonic nanoparticles, serve to enhance light absorption and charge transport for increased solar cell efficiency and material-to-cost efficiency. For example, perovskite solar cells have surpassed 25% efficiency in lab settings—equal to that of traditional silicon cells—even though they're lighter and cheaper to make. Nanostructured coatings also enhance the longevity and efficiency of solar panels by creating self-cleaning and anti-reflective surfaces, making the solar panels less prone to energy-bereaving dirt and dust and more energy productive.

In energy storage, nanotechnology has revitalized the battery industry by increasing energy density, fast charge and discharge rate, and long cycle life. Lithium-ion batteries with lower-dimensional (nanostructured) electrodes—such as silicon nanowires or graphene anodes—show excellent performances stemming from faster ion diffusion and lesser deterioration. These enhancements are important for HV applications such as EVs, renewable energy storage, and portable electronics. Furthermore, supercapacitors and solid-state batteries that are based on nano-engineered materials are an alternative to traditional storage with an improved safety profile and fast discharging performance. Hydrogen-based systems for fuel cell applications can minimize reliance on precious metals. For example, carbon-supported platinum nanoparticles are highly efficient while using less platinum, which helps reduce costs and improve economic viability. Outside of electricity generation and storage, nanogenerators based on piezoelectric and triboelectric effect can scavenge mechanical energy from the ambient vibrations, moving objects or human motion, offering decentralized power for sensors, medical devices, or IoT systems. Researchers are also working on thermoelectric nanomaterials that could turn waste heat from industry into electricity. Such technologies may play a major role in energy-efficient systems and circular industrial design. So with the world energy landscape moving toward renewables and distributed generation, nanotech becomes the enabler to outperform, lower cost, and make it reliable throughout the energy value chain (Oshilalu 2024).

10.5.3 Safety, Ethics, and Regulatory Aspects

Nanotechnology has significant contributions to offer in sustainability, the positive sides of which are clouded by its multidimensional toxicological, ethical, and regulatory (Gupta 2023). Nanoparticles are highly reactive and can act as anti-adhesion agents. Due to their minute size, they can penetrate biological membranes and serve as vehicles to deliver their active or potentially toxic cargo deep into the body, albeit toxic. Some nanoparticles have been shown to lead to oxidative stress, inflammation, or genotoxicity in organisms. Tracer exposure of ecosystems is another way that environmental exposure during the manufacture, use, or disposal of nano-enabled products may affect ecosystems, permeating aquatic life. The risk of such materials is not completely clear due, in part, to the diversity and the complex interactions of these nanomaterials in actual environment (Lead et al. 2018). Hence, risk assessment approaches should further develop to consider nanoparticle characterization, fate modeling, exposure routes, and toxicological assessment in realistic contexts. Moral issues also concern the use of nanotechnology most notably in surveillance, enhancement, and environmental fabrication. Applying nanosensors in agriculture or medicine poses privacy, ownership, and consent problems. In more developed countries, unequal access to nanotechnology could exacerbate social-economic imbalances, unless distribution and capacity building are considered. It could be that popular belief, colors by media-led narratives and the skepticism toward new technology encodings (including GMOs) will affect resistance, unless managed by open communication and public involvement. To address these concerns, a strong regulatory and governance framework is required. Regulatory agencies, such as the United States Environmental Protection Agency (US-EPA), the European Chemicals Agency (ECHA), and national food and drug agencies are in the process of establishing new guidelines specifically targeting the nano-regulatory aspects, but harmonization is still missing. Major issues are definitional, measurement, and labeling inconsistencies. It is essential that work be co-ordinated at international level through the agencies of the Organisation for Economic Co-operation and Development (OECD) and the International Organization for Standardization (ISO) so that common safety standards, certification schemes, and best practice can be put in place. Additionally, the notion of responsible research and innovation (RRI) is being promoted—projected as focusing on the anticipation of societal effects, reflection of scientists, inclusion of multiple perspectives, and responsiveness to public values (Owen and Pansera 2019).

10.6 Synergies Between Bioenergy and Nanotechnology

10.6.1 Integrative Case Studies

Integrative case studies from across the globe highlight how bioenergy and nanotechnology can collaboratively transform traditional processes into sustainable and circular systems (Klemeš et al. 2019). One such case is the use of nanoengineered catalytic systems in biodiesel manufacturing. Conventional transesterification technologies for turning oils or fats into biodiesel with a catalyst tend to have long reaction times and high energy consumption. However, by incorporating nanocatalysts like magnesium oxide, calcium oxide, and zeolite-based nanoparticles, the reaction efficiency was dramatically improved, processing temperatures minimized, and the reusability of the catalyst was made possible. In a study, the biodiesel production time was reduced to half and conversion efficiency was improved above 95% with addition of zinc oxide nanoparticles. In another case study, researchers of the Netherlands fabricated nano-walls/nano-pores for gas purification in biogas plants. They are promising membranes to selectively remove methane from carbon dioxide (and impurities) by a more efficient process and energy-saving process compared to existing methods and increasing energy content and storage potential of the produced bio-CNG (Kabeyi and Olanrewaju 2022). In India, nano biogas potential metabolizing enhancement-based integrated rural energy systems are piloted through waste management and clean cooking response. These technologies utilize nano-coated reactor walls to resist corrosion and promote microbial activity, increasing gas generation and extending use-life. These case studies illustrate the economic and technical feasibility of bioenergy systems enhanced by nanotechnology, especially in remote or off-grid settings. Another potential top-down application is biochar doped with nanoscale iron or silver particles for water treatment; this option kills two birds with one stone by addressing water quality and having a product that helps sequester carbon. These combined approaches showcase how bio-nano synergies can be directly related to multiple United Nations Sustainable Development Goals (SDGs) for clean energy, water and sanitation, responsible consumption, and climate action (Prajapati et al. 2024).

10.6.2 Hybrid Systems in Energy and Waste Management

These case studies are extended by the notion of hybrid systems in the domain of energy and waste management, which integrates all forms of applications into multifunctional platforms, as where the inputs and outputs are optimally managed. In these systems, organic wastes are utilized not only for energy production (e.g., by anaerobic digestion or gasification) but also as feedstock for producing high-value products (e.g., biochar, bio-based chemicals, and nanocomposites) (Ashokkumar et al. 2022). Nanotechnology has an influence on several points in this chain: as a process intensifier (e.g., nano-catalysts), a sensor technology (e.g., real-time monitoring of gas compositions), and a product quality aid (e.g., nano-modified bio-based materials). In urban biorefineries, well characterized nano-biosensors can be used to monitor the fermentation conditions, by on-line detection of contaminants or suboptimal microbial activities. This accuracy increases process reliability and product quality and conserves energy and raw materials. Nanomaterials enhance the combustion efficiency and lower emissions of NOx, Sox, and particulate matter in waste to energy (WTE) system by nano-additives and catalytic converter (Ghazaly and Abdulhameed 2023). In addition, heat exchangers and reactor surfaces are coated with nano-coating, which prevents fouling as well as thermal losses thus improving plant efficiency. For thermal gasification, novel nanotechnologies allow tar and contaminants to be captured in-process, which results in lower syngas tar content and limits the need for subsequent treatment. Integrated systems that link anaerobic digestion to algal-based carbon capture—where algae biomass can be used in biodiesel or nanomaterial production—are examples of a closed-loop concept. Thus, bioenergetics meets nanotechnology, facilitating approaches toward circular economy in which waste is minimized, energy is optimized, and environmental footprints are reduced at each stage of the life cycle (Saxena 2023).

10.6.3 Smart Sustainable Manufacturing

A field of considerable interest is smart sustainable manufacturing, in which digital technologies, bio-based materials, and nanoscale engineering are blended to produce responsive, resource-efficient industrial procedures. Real-time: The sensor-driven, automated, data-analyzed methodology of smart manufacturing is used to monitor and improve processes in real time. These systems become more accurate, responsive, and more capable when powered by nanotechnology. For instance, tiny sensors placed inside fermentation tanks offer real-time monitoring of temperature, pH, and nutrient levels, enabling autonomous process fine-tuning and predictive maintenance. Real-time control systems like these result in reduced energy consumption and material losses, as well as product quality stability. In addition, nanotechnology makes possible the development of high-performance biocomposites with natural fibers (e.g., hemp, jute, kenaf) reinforced with nanofillers, such as graphene oxide (GO), nanosilica, or carbon nanotubes (CNTs). They have high strength-to-weight ratios and are thermally stable, biodegradable and therefore fit for sustainable packaging, automotive parts, building materials, and other products. For example, biodegradable nano-plastics are now being derived from agricultural residues processed in biorefineries (instead of petroleum-based plastics) for a variety of applications. The smart factories operating on biomass process with nano-enabled equipment not only reduce their carbon and water footprint but also add coproducts of value that help in circular manufacturing paradigms.

The economic consequences of such bio-nano integration are potentially enormous. It improves resource utilization, cuts cost, and creates new market segments in green chemicals, sustainable construction, clean energy, and intelligent materials. What picked amounts to a combination of chicken and egg thinking, since it fosters innovative ecosystems where startups, universities, and industries can team up to work on cross-disciplinary solutions. Nations have also started investing in innovation hubs and demonstration facilities that integrate bioenergy, nanotechnology, and digital tools to develop prototypes for tomorrow's industrial systems, such as in Germany, the Netherlands, South Korea, and Singapore. These hubs are not just testing technologies but rather developing regulatory structures and safety protocols and workforce training programs that can help to establish that new aerospace ecosystem. But standardization, scalability, and the governance of risk still present challenges. Development and adoption of bioenergy and nanotechnology is an expensive and cross-disciplinary endeavor that needs to bring the communities from industry, academia, and different governments together. Uncertainties about the toxicity of nanoparticles, lifecycle emissions of hybrids and intellectual property right issues should be addressed through research, regulation, and public dialog. In addition, application of such technologies in the Global South is reliant on capacity development, access to finance and fair technology exchange. Stakeholders need to make a conscious effort to prioritize co-creation with local communities, open-access technology platforms, education around emerging green skills (Anthony Jr 2024).

10.7 Global Perspectives and Comparative Analysis

10.7.1 Leading Countries and Strategic Approaches

Leading countries in sustainable technology adoption such as Germany, Sweden, Brazil, South Korea, and the Netherlands—have implemented forward-thinking policies that integrate clean energy, innovation, and green manufacturing. Germany's Energiewende ("energy transition") is one of the most ambitious sustainability roadmaps, underpinned by strict regulatory regimes, public subsidies, and decentralized energy models. In Germany, where some 9500 biogas plants generate electricity and heat for industrial and farm use, there has been a massive investment in bioenergy. It has further backed nanotechnology R&D, notably for energy applications, by energy efficiency, battery technology, and new materials research through the High-Tech Strategy, as well as via the Nano-Initiative Action Plan. Through aggressive carbon tax measures and a massive investment in bioenergy-based district heating, Sweden has virtually decarbonized the residential heating sector. Its effectiveness depends in pushing for green taxation along innovation incentives and public ownership of strategic infrastructure (Fig. 10.4) (Bu et al. 2025). There is a different, however, just as effective, model presented by Brazil, based on the production of Bioethanol from sugarcane, that fuels a substantial fraction on the countries transport sector. From the 1970s, the Proálcool program has developed into one of the largest renewable fuel programs in the world with the support of obligatory blending policies, fiscal incentives, and public/private R&D partnerships (Stolf and de Oliveira 2020). In Asia, South Korea's Green Growth Strategy and National Nanotechnology Development Plan make the nation a frontrunner in merging advanced materials science with sustainability goals. For South Korea, it is investment in hydrogen, energy storage, and nano-enabled electronics that will get it to carbon neutrality. The Dutch is smaller but driving circular

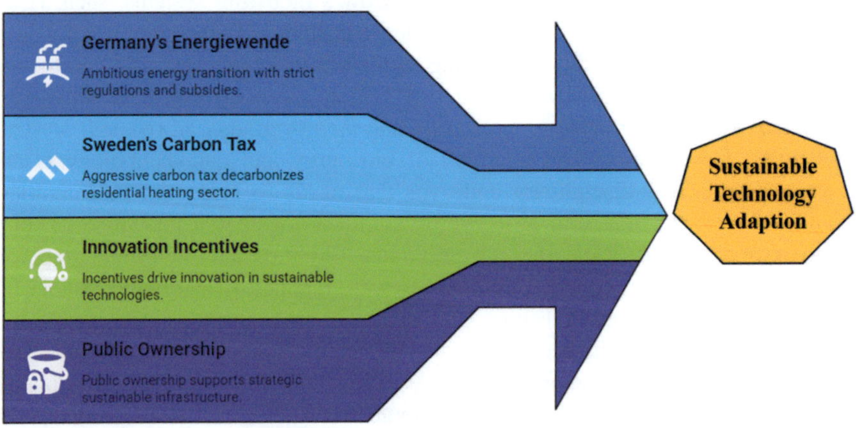

Fig. 10.4 Describe the pathways to sustainability

adoption in bio-based innovation hubs around the likes of Wageningen University, and nanotech clusters such as Eindhoven, reshaping agri-industrial and manufacturing processes. These exemplary countries offer a range of models—bioenergy-centric, technology-based, and policy-derived demonstrating different pathways to sustainability rooted by sound governance, infrastructure, and sound strategy.

10.7.2 Economic Models for Sustainability Transition

The economic models underpinning these sustainability shifts are as diverse as the countries but are all taking the agenda beyond the standard narrative of decarbonization, decentralization, and digitalization. The conventional linear "take-make-dispose" models are giving way to circular economic models that focus on recycling of resources, value addition to wastes, and lifecycle maximization. In Europe, the Circular Economy Action Plan, part of the Green Deal, provides for financial, regulatory, and innovation instruments that will support businesses in their shift to resource-efficient and low-carbon activity. Germany and the Scandinavian countries have developed so-called eco-industrial parks, in which the waste of a factory is the input of another one, to optimize the waste outputs and minimize the resulting environmental impact (Holländer et al. 2009).

In Latin America, Brazil's economy of ethanol works on the model of the Verti-limbed bio-refinery, simultaneously producing electricity, bioplastics, and fertilizers from sugarcane in addition to fuels, which guarantees the economic diversification and value to the product. In Asia, a state-directed industrial ecology approach has been taken in China that includes the creation of regional green developmental zones and massive new biomass utilization efforts; however, fears of environmentalism and centralized control continue to reign. South Korea's aspiration is an innovation-led economic transition that taps into nanotechnology and Industry 4.0 to produce high-value sustainable manufacturing and clean energy systems. In the meantime, countries such as India and Kenya are pioneering decentralized economic model's biogas cooperatives, solar-bio hybrid systems, nanotech-enabled local water purification companies that are aimed at inclusive growth and rural resilience. They draw equally on market mechanisms, state intervention, and international cooperation in their economic models. Instruments such as carbon pricing, feed-in tariffs, public purchasing of green goods and services, or environmental fiscal reforms have attempted to internalize environmental costs and direct investments toward sustainable technologies. And financing for sustainability via green bonds, E.S.G. investing, and impact funds has emerged as a significant economic lever. The World Bank, UNEP Finance Initiative, and regional development banks are increasingly targeting their portfolios at climate-smart infrastructure and technology deployment. But economic models also need to consider equity, which means making sure that transitions don't fall most heavily on already marginalized communities or worsen global technology inequality. Ensuring conducive finance, building skills, and just transition are important pillars of any sustainable economy that will work for us all (Schröder 2020).

10.7.3 Cross-Sector Collaborations and Innovation Hubs

A key enabling factor of sustainability transformation is the multiple cooperation arenas and innovation hubs that are emerging to bridge the worlds of science, industry, policy, and civil society. In Germany, Fraunhofer Institutes carry out applied research, linking research, and industrial application and assist SMEs and larger companies in implementing sustainable technologies, such as nanotech-based bioenergy systems. Sweden RISE (Research Institutes of Sweden) Founded in 2017, RISE is used as a platform to connect academic researchers, government bodies, and big Swedish and global businesses as it works to develop near-commercial scale systems like bio-based packaging materials or nano-enabled water treatments. Complementary in Brazil, the New York-based genome editing specialist in 2014, the São Paulo Research Foundation (FAPESP) signed up to co-fund autumn 2017 Grant life science industry today. Networks such as the Amsterdam Smart City Initiative and Greenport Venlo have turned the Netherlands into a magnet for circular innovation, blending urban planning, digital infrastructure, nanotechnology, and bio-based materials. Other examples include South Korea's Innopolis Daedeok and Ulsan MIRECO Smart Industrial Complexes, which promote intersectoral innovation in hydrogen, clean mobility, and smart manufacturing based on nanobio convergence technologies. In India, bottom-up innovation networks, like the National Innovation Foundation, are integrating indigenous wisdoms with twenty-first century nanoscience to address sustainable agriculture, water purification, and affordable bioenergy systems. These hubs are more than just technology incubators: they are in fact "policy labs," designing new models of governance, sustainability criteria, and of public engagement. Cross-sector collaboration enables nimble, real-time problem solving and facilitates breaking down sectoral silos that can limit the escalation of sustainable technologies. Collaboration globally, including through Mission Innovation, the Clean Energy Ministerial, and the Global Green Growth Institute (GGGI), complements and enhances the work of these hubs by facilitating knowledge sharing, standard setting, and investment across borders (Tagliapietra 2024).

10.8 Challenges and Future Directions

Despite the fast development of sustainable technologies, e.g., bioenergy and nanotechnology, their deployment at scale still encounters a number of problems ranging from technical and economic to social and regulatory aspects. Optimizing the efficiency, scalability, and the "fit" of these technologies with the environment is among their most important technical challenges. Bioenergy technology usually faces feedstock's heterogeneity and seasonal availability and the inconsistency in the resulting energy, which affects the reliability and economic aspect. In the meantime, while nanotechnology is one of the most promising technologies, it still suffers from difficulty in scaling up, especially to produce superior performance nanomaterials in an economical and environmentally friendly process. Moving up from laboratory scale devices to commercially viable systems necessitates progress in manufacturing techniques, material consistency, and performance predictability in the field. Economically, high installation costs and uncertain payback continue to be big challenges. While the long-term advantages of greener technology are well known—from cleaner air to cost savings—some industries, and governments, are wary because of near-term budget restrictions, and multiple competing demands for funds. In developing economies, issues of finance, low institutional capacity, and less developed markets would exacerbate these economic obstacles. Policy-enabling instruments such as green subsidies, tax exemptions, carbon taxes, and public–private partnerships are imperative to compensation for these deficiencies, but they must be effectively rolled out and have long-term stability to generate confidence for investors. On the social and ethical aspect, perception, safety, and equity are also taken into account in order to guide development path for sustainable technologies. Especially nanotechnology, which brings forward new physicochemical properties, causes concerns about health and environmental risks, toxicity, and environmental impact. Lack of clear risk communication, and regulatory oversight could slow progress through public resistance. Likewise, other large-scale bioenergy projects might result in land-use competition that compromises food security, biodiversity, or dispossesses local communities if not well planned. A just transition— ensuring that gains from economic and environment improvements accrue to all and are not undercut by social and equity costs—is central to inclusive and sustainable development. Finally, the future bioenergy and nanotechnology will move toward synergetic integration, life cycle consideration, and cross-disciplinary innovation. Hybrid systems that integrate nanotechnology-driven bioenergy processes (e.g., nano catalysts for faster biofuel generation, nano sensors for process optimization), however, have the potential to significantly enhance efficiency and scalability. Another key will be embedding sustainability into the design of technologies (eco-design), increasing circularity via waste valorization, and integrating digital tools such as AI and IOT in smart process control. Research opportunities will include green synthesis of nanomaterials, safer-by-design concepts, and real-time monitoring of environmental and health impacts. Lastly, a world-wide coordination and capacity building as well as harmonization of regulations will be necessary. Given that sustainability transcends physical and geographic boundaries, whoever wants to move forward in efforts should work together in research, funding, and standard setting to speed research and innovation, in a way that is safe and that doesn't leave anyone behind. Effective responses to these challenges, using collective action and innovation, can place bioenergy and nanotechnology at the center of a sustainable, low-carbon, and equitable global economy.

10.9 Conclusion

The present chapter has described the importance of bioenergy and nanotechnology for a sustainable industrial/environmental future. In a world where economies struggle between the demands for growth and ecological stewardship, these interchangeable, two technology systems from these interdisciplinary sources offer two efficient scalable solutions to decrease dependence on fossil fuels, optimize resource utilization, and move toward circular economic models. However, bioenergy as the way for valorization of organic waste and environmentally friendly decentralized energy is a game-changer in rural development, energy security, and low carbon industrialization. At the same time, nanotechnology improves the performance of materials and the efficiency of processes and helps us to control our environment with extra precision, promoting massive reductions in pollution, energy storage, and sustainable manufacturing. The synergetic of the bioenergy and the nanotechnology create opportunities of hybrid systems, smart manufacturing, and real-time environmental monitoring, transforming the way of how industries work and innovate. Examples from countries such as Germany, Sweden, Brazil, and South Korea show that strategic policy support, innovation ecosystems, and public engagement are essential to scale up these technologies in a manner that is effective and inclusive. Yet, challenges remain. Technical, regulatory, societal, ethical, and market challenges should be managed proactively, transparently, and collaboratively. Further development will require dedication to interdisciplinary research, responsible innovation, and inclusive policy making. Second, international collaboration in information sharing, standardization, and capacity-building is as important as the development of sustainable technologies themselves to achieve equal opportunities for the benefits of such technologies. With the global progress toward net-zero and resilient economies, the possibility of bioenergy and nanotechnology is converging as a vehicle for sustainable change across industries. Ultimately, the social, economic, and environmental promise of these technologies is as dependent on the science as it is on institutions, industries, and communities collaborating to co-create solutions that are fair, flexible, and prepared for life ahead. This chapter explains that with stewarding human designs and cooperative governance, bioenergy and nanotechnology are two foundations for a genuinely sustainable and regenerative world industry.

References

Adams P, Bridgwater T, Lea-Langton A, Ross A, Watson I (2018) Biomass conversion technologies. In: Greenhouse gas balances of bioenergy systems. Elsevier, pp 107–139

Aguilar R (2022) Co-designing planning support tools for stakeholder engagement in collaborative spatial planning processes. PhD thesis, University of Twente

Ahmed SF, Mofijur M, Ahmed B, Mehnaz T, Mehejabin F, Maliat D, Hoang AT, Shafiullah G (2022) Nanomaterials as a sustainable choice for treating wastewater. Environ Res 214:113807

Al-Rumaihi A, Shahbaz M, Mckay G, Mackey H, Al-Ansari T (2022) A review of pyrolysis technologies and feedstock: a blending approach for plastic and biomass towards optimum biochar yield. Renew Sust Energ Rev 167:112715
Anas M, Khattak WA, Fahad S, Alrawiq N, Alrawiq HS, Abdelsalam NR, Jaremko M, Quraishi UM (2024) Mitigating heavy metal pollution in agriculture: a multi-omics and nanotechnology approach to safeguard global wheat production. J Hazard Mater Adv 17:100584
Anas M, Falak A, Hassan S, Khattak WA, Saleem MH, Khan KA, Khalid A, Fahad S (2025) Microbial interactions and bacterial responses to metal stress in plants: mechanisms, adaptations, and applications for sustainable agriculture. J Crop Health 77(1):36
Anthony B Jr (2024) The role of community engagement in urban innovation towards the co-creation of smart sustainable cities. J Knowl Econ 15(1):1592–1624
Ashokkumar V, Flora G, Venkatkarthick R, SenthilKannan K, Kuppam C, Stephy GM, Kamyab H, Chen W-H, Thomas J, Ngamcharussrivichai C (2022) Advanced technologies on the sustainable approaches for conversion of organic waste to valuable bioproducts: emerging circular bioeconomy perspective. Fuel 324:124313
Barot S (2022) Biomass and bioenergy: resources, conversion and application. In: Renewable energy for sustainable growth assessment. Wiley, pp 243–262
Baumber A (2017) Enhancing ecosystem services through targeted bioenergy support policies. Ecosyst Serv 26:98–110
Behera BK, Varma A (2019) Bioenergy for sustainability and security. Springer
Bu Z, Liu J, Zhang X (2025) Collaborative government-public efforts in driving green technology innovation for environmental governance in PPP projects: a study based on prospect theory. Kybernetes 54(3):1684–1715
Chopra J, Rangarajan V, Rathnasamy S, Dey P (2024) Life cycle assessment as a key decision tool for emerging pretreatment technologies of biomass-to-biofuel: unveiling challenges, advances, and future potential. Bioenergy Res 17(2):857–876
Das HS, Basak A, Maity S (2025) Materials science and nanotechnology. In: Innovations in energy efficient construction through sustainable materials. IGI Global, pp 175–206
David E, Niculescu V-C (2021) Volatile organic compounds (VOCs) as environmental pollutants: occurrence and mitigation using nanomaterials. Int J Environ Res Public Health 18(24):13147
Davis JK, George J, Balachandran M (2024) Integrating dye-sensitized solar cells and supercapacitors: portable powerpacks for future energy applications. J Mater Sci 59:1–28
Degner M, Moser S, Lewalter D (2022) Digital media in institutional informal learning places: a systematic literature review. Comput Educ Open 3:100068
Erhueh OV, Nwakile C, Akano OA, Esiri AE, Hanson E (2024) Carbon capture and sustainability in LNG projects: engineering lessons for a greener future. Glob J Res Sci Technol 2(2):38–64
Eze OI, Chibuzor CN, Okafor JC, Osita JI (2024) Leveraging technology to bridge the policy innovation gap in developing countries: enhancing service delivery and public engagement. West Afr J Interdiscip Res (ISSN: 3027-1878) 2(2):1–73
Fioranelli A, Bizzo WA (2023) Generation of surplus electricity in sugarcane mills from sugarcane bagasse and straw: challenges, failures and opportunities. Renew Sust Energ Rev 186:113647
Fisher, J. A. (2021). Enhancing 'Human nature': the human enhancement debate in US bioethics. Columbia University
García R, Gil MV, González-Vázquez MP, Rubiera F, Pevida C (2019) Biomass pelletization: contribution to renewable power generation scenarios. In: Production of materials from sustainable biomass resources. Springer, pp 269–294
Gebreegziabher Z, Naik L, Melamu R, Balana BB (2014) Prospects and challenges for urban application of biogas installations in sub-Saharan Africa. Biomass Bioenergy 70:130–140
Ghazaly NM, Abdulhameed AN (2023) Performance and emissions of nanoadditives in diesel engine: a. J Ilmiah Tek Elektro Komput Dan Inform (JITEKI) 9(4):997–1008
Gorus MS, Aslan M (2019) Impacts of economic indicators on environmental degradation: evidence from MENA countries. Renew Sust Energ Rev 103:259–268
Gupta P (2023) Nanotoxicology in nanobiomedicine. Springer

Hariram N, Mekha K, Suganthan V, Sudhakar K (2023) Sustainalism: an integrated socio-economic-environmental model to address sustainable development and sustainability. Sustainability 15(13):10682

Holländer R, Chunyou W, Ning D (2009) Sustainable development of industrial parks. University of Leipzig, Leipzig

Jenssen T, König A, Eltrop L (2014) Bioenergy villages in Germany: bringing a low carbon energy supply for rural areas into practice. Renew Energy 61:74–80

Junginger HM, Mai-Moulin T, Daioglou V, Fritsche U, Guisson R, Hennig C, Thrän D, Heinimö J, Hess JR, Lamers P (2019) The future of biomass and bioenergy deployment and trade: a synthesis of 15 years IEA bioenergy task 40 on sustainable bioenergy trade. Biofuels Bioprod Biorefin 13(2):247–266

Kabeyi MJB, Olanrewaju OA (2022) Biogas production and applications in the sustainable energy transition. J Energy 2022(1):8750221

Kiani M (2025) Nanomaterials unveiled: catalyzing the future of technological innovation. Available at SSRN 5086632

Klemeš JJ, Varbanov PS, Walmsley TG, Foley A (2019) Process integration and circular economy for renewable and sustainable energy systems. Renew Sust Energ Rev 116:109435

Kousar S, Ahmad D, Khattak WA, Anwar MM, Majeed M, Shah GM, Muhammad M (2025) Environmental resilience: navigation of lead and beneficial elements in soil. In: Beneficial elements for remediation of heavy metals in polluted soil. Elsevier, pp 221–240

Koutoudjian G, Diniz L, Cespedes R, UNDP VG (2021) About IRENA The International Renewable Energy Agency (IRENA) is an intergovernmental organisation that supports countries in their transition to a sustainable energy future, and serves as the principal platform for international co-operation, a centre of excellence, and a repository of policy, technology, resource and financial knowledge on renewable energy. The IRENA Promotes, 24

Larki I, Zahedi A, Asadi M, Forootan MM, Farajollahi M, Ahmadi R, Ahmadi A (2023) Mitigation approaches and techniques for combustion power plants flue gas emissions: a comprehensive review. Sci Total Environ 903:166108

Lead JR, Batley GE, Alvarez PJ, Croteau M, Handy RD, McLaughlin MJ, Judy JD, Schirmer K (2018) Nanomaterials in the environment: behavior, fate, bioavailability, and effects—an updated review. Environ Toxicol Chem 37(8):2029–2063

Malik R, Patil S (2020) Nanotechnology: regulatory outlook on nanomaterials and nanomedicines in United States, Europe and India. Appl Clin Res Clin Trials Regul Aff 7(3):225–236

Mangal P, Lakshmi D, Nagpal N (2025) Cross-sectoral collaborations for advancing renewable energy and conservation goals. In: Digital innovations for renewable energy and conservation. IGI Global, pp 47–74

Manikandan S, Vickram S, Deena SR, Subbaiya R, Karmegam N (2024) Critical review on fostering sustainable progress: an in-depth evaluation of cleaner production methodologies and pioneering innovations in industrial processes. J Clean Prod 452:142207

Matus KJ, Xiao X, Zimmerman JB (2012) Green chemistry and green engineering in China: drivers, policies and barriers to innovation. J Clean Prod 32:193–203

Munagala M, Shastri Y, Nagarajan S, Ranade V (2022) Production of bio-CNG from sugarcane bagasse: commercialization potential assessment in Indian context. Ind Crops Prod 188:115590

Nuhu SK, Emmanuel EO, Manan ZA (2023) Global challenges to sustainable renewable energy expansion: the funding and incentives structures. Chem Eng Trans 106:553–558

Oshilalu AZ (2024) Sustainability meets scalability: transforming energy infrastructure projects into economic catalysts through supply chain innovation. Int J Res Publ Rev 5(12):762–779

Owen R, Pansera M (2019) Responsible innovation and responsible research and innovation. Edward Elgar Publishing

Prabhakar A (2025) A sustainable and inclusive economic development: a global imperative: a global imperative. J Recycl Econ Sustain Policy 4(1):1–16

Prajapati D, Jabborova D, Saharan BS, Singh N, Patani A, Singh S, Joshi C (2024) Bionanotechnology: a paradigm for advancing environmental sustainability. Indian J Microbiol 65(1):306–332

Prasad RD, Prasad NR, Prasad N, Prasad RS, Prasad SR, Shrivastav OP, Prasad RB, Prasad RA, Prasad RG, Saxena P (2025) A review on aspects of nanotechnology in environmental science and engineering. ES General 7:1397

Rane N, Paramesha M, Choudhary S, Rane J (2024) Business intelligence and artificial intelligence for sustainable development: integrating internet of things, machine learning, and big data analytics for enhanced sustainability. In: Machine learning, and big data analytics for enhanced sustainability, 20 May 2024

Saxena N (2023) Bio-nanotechnology in waste to energy conversion in a circular economy approach for better sustainability. In: Bionanotechnology towards green energy. CRC Press, pp 253–274

Schröder P (2020) Promoting a just transition to an inclusive circular economy. Royal Institute of International Affairs, London

Scott AJ (2012) A world in emergence: cities and regions in the 21st century. In: A world in emergence. Edward Elgar Publishing

Silveira S, Johnson FX (2016) Navigating the transition to sustainable bioenergy in Sweden and Brazil: lessons learned in a European and international context. Energy Res Soc Sci 13:180–193

Solomon NO, Simpa P, Adenekan OA, Obasi SC (2024) Sustainable nanomaterials' role in green supply chains and environmental sustainability. Eng Sci Technol J 5(5):1678–1694

Srivastava N, Mishra G (2022) Nanotechnology for environmental pollution detection and remedies. In: Miniaturized analytical devices: materials and technology. Wiley, pp 279–294

Stolf R, de Oliveira APR (2020) The success of the Brazilian alcohol program (PROÁLCOOL)—a decade-by-decade brief history of ethanol in Brazil. Eng Agric 40(2):243–248

Tagliapietra S (2024) The European Union's global gateway: an institutional and economic overview. World Econ 47(4):1326–1335

Thanigaivel S, Priya A, Dutta K, Rajendran S, Sekar K, Jalil A, Soto-Moscoso M (2022) Role of nanotechnology for the conversion of lignocellulosic biomass into biopotent energy: a biorefinery approach for waste to value-added products. Fuel 322:124236

Tong C, Tong C (2019) Biomass for bioenergy. In: Introduction to materials for advanced energy systems. Springer, pp 503–586

Vasantha V, Sharvari S, Alfia N, Praveen N (2021) Application of nanotechnology toward improved production of sustainable bioenergy. In: Nanomaterials. Elsevier, pp 445–479

Verma J, Shah YT, Sundaramurthy S, Arisutha S (2024) Insights into next-generation biofuel strategies and policy considerations in India. In: Clean energy transition-via-biomass resource utilization: a way to mitigate climate change. Springer, pp 251–310

Wong KV, Paddon PA (2017) Nanotechnology impact on the automotive industry. In: Nanotechnology and energy. Jenny Stanford Publishing, pp 157–208

Yrjälä K, Ramakrishnan M, Salo E (2022) Agricultural waste streams as resource in circular economy for biochar production towards carbon neutrality. Curr Opin Environ Sci Health 26: 100339

Zabaniotou A (2018) Redesigning a bioenergy sector in EU in the transition to circular waste-based bioeconomy—a multidisciplinary review. J Clean Prod 177:197–206

Zhang Y, Duan L, Esmaeili H (2022) A review on biodiesel production using various heterogeneous nanocatalysts: operation mechanisms and performances. Biomass Bioenergy 158:106356

Chapter 11
Geo-Environmental Consequences of Bioenergy Production Compared to Fossil Fuels

Fredrick Kayusi and Petros Chavula

Abstract Bioenergy has emerged as a promising alternative to fossil fuels due to its potential for renewable, carbon-neutral energy production. However, the geo-environmental implications of large-scale biomass cultivation require thorough evaluation. This chapter employs a systematic review methodology aligned with Preferred Reporting Items for Systematic Reviews and Meta-Analyses (PRISMA) guidelines to compare the environmental performance of bioenergy versus conventional fossil fuels, focusing on indicators such as soil quality, biodiversity, water resources, carbon sequestration, and phytoremediation potential. An initial pool of 7764 records was screened across major academic databases, yielding 46 peer-reviewed studies for qualitative synthesis. Results indicate that bioenergy systems sourced from residues and perennial feedstocks can reduce lifecycle greenhouse gas emissions and improve soil organic matter, yet they may also intensify water use and nutrient runoff if managed unsustainably. Land-use changes for dedicated energy crops can threaten biodiversity and alter hydrological cycles, though integrated practices—such as co-cultivation on marginal lands and adoption of conservation agriculture—mitigate these risks. Comparative analysis reveals that fossil fuel extraction and combustion produce higher direct CO_2, CH_4, and toxic pollutant emissions, alongside severe land degradation from mining and drilling activities. The chapter concludes that bioenergy's environmental benefits are contingent on feedstock choice, site management, and policy frameworks that internalise indirect land-use changes and ecosystem service trade-offs. Strategic deployment of second-

F. Kayusi (✉)
Department of Environmental Studies, Geography and Planning, Maasai Mara University, Narok, Kenya

Department of Environmental and Earth Sciences, Pwani University, Kilifi, Kenya
e-mail: mg22pu3605021@pu.ac.ke

P. Chavula
Department of Agricultural Economics and Extension, School of Agricultural Sciences, University of Zambia, Lusaka, Zambia

Africa Centre of Excellence for Climate-Smart Agriculture and Biodiversity Conservation, Haramaya University, Dire Dawa, Ethiopia

N. Srivastava (ed.), *Sustainable Nanostrategies for Biorefineries*, Clean Energy Production Technologies, https://doi.org/10.1007/978-981-95-5875-9_11

generation biofuels, combined with robust sustainability criteria, is essential to realise bioenergy's role in a low-carbon energy transition without compromising food security or ecological integrity.

Keywords Bioenergy · Fossil Fuels · Geo-environmental Impacts · Systematic Review · Land-Use Change · Sustainability

11.1 Introduction

In the global quest for sustainable energy transitions, bioenergy has emerged as a compelling alternative to fossil fuels. As concerns over climate change, energy insecurity, and fossil fuel depletion intensify, bioenergy promises renewable and potentially carbon-neutral solutions (Gandy 2018; Memon et al. 2024). However, its environmental and socio-economic consequences—particularly those related to land use—require critical examination. Bioenergy production involves the cultivation of biomass, often on agricultural or marginal lands, raising questions about its broader geo-environmental implications. The intensification of biomass cultivation has the potential to disrupt soil health, degrade biodiversity, compete with food production, and alter hydrological systems (Gunjyal et al. 2023; Raihan 2023). Furthermore, land-use change driven by bioenergy expansion can result in deforestation, peatland drainage, and ecosystem fragmentation, inadvertently increasing greenhouse gas emissions rather than mitigating them (Fig. 11.1).

Bioenergy production on productive agricultural land, while offering higher yields, can threaten food security, especially in regions already vulnerable to hunger and poverty. In contrast, targeting marginal or degraded lands may reduce food competition but introduces limitations in terms of yield potential, soil fertility, and water scarcity. This trade-off highlights the need for nuanced land-use planning and an integrated sustainability framework (Hossain et al. 2019; Yüksel et al. 2023). Additionally, early modelling exercises indicate that even when marginal lands are prioritised, the resulting pressure on natural ecosystems is significant. This suggests

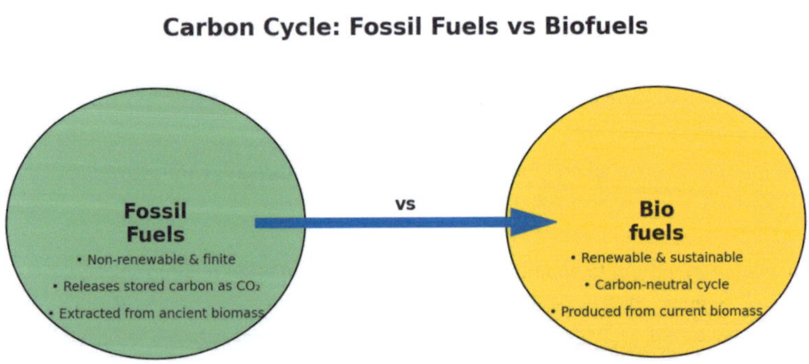

Fig. 11.1 Comparison of the carbon cycle between fossil fuels and biofuels

that bioenergy policies must account for indirect land-use changes and cascading ecological effects to ensure that unintended consequences do not offset environmental gains.

Emerging approaches propose the use of second-generation biofuels, relying on non-food feedstocks such as agricultural residues, perennial grasses, and forest biomass. These feedstocks, particularly drought-resistant perennials like switchgrass and miscanthus, offer ecological advantages including improved soil structure, enhanced nutrient cycling, and higher carbon sequestration potential (Achi et al. 2024; Rane et al. 2024). When managed responsibly, such systems could align with conservation agriculture, thereby reducing the environmental footprint of energy production. Moreover, co-locating food and fuel production through agroecological practices or polyculture systems may offer a viable pathway to mitigate land-use conflict, increase resilience, and diversify rural income streams (Marchiori et al. 2023b; Mutascu 2025).

Nonetheless, a comprehensive assessment of bioenergy systems must include not only feedstock productivity and emissions but also the impacts on ecosystem services, landscape integrity, and long-term soil health. This is particularly urgent in regions such as sub-Saharan Africa, where bioenergy is being promoted as a pathway to energy independence and economic development (Jaiswal et al. 2024; Oswald 2022). In these contexts, environmental assessments must consider local biophysical conditions, land tenure dynamics, and the sociopolitical drivers of land allocation. A further dimension is the competition between fossil fuels and biofuels in energy markets, where fossil fuels continue to dominate due to lower costs and existing infrastructure, despite their environmental drawbacks (Table 11.1).

In light of these complexities, this chapter critically examines the geo-environmental consequences of bioenergy production compared to fossil fuels. It explores the trade-offs associated with different land-use scenarios, assesses the sustainability of various feedstock options, and investigates their impacts on carbon emissions, biodiversity, and soil quality. It also evaluates the role of emerging technologies and policy interventions in enhancing the environmental performance of bioenergy systems. Ultimately, this chapter aims to provide a grounded perspective on how bioenergy can contribute to climate goals without undermining ecological and food system integrity. Through this analysis, it seeks to inform sustainable land-use and energy policy frameworks that balance development with planetary boundaries.

11.2 Materials and Methods

11.2.1 Overview and Purpose

This chapter details a systematic review methodology used to evaluate the environmental impacts of bioenergy production compared to fossil fuels. It relates this approach to advanced forensic methods in data extraction and analysis to ensure

Table 11.1 Geo-environmental consequences of bioenergy production compared to fossil fuels

Impact category	Bioenergy production	Fossil fuels
Land-use change	Often requires dedicated cropland or plantations; can lead to deforestation or indirect land-use change if not managed with sustainability safeguards	Extraction (mining, drilling) fragments landscapes, causes subsidence, and may require large surface disturbances (e.g. strip mining, well pads)
Greenhouse gases	Lifecycle CO_2 emissions can be low or even negative when residue-based or waste feedstocks are used; however, intensive cultivation can emit N_2O from soils	High direct CO_2 and CH_4 emissions in combustion; upstream emissions from extraction, refining, and transport add substantially to the climate burden
Water use and quality	Irrigation demands for energy crops can be significant; fertiliser and pesticide runoff may degrade surface and groundwater quality	Water use in extraction and refining can be heavy; produced-water disposal often contains toxic contaminants, risking aquifer pollution
Soil health	Perennial bioenergy crops can improve soil structure and organic matter, but monocultures can deplete nutrients and increase erosion if mismanaged	Oil and gas spills, coal ash disposal, acid mine drainage severely degrade soil and river sediments, often creating long-term contaminated zones
Biodiversity	Conversion to energy plantations can reduce habitat heterogeneity; mixed-species systems and buffer zones can mitigate impacts	Habitat destruction from mining, drilling infrastructure, and associated transport corridors fragments ecosystems and drives local species loss
Air pollution	Combustion of biomass can emit particulates, NO_x and VOCs, but modern controls and clean feedstocks minimise local air-quality impacts	Combustion and refining release SO_2, NO_x, particulates, VOCs, and heavy metals; contributes to smog, acid rain, and respiratory illnesses
Geo-engineering risks	Large-scale carbon capture and storage paired with bioenergy (BECCS) poses seismic and leakage risks in storage reservoirs	CCS on fossil plants shares similar storage risks, plus oil/gas wells can create subsurface pathways for leakage if poorly sealed

Notes:

- Bioenergy impacts vary widely by feedstock type (e.g., residues vs purpose-grown crops), site history, and management practices
- Fossil fuels impacts are generally concentrated around extraction and combustion, but cumulative effects are global (climate change) and local (pollution, land degradation)

transparency, reproducibility, and scientific rigour. The main aim is to examine how bioenergy systems provide environmental advantages over traditional fossil-based systems, with a focus on soil quality, biodiversity, water resources, carbon sequestration, and phytoremediation.

11.2.2 Search Strategy and Eligibility Criteria

A comprehensive search strategy was developed in accordance with PRISMA guidelines. Major databases—Google Scholar, ScienceDirect, SpringerLink, and

IEEE Xplore—were searched using combinations of keywords such as "bioenergy," "fossil fuels," "environmental impact," "soil quality," "carbon emissions," "biodiversity," and "sustainability." The initial search retrieved 7764 articles spanning the period from 1937 to 2021, with a particular focus on the sub-Saharan African context.

Inclusion criteria were:

- Empirical evidence on geo-environmental impacts of bioenergy or fossil fuels.
- Articles focused on indicators, models, or dimensions of environmental performance.
- Alignment with the research objective and questions.
- Publications in any language within the timeframe of interest.

Exclusion criteria included:

- Conceptual reviews lacking empirical findings.
- Studies with geographically irrelevant scopes.
- Articles with fewer than 0.01 citations per year.
- Technical reports lacking peer review.

After screening, 246 articles were selected for content analysis.

11.2.3 Screening and Data Extraction

The screening process involved the elimination of duplicates, followed by title and abstract reviews. Studies were selected for full-text analysis based on their relevance to geo-environmental indicators of energy systems. Only studies that provided measurable environmental outcomes or comparative analysis between bioenergy and fossil fuels were retained.

From the initial 7764 records, 612 duplicates were removed. A further 1597 records were excluded during abstract screening. Full-text assessment of 207 publications led to 46 studies being included in the qualitative synthesis (Fig. 11.2).

11.2.4 Data Extraction Process

The data extraction process collected:

- Study metadata (title, authors, year, and source).
- Environmental focus (soil, carbon, water, biodiversity, etc.)
- Type of energy system studied (bioenergy/fossil fuel).
- Metrics for impact assessment.
- Region or geographic scope.

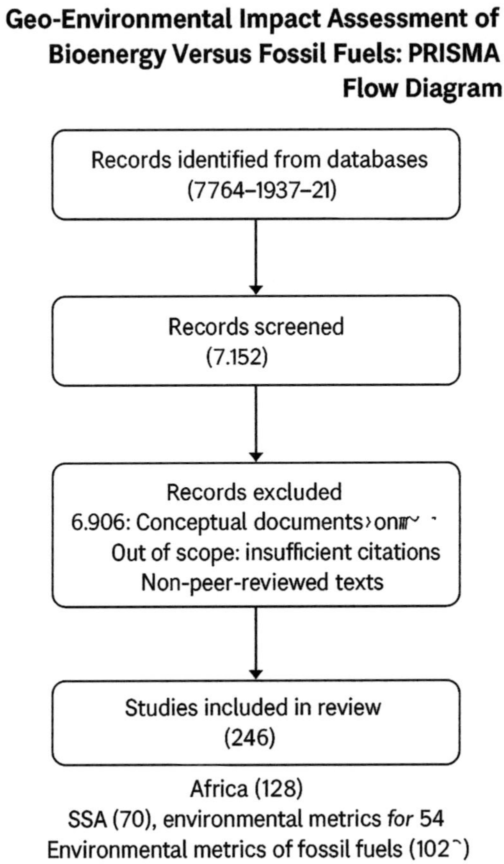

Fig. 11.2 PRISMA model

The analysis emphasised sub-Saharan Africa while drawing comparative lessons from global studies. A standardised rubric guided the selection to ensure consistency.

11.2.5 Data Synthesis and Analysis

Two independent reviewers conducted the initial screening and data extraction to minimise bias. The studies retained were assessed across environmental dimensions, and results were grouped by focus areas such as:

- Carbon Sequestration.
- Water and Mineral Use.
- Soil Quality.
- Biodiversity Impacts.
- Land Use and Phytoremediation Potential.

11.3 Synthesis of the Findings

11.3.1 Fossil Fuels: An Environmental Perspective

The UN Declaration on Bioenergy supports reducing fossil fuel use, promotes ecologically sustainable biofuel practices, and emphasises the preservation of food and water resources. Burning fuel not only pollutes the atmosphere but also leaves harmful residues on the earth. Humanity has faced significant risks from biofuel practices for over one and a half centuries. Wastes such as eggs, fruit scraps, cotton seed hulls, date stones, sugarcane trash, leaf litter, animal wastes, and carcasses are still utilised in biofuel production, often resulting in large amounts of waste (Luo et al. 2025; Rasheed et al. 2025; Tohm and Lingwall 2025). Developing methods to refine these bio-resources into higher-quality biofuel would be a major step forward in securing energy sustainability. Alongside production, the effective utilisation of bio-resources is equally critical for sustainable energy use. The planet's diverse climate conditions influence where different plants grow and thus where certain bio-resources are available. This explains why different countries report varied uses of bio-resources: for example, oil palm shells in Indonesia, Jatropha seeds as oil in India, and bamboo used as briquettes in China.

Comparative analysis and collaborative techno-financial-educational initiatives are essential for biofuel research, pilot testing, utilisation, and policy development among nations to promote sustainable biofuel production. Compared to fossil fuels, biofuels reduce exhaust emissions and greenhouse gases because they lack sulphur and nitrogen. The flue gases from direct combustion can be further processed, where bio-residue gases after hydrogenation and combustion purging leave no solid deposits in benzene trimethyl phenolic catalysts. Toxic elementary nitrogen (used for nitric acid), sulphur suspensions (for sulfuric acid), and volatile gases (for hydrazine) are oxidised into HNO3, H2SO4, and H2O, respectively, while releasing free hydrogen, which again forms ionic ammonia suitable for AMES grade. Availability: Over 60 per cent of vast bio-resources can be collected through fuel densification, sludge conversion, bio-banking, tube packing, chemical salt characterisation, coating, motor use, post-rice biospring, washing and drying pulverisation, foam preparation on perovskite recycling, reductive atmospheric plasma treatment, combination with wood logs, river grasses, and other sources. More than two-thirds of global energy consumption relies on concentrated fossil fuels, subjecting industrial operations to significant pressures. For these reasons, ensuring a sustainable energy supply remains a considerable challenge (Huang et al. 2021; Schulte et al. 2019).

11.3.2 Comparative Analysis of Bioenergy and Fossil Fuels

The demand for energy is growing throughout the world, primarily driven by the growth of developing nations. Fossil fuel reserves, however, are limited and are concentrated in politically unstable parts of the world (Liu et al. 2025; Reix et al. 2021). As a result, there is interest in developing alternative energy sources, including bioenergy. At the same time, use of fossil fuels contributes to anthropogenic greenhouse gas (GHG) emissions, which are causing climate change and associated adverse consequences, and so there is also interest in energy sources that will not exacerbate the climate problem.

Modern bioenergy is defined as energy derived from biological resources, other than the burning of wood and agricultural waste. There is considerable interest and increasing investments in bioenergy, primarily of the modern variety, as it is believed that bioenergy can supply renewable energy with greater net energy production and lower net GHG emissions than fossil fuels. Some studies show that bioethanol from sugarcane, for example, can supply over 8 times more energy than is consumed to produce it, and can provide energy with only 1% of the GHG emissions of gasoline (Liu et al. 2025; Negm and Shareef 2019; Rasheed et al. 2025; Taoufik and Fekri 2023). These projections, however, depend critically on several different factors, such as land use, fertiliser, and fossil fuel inputs, which vary widely, even within a single country. The robustness of the primary conclusions varies, but in general, information on land use changes, indirect GHG emissions, local air quality, and globalisation issues, such as environmental justice, is lacking. Several key factors that have been neglected in comparing bioenergy to fossil fuel energy systems are also mentioned. Bioenergy is being promoted as an environmentally sustainable energy source that can replace fossil fuels and, at the same time, contribute to environmental sustainability.

11.3.3 Geo-Environmental Impacts of Bioenergy Production

The rising demand for energy alongside efforts to cut CO_2 emissions has put bioenergy at the heart of international discussion. Bioenergy involves converting biomass—capturing energy from sunlight—into usable fuel. The biomass feedstock can come from natural or managed ecosystems. Both sources can be used to produce solid, liquid, or gaseous fuels. However, managing ecosystems can significantly affect soils, water resources, biodiversity, and climate. Therefore, it is vital to evaluate the geo-environmental impacts of modern bioenergy production compared to fossil fuels. After the surge in interest in bioenergy following the oil price peak in May 2008, a systematic analysis of the geo-environmental effects of bioenergy versus fossil fuels was conducted.

Soils can store and emit GHGs and, therefore, play a crucial role in climate change mitigation. A common belief is that biofuels derived from biomass residues have a net zero global warming GHG emissions balance. However, neutral GHG emissions are only theoretically achievable or accurate if the major sources of CO_2 in residues remain unaffected during biofuel production and combustion. Nevertheless, the GHG emissions can increase due to residues on aisles or dolines, and the damage to the carbon stock caused by fire on agricultural biofuel plantations may exceed the

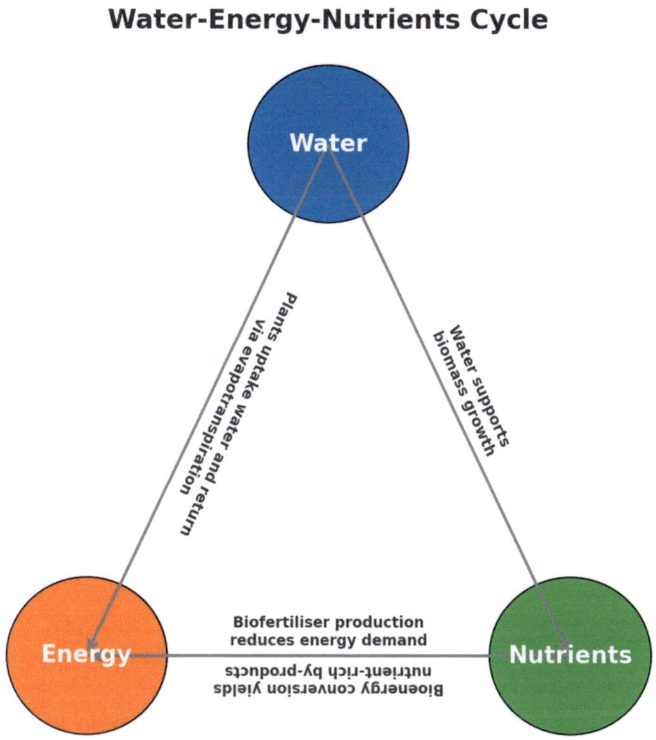

Fig. 11.3 Water–energy–nutrients cycle in bioenergy systems

fossil fuel substitution benefits. Managed ecosystems can store and release soil CO_2, nitrous oxide, and methane, while also providing heat and electricity (Fig. 11.3).

Freshwater is a rare resource of high environmental value. Natural precipitation is guiding in the computation of the environmental water flows for the maintenance of natural and managed ecosystems. The cumulative balance at the biosphere–atmosphere interface defines the land use impacts on the hydrological cycle. It is expected that biofuel plantations, and agriculture in general, increase the demand for water and thus reduce the effect of increased soil moisture. Water quality might also be affected by a change of land cover. Water quality and salinisation increase due to overexploitation and cultivation close to the aquifers.

11.3.4 Land Use Changes

Global energy production and consumption will increase in the next decades. Additional energy demand can be supplied by various options in the short term. The use of fossil fuels, or a combination of fossil fuels and nuclear power, appears as the “cheap” alternative. Unfortunately, this would further enhance a global environmental disaster, which is expected to raise greenhouse gas concentrations to levels potentially dangerous for societies and natural systems (Kumar et al. 2020).

Therefore, taking a global approach to finding and implementing a sustainable energy supply requires attention to a wide range of scientific, technological, economic, and social considerations. Especially promising is bioenergy production from agricultural and forest biomass, as the biomass producer can largely buffer the energy cycle. While fossil fuels are IR-active and may increase greenhouse gas levels, agricultural and forestry practices can limit emissions under certain conditions. This contribution examines the geo-environmental impacts of bioenergy production from agricultural and forest biomass on a global scale (Whittaker et al. 2023).

11.3.5 Water Resource Management

Ground and surface water are critical elements to consider in biofuel assessments because biofuel crop cultivation can be more water-intensive than fossil fuel plants, resulting in potentially large increases in water use and water-related impacts (Alattabi et al. 2025; Rasheed et al. 2025; Roy et al. 2023). Water consumption associated with biofuel production and use stems from three major processes: water needed for crop cultivation, water needed for fuel processing, and water released as effluent from processing plants. Each of these sets of impacts can be large, especially if biofuel feedstocks have high irrigation requirements or are processed in large plants. Given the generally limited availability of water in many regions and aquifers, even modest increases in water consumption can result in disproportionately large increases in water-related impacts (Fig. 11.4).

Less obvious but equally important are the water quality impacts associated with land use changes, crop cultivation, and processing plant effluent. Lessened quality impacts may originate from increased runoff from biofuel feedstocks compared to fossil fuel counterparts, increased use of nitrogenous fertiliser on crops, or increased water temperatures due to cooling water discharges. Each of these sets of impacts can severely impact water supply quality by increasing costs of treatment or directly impairing aquatic ecosystems. In principal, biofuel conversions for an equivalent carbon intensity may aggravate existing water supply-related problems, and research is needed to quantify these risks (Sharma 2024).

Diverse approaches can be adopted to mitigate the potential impacts resulting from emerging biofuels on water resources, including efficiency improvements, systemic changes, and broad-scale regulation methods. Efficiency improvements form the most straightforward approach and have been extensively discussed in previous works. Some of these efficiencies may arise from research and development advances focused on: developing plants with hardier water profiles; shifting emissions discharge to less critical locations and/or cooler temperatures; identifying potential feedstocks that are less water intensive; increasing retention of nitrogenous runoff; directly capturing carbon or shredding biomass emissions before release to the atmosphere; and improving agricultural efficiency or yield.

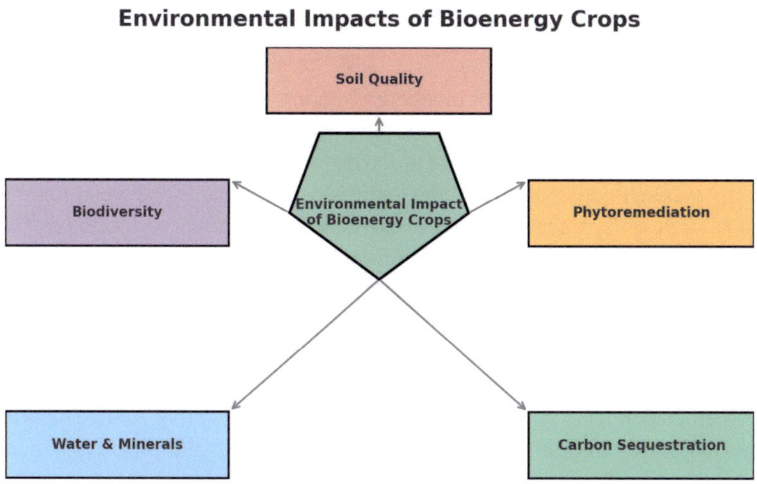

Fig. 11.4 Environmental impacts of bioenergy crops

11.3.6 Biodiversity Effects

The development of feedstock is a key factor for successful bioenergy production. Currently, palm oil proposals that pledge forest protection must be complemented with strict downstream safeguards controlling the output of palm oil trade. Newly planted oil palm tends to be spatially contiguous within an area dominated by oil palm. Consequently, mainstreaming biodiversity needs to become part of the five key risks that currently constitute standards for responsible investment in palm oil (Alattabi et al. 2025; Marchiori et al. 2023a; Whittaker et al. 2023). Comparisons of biofuel life-cycle environmental effects with their fossil fuel counterparts must take into account emissions attributable to energy production. Primary species threshold responses to this net change are revealed to exacerbate a global climate-crop biofuel threat with potentially severe ramifications.

This new approach anticipates a significant increase in biofuel land revival driven by land prices until 2100. Specifically, food crop prices are expected to rise sharply at decadal intervals if waste and residue limits are relaxed; this will lead to land expansion that increasingly shifts away from the dominant energy maize towards oil palm. This can largely be mitigated by implementing interventions to maintain palm oil prices. Additionally, biodiversity loss can be minimised by expanding oil palm plantations on land formerly used for sugarcane, while directly controlling deforestation. Forest lands that are less suitable or unattractive for agriculture are identified as priority areas for oil palm development (Roy et al. 2023). Within the framework of ‘more crop per drop’, dual-irrigation management improves water use efficiency, resulting in a 50% reduction in groundwater extraction. Ultimately, projections of global crop expansion should consider the potential for paddy rice to contribute to extensive land conversions, which are highly relevant.

11.3.7 Market Dynamics of Bioenergy Production

Where and how will the wood for bioenergy be sourced? The production of wood for bioenergy will most likely involve existing wood supply chains and markets in regions that are major net exporters of wood energy. It is also evident that sustainable bioenergy production economic models will vary significantly between regions due to factors such as traditions. The socio-economic consequences (opportunities and risks) of bioenergy production are therefore unique and must be examined within carefully defined regional contexts. Simultaneously, there are many general socio-economic concepts and paradigms that should be considered in bioenergy research and modelling. New bioenergy production models will be developed. Key questions include how bioenergy markets will be affected by the costs and prices of fossil fuels, whether demands can be satisfied using only the existing timber market, and whether there are opportunities to broaden supply chains and bioenergy industries (Bastola et al. 2019; Liu et al. 2025; Roohidehkordi and Krol 2021).

Use of such fuels as wood pellets and energy logs may expand the market for low-grade timber from thinning and low-quality stands. For end-users not connected to natural gas networks, the installation and conversion costs of heating systems can potentially be compensated by relatively cheap wood fuels, which is particularly applicable to larger plants. The impact on energy consumers should be like sizeable and sudden increases of fossil energy prices when woody fuels become comparatively more economical, and an economically ideal situation is found. The use of primary energy for bioenergy conversion into electricity is likely to be above the reasonable limits of techno-economical competitiveness.

There are some aspects, however, that have not been treated carefully or sufficiently at all. These include the influence of the policy frameworks that facilitate the penetration of renewable sources into energy markets. Policy is most important in driving and investing in developing countries (Ahmed et al. 2021). Awareness must be provided of the potential importance of these markets, including risks. It is also important to consider the influences of biofuel prices and potential restrictions on existing raw material supply chains.

11.3.8 Current Market Trends

A rapid expansion of bioenergy projects is taking place around the world despite continual uncertainty concerning the sustainability and climate benefit of many practices. Estimates of current world bioenergy production often differ based on assumptions about biomass resource availability, technology costs, and production parameters. Most bioenergy projects today come from technologies widely deployed and already commercialised, and they generally concern the most widely practised biomass sources—crops and residues, wood, and wastes. The recent surge in interest in algae bioenergy is yet to be backed by a visible commercial scale; nevertheless, niche markets for specific products using algae are developing. Compiling and interpreting production, resource, and cost data estimates for bioenergy technologies, networks, markets, and trade patterns globally allows understanding the future global bioenergy growth potential. More than 90% of bioenergy today is produced from biomass sources traditionally used for thermal energy generation, while biofuels production is rapidly expanding (Whittaker et al. 2023).

The current focus of many analyses and predictions remains on solid biomass, mainly because it is cheaper, even though it has a higher emissions factor than other biofuels. There is an expected rapid displacement of solid biomass by other biofuels, but estimates disagree on the pace, future share of different bioenergy sources, and how their markets are interconnected. However, estimates of current global biofuel production are more consistent, indicating that price-driven development of biofuels and their crop sources has begun (Kayusi et al. 2024a, b). Lower prices, improved conversion methods, and livestock waste management could speed up this process in the next decade. Biodiesel is mainly produced from oilseed crops; therefore, the production of the related meal continues to support the sector's viability. Since waste materials significantly reduce the cost of biodiesel production, more effort is directed towards utilising non-food crops and waste rather than improving oil extraction efficiency.

11.3.9 Economic Viability

Although biofuels like corn ethanol have well-known economic benefits, they nevertheless cause several geo-environmental impacts, including but not limited to increased use of fertilisers, pesticides, and water; enhanced soil erosion and runoff; land use change and spreading degradation; air pollution; socio-economic dislocation; and the effects of energy and litter management acreage requirements on biologically-based projects compared to fossil fuel systems. While there are both advantages and disadvantages to biofuels and the risks to food security are genuine, current concerns seem to focus more on the geo-environmental implications of bioprojects (which are also politicised) than on comparable issues related to fossil fuels (Huang et al. 2021; Schulte et al. 2019).

Therefore, a balanced analysis of each, in terms of their respective geo-environmental metrics, is necessary to inform policy decisions on funding or implementing alternative energy systems that maintain, or potentially improve, food security compared to oil. In comparison, biofuels are mainly classified as net-import-fuel (NIF) cropping systems designed to maximise biomatter yield per hectare-year, driven by goals of oil energy independence, sustainability, and security. Petrofuel systems, on the other hand, are predominantly net-export-fuel (NEF) land-not-in-crop systems aiming to optimise land profit through rent or waste transference in agriculture, generating cheap, exportable bioenergy. Proposed geo-environmental savings are standardised against US corn fuel-cycle production data (from 2000 to 2006) and/or current crop best-management-practice (BMP) scenarios.

Results vary depending on the bioproduction system chosen but are consistent across comparisons when all net-energy-gain (NEG) results for both bioenergy and petrofuel are assessed in kind. In conclusion, biofuels appear to be considerably less damaging to the geo-environment than fossil fuels (Luo et al. 2025; Tohm and Lingwall 2025).

11.3.10 Government Policies and Incentives

Prioritisation of the enactment of policies prescribing biofuel consumption may lead to massive increases in the price of food and inflation. Political will to support the bioenergy sector should couple a carbon tax with conservation instruments such as Payment for Ecosystem Services (PES). Regardless of the nature and magnitude of the subsidy schemes, only a negligible direct effect on bioenergy consumption is to be expected. Several potential remedies to mitigate the induced environmental consequences of the prescribed biofuel mandates are suggested.

While Brazil has been one of the most successful countries in adapting to biofuels and has the highest CS index, a correct object (ethanol) and correct means (sweet-sugarcane) also seem to be key. Furthermore, this analysis also identifies some areas to improve the implementation of Brazil's experiences by other countries. For example, while the Brazilian ethanol pricing policy is feasible and has many advantages, it is nonetheless contested, and a CBOT-style mitigation strategy is recommended. The Brazilian ethanol pricing policy is not fully market-oriented. Efficient allocation of land resources for sugarcane plantation is primarily due to the quality of the imported sugarcane varieties, the level of infrastructure investment, and the climate conditions. This implies that those countries should pay attention to these key aspects prior to industry development. Since the Brazilian sugarcane sector has adopted a relatively more market-oriented composition compared with organic production, it is worthy of further replication. Furthermore, Brazilian sugarcane development has a lengthy time frame and close cooperation with financial institutions, indicating a more long-term market-oriented environmental policy approach.

11.3.11 Technological Innovations in Bioenergy

Modern bioenergy is anticipated to become an increasingly environmentally friendly energy source utilising biological materials, such as crops and residue. This is perceived as a promising option for counter balancing increasing greenhouse gas emissions, attributable to pronounced changes in energy supply, use, and transformation, engendered as an outcome of the use of fossil fuels for energy supply. Despite greenhouse gas exchange being a natural process, prevailing agricultural practices for volatile organic carbon pools (land use) in the terrestrial environment are climatically relevant and must be assessed. Bioenergy is a possible auxiliary energy-supply technology to substitute fossil fuels, derived from two general sources: dedicated and residual production of biomass and biological products by agricultural and forestry processes. Such biomass is characterised by both specific traits (availability, location, transportability, seasonal volatility, and production) which impact the conversion technologies employed and specific constraints on its bioenergy potential (quality, deliverability, transportable energy concentration, and cost). Globally, it has a lesser market share, but it represents the largest component of renewable energy consumed in the world, 9.5% in 2008. Since the late 2000s, bioenergy has been gaining ever increasing attention as a means of counterbalancing climate change, energy security, and sustainable development in developing countries and regions (Marchiori et al. 2023b). There may be a potential competition for land and water between bioenergy and food crops. Another question is whether biomass for energy use can be produced in a sustainable manner. There may therefore be a trade-off between policies to increase bioenergy production and those aimed at increasing sustainability in agriculture, through farmers switching from conventional to organic farming. Generally, bioenergy is becoming an ever more important option in the arsenal of measures that may be adopted to mitigate climate change. This development has raised discussions of potential problems regarding the sustainable production of biomass for energy use.

11.3.12 Advancements in Biofuel Production

The increasing concern over fossil fuel-based energy security and greenhouse gas emissions has led to the search for alternative sources of energy that are environmentally suitable. Bio-fuels from biomass are considered as a solution to overcome both energy security and environmental sustainability. Biomass (non-fossilised, biodegradable organic materials, including agricultural products and residues, forestry, residues and industrial wastes, animal manures, urban wastes, etc.) is abundantly available on the earth. It is carbon neutral in production and utilisation in theory (Schmer et al. 2014). Biomass can be processed into bio-fuels through thermochemical and biochemical processes, and it is thought that thermochemical bio-refinery is capable of inter-converting all types of biomass to fuels as a result of a series of closely linked processes. Bio-energy is environmentally friendly because of the potentially highest feedstock energy conversion efficiency into bio-fuels when compared to fossil fuels in a well-designed bio-refinery (Yüksel et al. 2023).

It has also been accepted that fossil fuels should remain the dominant energy sources for industrial development until about 50% of all non-conventional oil reserves are processed, since fossil fuels exist almost everywhere near the earth's surface. The environmental concern of fossil fuel-based energy came only since the inundation of the earth with the post-combustion greenhouse gases. However, as fossil fuels are still searched for, both the primary and intermolecular selective conversion pathways for fossil fuels to useful chemicals and fuels have long been revealed without a full coverage of cobalt-catalysed Fischer-Tropsch synthesis.

Bio-fuels are renewable, biodegradable, and environmentally friendlier than conventional fossil fuels. They have been considered the best substitute alternatives of fossil fuels for sustainable development since bio-fuels produced from biomass are carbon-neutral in practice. Similar to fossil fuels, bio-fuels can be also classified according to the processing routes. Bio-fuels are classified into gas-phase (synthesis gas, methanol, DME) and liquid-phase (bio-dimethyl ether, bio-diesel, and bio-ethanol) products according to the physical states (Mutascu 2025). Bio-fuels may involve the production of transient intermediate fuel gases, such as volatiles released from pyrolysis products and syngas for Fischer-Tropsch synthesis.

The bio-fuels may be industrially converted to bio-gasoline, bioXTL (i.e. bio-fuel, bio-petrol, bio-jet fuel, etc.) or bio-carbon—a solid fuel in graphite/activated carbon forms. Their feedstocks commonly classify bio-fuels as first generation, second generation, and third generation bio-fuels. First generation bio-fuels include bio-ethanol from corn and cane sugar, bio-diesel from soy and palm oils, and bio-gasoline from 5-carbon sugars through biological conversions. Second-generation bio-fuels include the conversion of a diverse range of lignocellulosic biomass into bio-fuels and solid bio-fuels for combustion. However, commercially viable bio-refineries for producing caprolactam, polyphenols, etc. from wall-free algae residues and chemical-grade ethanol from non-food sugars are unavailable yet (Rane et al. 2024).

11.3.13 Waste-to-Energy Technologies

Waste-to-energy (WtE) technologies have been developed for over 30 years to process municipal solid waste (MSW) in an environmentally benign manner while recovering energy from waste (Gandy 2018; Gunjyal et al. 2023; Memon et al. 2024; Raihan 2023). The waste-to-energy (WtE) technologies, including pyrolysis, gasification, and incineration systems, are analysed and evaluated in terms of their environmental performance. The following preliminary conclusions are drawn: the source of the WtE systems significantly affects the direct emissions to air during operation. Major pollutants include CO2, NOx, SO2, HCl, and HF, all of which originate from fossil fuels in MSW. Despite the difference in the gCO2/kWh emission calculation method and the additional contributions from indigestible inorganic elements in the biomass feedstock, gasification systems using gas turbine or gas turbine combined cycle and incineration systems tend to have the highest gCO2/kWh values.

With the expansion of waste treatment amounts, innovative technologies capable of handling multiple kinds of waste or of converting waste into recyclables or fuels tend to be preferred. For new treatment technologies utilising WtE processes at a large scale, such as gasification, incineration and pyrolysis, technical performance is usually of primary concern before environmental life cycle benefit analysis (Chavula and Kayusi 2025). However, as scrutiny of environmental incidents and emissions from waste treatment technologies has been increasingly raised and the implementation of stricter environmental regulations is envisaged, there is a strong need to evaluate the environmental performance of emerging or modified waste treatment technologies.

11.3.14 Successful Bioenergy Projects

In modern society, the energy demand is constantly growing. So far, the ever-increasing use of fossil energy (coal, oil, and natural gas) has supplied the much-desired energy, but it comes at a cost. Burning fossil fuel releases greenhouse gases (GHG), and important products of combustion such as sulfur oxides, nitrogen oxides, and soot. These products of combustion contribute significantly to anthropogenic climate change, biodiversity loss, forest degradation, acidification, and human health problems, including respiratory diseases and mortality (Luo et al. 2025; Yüksel et al. 2023). Coal reserves are rapidly dwindling, oil crops highly depend on limited geological resources, and the potential of natural gas also has its limits. These facts have made modern bioenergy (fuels made from biomass) a promising option for international climate policy.

All over the world, land is being cultivated for the production of biomass that will serve as a renewable energy resource. Palm oil from native forests, rapeseed from moderate cereal cropping areas, sugarcane from arid coastal areas, and wood pellets from Scandinavian and Siberian boreal forests are only examples of this growing biobased economic sector of international significance that will be examined in this chapter. For each bioenergy project described, the provision of bioenergy, any changes in land cover required, and possible geo-environmental consequences such as GHG emissions, biodiversity loss, changes in water quantity, and pollution of water bodies are briefly outlined (Kayusi 2025b; Kayusi et al. 2025). Each of these issues is recapitulated by selecting one or more case studies that presently exemplify these geo-environmental consequences.

Bioenergy projects producing palm oil, biodiesel, waste palm oil, and fibre. Elaborate land cover changes associated with this flourishing industry in Southeast Asia consist of high GHG emissions, biodiversity loss, and soil subsidence (Kumar et al. 2020; Whittaker et al. 2023). The Australian solar biofuel project has experienced technology problems and would need land cover. Wind energy projects under consideration in China, however, have thrived and are urgently needed to reduce GHG emissions. Wood pelleting, sawdust production, and bioheat projects are being prepared in Scotland. However, forest biodiversity would be threatened by an expanded annual felling of 2 million extra m^3. Finally, black liquor by steam gasification to an AeroMethanol was first successful in Sweden but was terminated in South Africa.

11.3.15 Challenges Faced in Implementation

Commercialisation and implementation of biofuel technologies face challenges in the advancement of next-generation technologies. These hurdles include cost, feedstock supply, technology risk, product distribution, and infrastructure adaptations for distributing biofuels in the existing oil pipeline system, regulatory and policy uncertainty, and long timelines for market acceptance. However, while significant expansion of biofuels is expected over the next few decades, it is believed that they will compose only a small fraction of total transportation fuel consumption because of costs, availability of extensive alternative fuel supply options, and potential feedstock sustainability issues in the longer term. However, biofuels can be considered as an environmental bridge until more sustainable alternative fuels become commercially available (Roy et al. 2023; Taoufik and Fekri 2023).

Among the countries seen as top biofuel producers, the U.S. and Brazil mainly use plant oils and sugar from the damaging and controversial monoculture crops. EU countries mostly rely on rendered tallow and cooking oils from food waste. A cautious, systemic approach to managing the risks of land-use change for biodiversity and climate involves phasing out the allowance of indirect land-use change-related emissions through a cap on sustainable biofuels used in transport (Kayusi et al. 2024a). The supply chain of crop-based biofuels is highly vulnerable to land-use change and market fluctuations. Limiting crop use for biofuels is essential to prevent rising food prices and to decrease nutrient runoff into aquatic ecosystems. It still needs to be decided whether to keep the first-generation biofuels within a low GHG cap or to remove them from the EU burden-sharing agreement (Liu et al. 2025; Sharma 2024).

Positive performance indicators include the sustainable sourcing of biomass. Negative performance indicators include the penetration of uncultivated natural land into the food-energy supply chains and consequential carbon sink loss through land conversion, in combination with massive deforestation in the tropics, nutrient loading, and linked welfare loss for populations dependent on forest resources.

11.3.16 Public Perception and Acceptance of Bioenergy

Bioenergy has received wide-ranging attention concerning its suitability to mitigate climate change. However, the recent rapid expansion of the bioenergy sector has sparked a dilemma. This dilemma is framed as a question of scale and aims. Two scales of discussion emerge. A larger scale refers to commercialised bioenergy usage in industrialised regions; it has attracted wide scientific and public attention due to a more developed bioenergy market (Yüksel et al. 2023; Achi et al. 2024). A smaller scale refers to small-scale, local types of bioenergy production; polls indicate a favourable attitude towards bioenergy's contribution to rural development. Local case studies centred on local contexts illustrate many benefits and sustainability issues of bioenergy. However, whether and how bioenergy contributes to a region's sustainable rural development is poorly understood. Bioenergy describes renewable energy derived from the conversion of lignocellulosic biomass. Countries with an energy shortage or large remote areas take a keen interest (Kayusi 2025a). Local biomass is a form of "sustainable" energy that countries have combined to mitigate global warming and greenhouse gas emissions. In East Asia and Europe, many efforts have been made to promote the broader use of bioenergy.

There are only a few hundred studies focused on bioenergy policy adoption. The most common policy instrument is financial support, which is popular among both advanced and developing governments (Taoufik and Fekri 2023). However, the subsidy scale varies greatly. Almost all countries use multiple combinations of policy instruments, reflecting the complexity and varying priorities of bioenergy technologies in different countries. The commonly used instruments vary, with awareness initiatives most common for advanced countries. For developing countries, the demand and capacity for specific instruments depend on the country. Research and development investment tends to focus on advanced economies, with high adoption levels. In the developing world, public-private cooperation is a favoured option, and some have specific instruments targeting biomass and biogas, as well as their technological options. In Europe, competitiveness policies generally receive the widest support, followed by sustainability policies. For each bioenergy technology, general economic policies come first, while competition policies receive less attention.

11.4 Conclusion and Recommendations

Most bioenergy resources offer a limited GHG reduction potential compared to fossil fuels. Wastewood and agricultural residues are widely considered acceptable bioenergy resources by certification schemes. These feedstocks are, however, of limited supply, and there are concerns about their sustainability. Residues have been dealt with in legislation as waste, which should be reduced. High demand for agricultural residues and their limited supply can lead to undesirable effects such as an increase in soil depletion and erosion rates. Wasteful practices in agriculture, such as burning straw, have been observed and seem to be a significant barrier to sustainable bioenergy production. Concerns around sustainability should not only be addressed in a local and regional context, but also on a global level. Some consensus has emerged around more responsible global biofuel policies, which must consider the influences on commodity prices, investments in agricultural land, and further market distortions. It is improbable that bioenergy produced from non-traditional feedstock sources will be a valuable and available option for widespread, economically priced energy production shortly. During the time needed for development, fossil fuel consumption will increase, leading to an increase in carbon emissions, and trapping the global temperature increase at dangerously high levels.

Sustainable practices on conventional agricultural land can result in minimal energy, and optimally less than 0.5% of the present-day global primary energy supply. On grass and shrublands, the global total biophysical potential of bioenergy from herbaceous biomass is estimated at only 2% of the present-day levels of global consumption of fossil fuels. It is also recognised that a maximum production potential for dedicated energy crops is on non-arable and marginal agricultural land, which may not be sufficient to meet future energy needs. Concerns about the sustainability of biomass harvest on large plantations held by industrialised countries should be addressed. Assessments of the bioenergy expansion potential require global approaches to estimate biomass supply and compare it to fossil fuel supply options.

References

Achi CG, Snyman J, Ndambuki JM, Kupolati WK (2024) Advanced waste-to-energy technologies: a review on pathway to sustainable energy recovery in a circular economy. Nat Environ Pollut Technol 23:1239–1259

Ahmed I, Zia MA, Afzal H, Ahmed S, Ahmad M, Akram Z, Sher F, Iqbal HMN (2021) Socio-economic and environmental impacts of biomass valorisation: a strategic drive for sustainable bioeconomy. Sustainability 13(8):4200

Alattabi IA, Almudhafar SM, Jafar I, Al-Naffakh J, Almayahi BA (2025) Environmental and geo-life assessment to produce alternative fuel from plastic waste and used motor oil: a sustainable energy approach for Iraq

Bastola S, Lim KJ, Yang JE, Shin Y, Jung Y (2019) Landuse and landcover change and the impacts on soil carbon storage on the Bagmati Basin of Nepal. J Korean GEO-Environ Soc 20(12): 33–39

Chavula P, Kayusi F (2025) Navigating the Paradox: Climate Change, Cutting-Edge Technologies, and Groundwater Sustainability Navegando el Paradoxo: Cambio Climático, Tecnologías de Vanguardia y Sostenibilidad del Agua Subterránea. https://doi.org/10.62486/latia202583

Gandy M (2018) Cities in deep time: bio-diversity, metabolic rift, and the urban question. City 22(1):96–105

Gunjyal N, Rani S, Asgari Lajayer B, Senapathi V, Astatkie T (2023) A review of the effects of environmental hazards on humans, their remediation for sustainable development, and risk assessment. Environ Monit Assess 195(6):795

Hossain MZ, Anawar HM, Santa-Regina I, Akter F (2019) Revegetation of energy crops on acidic and alkaline toxic metal-rich mining waste and soil: carbon sequestration, energy production and waste management. In: Sustainable and economic waste management. CRC Press, pp 127–144

Huang H, Reddy NG, Huang X, Chen P, Wang P, Zhang Y, Huang Y, Lin P, Garg A (2021) Effects of pyrolysis temperature, feedstock type and compaction on water retention of biochar amended soil. Sci Rep 11(1):7419

Jaiswal KK, Chowdhury CR, Yadav D, Dutta S, Banerjee I, Jaiswal KS, Ramaswamy AP, Mrinal, Sangmesh B, Jaiswal AK (2024) Eco-friendly approaches for converting organic waste to bioenergy for sustainable development. In: Applied biotechnology and bioinformatics: agriculture, pharmaceutical research and environment, pp 269–302

Kayusi F (2025a) Article Context and Technological Integration: AI ' s Role in Climate Change Research Contexto del Artículo e Integración Tecnológica: El Papel de la IA en la Investigación sobre el Cambio Climático. https://doi.org/10.62486/latia202585

Kayusi F (2025b) Enhancing Urban Green Spaces: AI-Driven Insights for Biodiversity Conservation and Ecosystem Services Mejora de los espacios verdes urbanos: Inteligencia Artificial para la conservación de la biodiversidad y los servicios ecosistémicos. https://doi.org/10.62486/latia202587

Kayusi F, Chavula P, Kengere D, Mbunde J (2024a) Policy influence on genetic modification: innovations in enhancing crop nutrition. XXX(Xx). https://doi.org/10.9734/AJARR/2024/XXXXX

Kayusi F, Kasulla S, Malik SJ, Wasike JA, Lungu G, Mambwe H, Chavula P (2024b) Advanced AI, machine learning and deep learning techniques for climate change studies: a review. South Asian Res J Agri Fish 6(6):101–110

Kayusi F, Wasike J, Chavula P (2025) The role of mulching in reducing greenhouse gas emissions and enhancing soil health among smallholder farmers in Zambia, Malawi, Kenya, and Tanzania: an AI-driven approach. LatIA 3:75

Kumar A, Shekhar S, Tirkey AS, Krishna AP (2020) Geo-environmental Hazard vulnerability and risk assessment over South Karanpura Coalfield Region of India. In: Sustainable development practices using geoinformatics, pp 23–45

Liu Y, Wang Z, Liu M, Zhang T, Liu S, Lu K (2025) Characterization, sources and risk assessment of organochlorine pesticides (OCPs), organophosphorus pesticides (OPPs) and polycyclic aromatic hydrocarbons (PAHs) in agricultural soils from Jilin Province in Northeast China. Stoch Env Res Risk A 39:1741–1755

Luo M, Yuan M, Ji C, Gao J, Shi Z (2025) Mycorrhizal types determined the response of yield of woody bioenergy crops to environmental factors. Int Microbiol 28(5):1057–1069

Marchiori L, Morais MV, Studart A, Albuquerque A, Andrade Pais L, Ferreira Gomes L, Cavaleiro V (2023a) Energy harvesting opportunities in Geoenvironmental engineering. Energies 17(1): 215

Marchiori L, Morais MV, Studart A, Albuquerque A, Andrade Pais L, Ferreira Gomes L, Cavaleiro V (2023b) Energy harvesting opportunities in Geoenvironmental engineering. Energies 2024(17):215

Memon AR, Makauskas P, Kaminskaite-Baranauskiene I, Pal M (2024) Unlocking geothermal energy: a thorough literature review of Lithuanian geothermal complexes and their production potential. Energies 17(7):1576

Mutascu M (2025) Impact of pole motion on CO2 emissions through global warming. Geol J 60: 2714–2739

Negm AM, Shareef N (2019) Introduction to the "waste management in MENA regions". In: Waste management in MENA regions. Springer, pp 1–11

Oswald Y (2022) France: international resources for a sustainable, inclusive, and innovative future? In: A triple bottom line analysis of global consumption. Jenny Stanford Publishing, pp 131–139

Raihan A (2023) A review of climate change mitigation and agriculture sustainability through soil carbon sequestration. J Agri Sustainab Environ 2:23–56. ISSN, 2997, 271X

Rane NL, Günen MA, Mallick SK, Rane J, Pande CB, Giduturi M, Bhutto JK, Yadav KK, Tolche AD, Alreshidi MA (2024) GIS-based multi-influencing factor (MIF) application for optimal site selection of solar photovoltaic power plant in Nashik, India. Environ Sci Eur 36(1):5

Rasheed F, Kayani WK, Asghar MU, Farrukh A, Gul S, Khan I, Nazli N (2025) Unlocking the potential of biowaste: sustainable strategies to produce valuable industrial products. J Environ Qual 54(2):319–334

Reix A, Zmamou H, Lenormand H, Leblanc N (2021) Valorization of sunflower bark in agropellets. Geo-Environ Eng GEE2021:ESITC-UniCaen

Roohidehkordi I, Krol MM (2021) Applicability of ground source heat pumps as a bioremediation-enhancing technology for monoaromatic hydrocarbon contaminants. Sci Total Environ 778: 146235

Roy P, Pal SC, Chakrabortty R, Saha A, Chowdhuri I (2023) RETRACTED: a systematic review on climate change and geo-environmental factors induced land degradation: processes, policy-practice gap and its management strategies. Geol J 58(9):3487–3514

Schmer MR, Vogel KP, Varvel GE, Follett RF, Mitchell RB, Jin VL (2014) Energy Potential and Greenhouse Gas Emissions from Bioenergy Cropping Systems on Marginally Productive Cropland. PLoS ONE 9(3): e89501. https://doi.org/10.1371/journal.pone.0089501

Schulte RPO, O'Sullivan L, Vrebos D, Bampa F, Jones A, Staes J (2019) Demands on land: mapping competing societal expectations for the functionality of agricultural soils in Europe. Environ Sci Pol 100:113–125

Sharma P (2024) Change in geo-environmental conditions and crop productivity. In: Advances in food security and sustainability, vol 9. Elsevier, pp 33–57

Taoufik M, Fekri A (2023) A GIS-based multi-criteria decision-making approach for site suitability analysis of solar-powered hydrogen production in the Souss-Massa region, Morocco. Renewab Energ Focus 46:385–401

Tohm, C., & Lingwall, B. N. (2025) Use of sawmill residuals as a geotechnical amendment for reclamation and site restoration. In Geo-EnvironMeet 2025, pp 207–215

Whittaker RJ, Fernández-Palacios JM, Matthews TJ (2023) Island biogeography: geo-environmental dynamics, ecology, evolution, human impact, and conservation. Oxford University Press

Yüksel S, Dinçer H, Uluer GS (2023) Evaluating possible ways to decrease negative environmental impact of nuclear energy projects. In: Economic, environmental and health consequences of conservation capital: a global perspective. Springer, pp 55–68

Chapter 12
Socioeconomic Balance Through Bioenergy Sector and the Role of Nanotechnology: Present-Time Scenario

Gursharan Singh, Shailendra Kumar Arya, and Jatinder Singh

Abstract In current scenario, the bioenergy is purely gesture towards the synthesis of first-generation alcohol from sugarcane, rice, sorghum, molasses and many other starch-based sources. In the last few decades, demand of ethanol is hugely increased especially in USA, Brazil, China and India. First generation of ethanol production is traditional and easiest method to generate the alcohols at large scale. The demand for alcohols all over the world increased the rural as well as suburban employment. As a consequence, socio-economic status of the related population has improved. On the other hand, there is also amplified use of land for the growth of sugarcane and other energy crops for production of ethanol, which creates a competition with food production for the billions of people. It's also recognised that due to the rise in distilleries, environmental pollution also increased, especially contamination of groundwater and decreased the soil fertility. In the same time, nanobiotechnology showed its positive influence in bioenergy sector, at least in laboratory-scale experiments. In the coming future, it may be helpful for the synthesis of second-generation of biofuels such as bioethanol from lignocellulosic biomass (LCB), especially belonging to different types of straws and forest waste. Nanobiotechnology can play the major role in the deconstruction of LCB that may be through immobilisation of enzymes, etc.

Keywords Bioenergy · Nanotechnology · Biofuels · Biotechnology and socio-economic

G. Singh (✉)
School of Allied Medical Sciences, Lovely Professional University, Phagwara, India
e-mail: gursharan.24447@lpu.co.in

S. K. Arya
UIET, Biotechnology Branch, Panjab University, Chandigarh, India

J. Singh
Department of Horticulture, SAGR, Lovely Professional University, Phagwara, India

N. Srivastava (ed.), *Sustainable Nanostrategies for Biorefineries*, Clean Energy Production Technologies, https://doi.org/10.1007/978-981-95-5875-9_12

12.1 Introduction

In developing countries, migration of rural population to urban areas is general process for the search of better livelihood and easy employment opportunities. But, in the last few years, reverse migration has been noticed people rushed towards villages during and after the COVID-19 spread. The most devastating effects of COVID-19 were the loss of occupations of the large population especially that was lived at the bottom of socio-economic level in cities (González-Leonardo et al. 2022). In this situation of reverse migration, rural populations have to make themselves self-sufficient, at least from the energy point of view, and they can rear their own livestock for self-consumption of milk and other products. But this is not enough for the long run; many recent opportunities are arising regarding the generation of employments related to bioenergy sector. The alcohol-producing business has shown a huge potential in recent years to enhance the GDP of nations, which plays an important role in creating the new employment opportunities as well as reduced CO_2 discharges as per the requirements of SDG 9–Industry, Innovation and Infrastructure. Figure 12.1 depicts the positive impact and contribution of bioethanol industry in achieving different sustainable development goals.

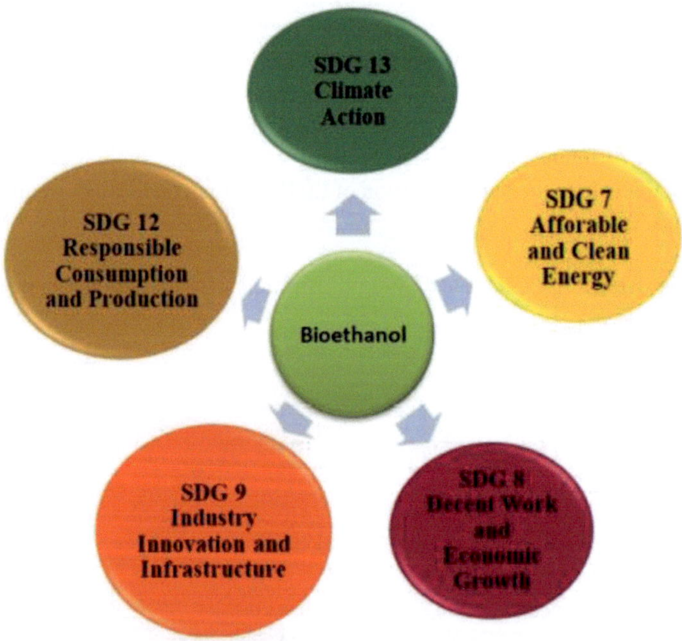

Fig. 12.1 Sustainable development goals (SDG) 7, 8,9,12 and 13 that fulfilled/support by the growth of bioethanol producing industries (Devi et al. 2023). SDGs are designed as to achieve the socio-economic growth and help to alleviate the poverty

According to NITI (National Institution for Transforming India) Aayog, the country will achieve the goal of less carbon generation through blending of ethanol up to 20% in petrol; as a result, it will strengthen the energy sector and also support the local enterprises and motivate the farmers to grow the ethanol-related cash crops. Net shell of this step will boost the economy of nation and create the local employment related to distilleries and farming sector. It also seems like generating new opportunities for sustainable agriculture and breaking the idea of monotonous cropping arrangements (https://www.niti.gov.in/). In recent literature, it's clearly discussed that molasses (left over after the synthesis of sucrose) plays a major role for the production of ethanol because this is well-preferred raw material by *Saccharomyces cerevisiae* (Arshad et al. 2019). More sugar mills may produce more molasses, while in India the number of sugar mills is declining gradually. According to National Federation of Cooperative Sugar Factories (NFCSF), in the season of 2024–25, 454 sugar mills were operational across the country while this number was 505 in the previous season. In the race for more production of ethanol by using an excess of sugar cane and food grains, insecurities may be created in future due to the competition for using the land for growing sugar cane or staple crops for the huge population of humans. Farmers will be inclined more towards the growing of sugar cane if they pay more in comparison to the other crops. Singh et al. (2023) reviewed the stern socio-economic issues discussed by many researchers also; in the case of Brazil, expansion of agricultural area swiftly only for sugarcane farming, and moreover, the possession is only in the hands of rich landholders. This type of outsized agribusiness owner has diminutive or no respect for environmental distresses; as a result, there is too much air smog blow-out from the burning of sugarcane waste after the harvesting of crops that ruined the soil and aquatic environment.

In the last few decades, there has been extensive research going on regarding the generation of cellulosic ethanol from lignocellulosic biomass (LCB) especially from the rice straw, wheat straw, corn straw and diverse type of grasses. This type of alcohol production from LCB is called second-generation ethanol synthesis. In case of second generation of alcohol production, recalcitrance of the lignin present in LCB is major hurdle for the extraction of cellulose. Nanoparticles (NPs) can play the significant role in pretreatment of LCB and further enzymes like cellulases can be immobilised on NPs for mobilising the conversion of glucose from cellulose (Cherian et al. 2015). NPs also can reduce the many inhibitors of the fermentation. Production of cellulosic ethanol is really a dream project of many nations, but still this research is only at pilot scale due its cost intensive procedures (Sharma et al. 2020; Singh et al. 2024).

Nanobiotechnology procedures recently showed the substantial role in case of breaking the recalcitrance and pretreatment of LCB (Kaur et al. 2019). The tiny size of NPs can easily pierce deeply to the cell wall of LCB and interact with cellulose, hemicellulose and lignin; as a result, all biorefinery materials are usually extracted easily. Further, the application of magnetic NPs can easily be done in biorefineries and also recovered for reusability. Nanobiotechnology reported beneficial for the biorefineries especially in pretreatment and during fermentation, but the exact mechanism of how NPs work is not clear. Production of second-generation ethanol from LCB by use of nanobiotechnology needs more comprehensive studies as a prerequisite to explore the efficiency of the NPs (Arora et al. 2020).

12.2 Life Cycle Assessment (LCA) of Alcohol Production from Starch and Molasses (First-Generation Biofuel)

LCA of alcohol production from molasses and starch is important aspect to study the effect on socio-economic benefits to the directly linked people. When starting from the sugar cane crop that grows in the fields requires water, fertilisers, herbicides, etc. Thereafter sugar cane is transported to the sugar mills, and finally juice is extracted from it, then crystallised and convert into the edible sugar (sucrose). The waste liquid that is not converted into the sugar crystals is called as molasses. This molasses is finally transported to distilleries for direct fermentation by yeast and finally produces ethanol. In this entire process, lot of rural and urban population is directly/indirectly economically benefitted. Same process implies on the sugar production from grains, but in this case grains like rice, sorghum, barley etc. first crushed in to the powder form then added to water and exposed to starch hydrolysing enzymes for conversion of this to glucose. Numerous people attached to this process from growing the grains, its transportation to distilleries and finally conversion to ethanol, large number of populations benefitted from this process.

12.2.1 Life Cycle Assessment (LCA) of Alcohol Production from LCB (Second Generation Biofuel)

Alcohol production from LCB is dream project of developed and many developing countries, but its cost intensive production of ethanol is still big challenge. Advantages of the second generation ethanol production are: there is no threat to the food security as compared to first-generation alcohol synthesised from the fermentation of starch and sugar cane juice. There are merely wasted raw materials required for the synthesis of second-generation ethanol such as rice straw, wheat straw, different types of peels, and forest waste debris. On the other hand, big draw backs like recalcitrance of lignin that binds tightly to the cellulose and hemicelluloses in LCB. To break down the rigidity of the LCB, its pretreatment is necessary step in which involves the heating of LCB at high temperature in alkaline conditions. Thereafter enzymatic hydrolysis of cellulose by use of cellulases is an immense challenge, as the cost of the enzymes is high and the crystallinity of the extracted cellulose from LCB is resistant to convert to glucose. Further fermentation of glucose extracted from cellulose is not easy process because numerous fermentation inhibitors retard the growth and functioning of yeast. Although production of ethanol from LCB is

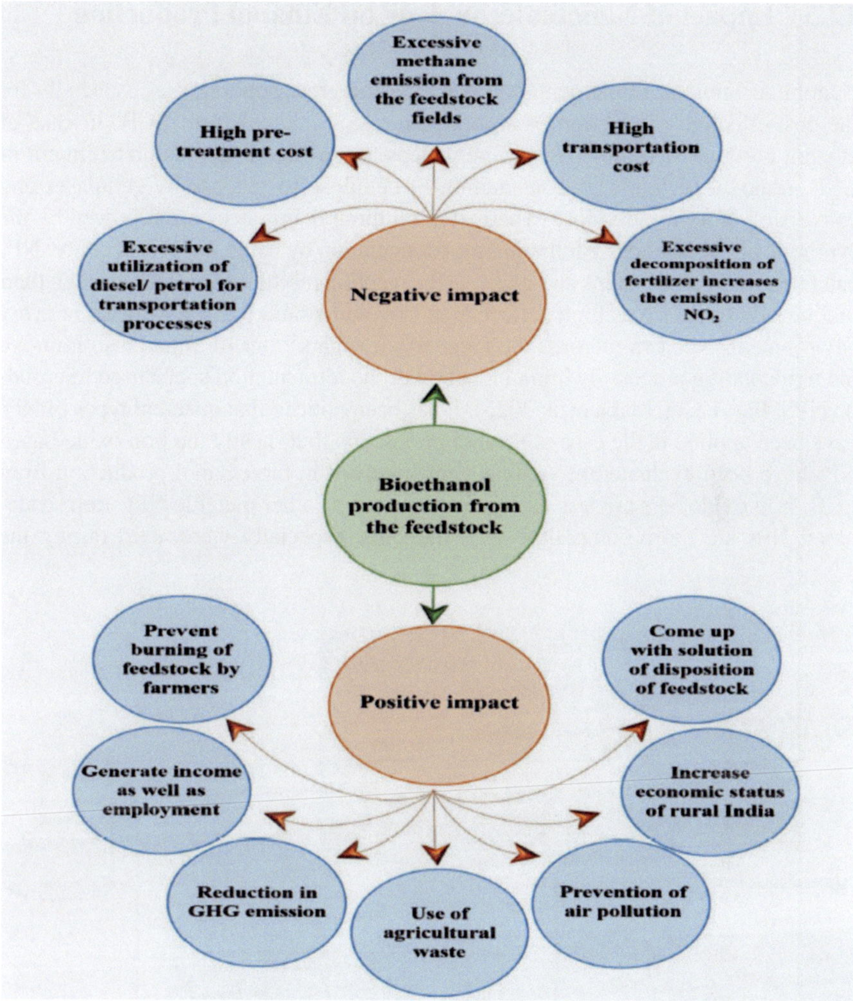

Fig. 12.2 Positive and negative aspects of the alcohol production from LCB (Shukla et al. 2023)

still under research in many countries but due to this technology agro-waste like rice straw, corn straw and many types of grasses can be managed easily (Kazmi et al. 2025). Further it seems, this technology will take time to be implemented as comparison to the setup of first-generation production of ethanol. There are many advantages and drawbacks of second-generation ethanol production from LCB feed stock Fig. 12.2.

12.3 Impact of Nanobiotechnology on Ethanol Production

Nanobiotechnology implementation is one of the great opportunities especially for the cost-effective production of second generation ethanol from LCB. In case of ethanol production from LCB, different steps are involved such as pretreatment of LCB, removal of lignin, deconstruction of cellulose to glucose by cellulases and conversion of xylan to xylose. Thereafter, all these major steps are followed by the synthesis of the ethanol by initiating the fermentation by using the yeast culture. NPs can interfere in all the steps and improve the reaction mixture outcomes through their unique structure such as high surface area-to-volume ratio for performance in every environment. NPs can increase the recovery of sugars from LCB and also improve the fermentation process by immobilisation of the fermentation specialised microbes on NPS, Fig. 12.3 (Ndaba et al. 2025). It has been noticed that different types of NPs have been applied in the case of ethanol production, but mostly the iron oxide-based NPs have been evaluated at various steps involved in the ethanol production from LCB. Iron oxide NPs are less toxic in comparison to other metallic NPs. Iron oxide-based NPs are more compatible with the cells, especially when used during the

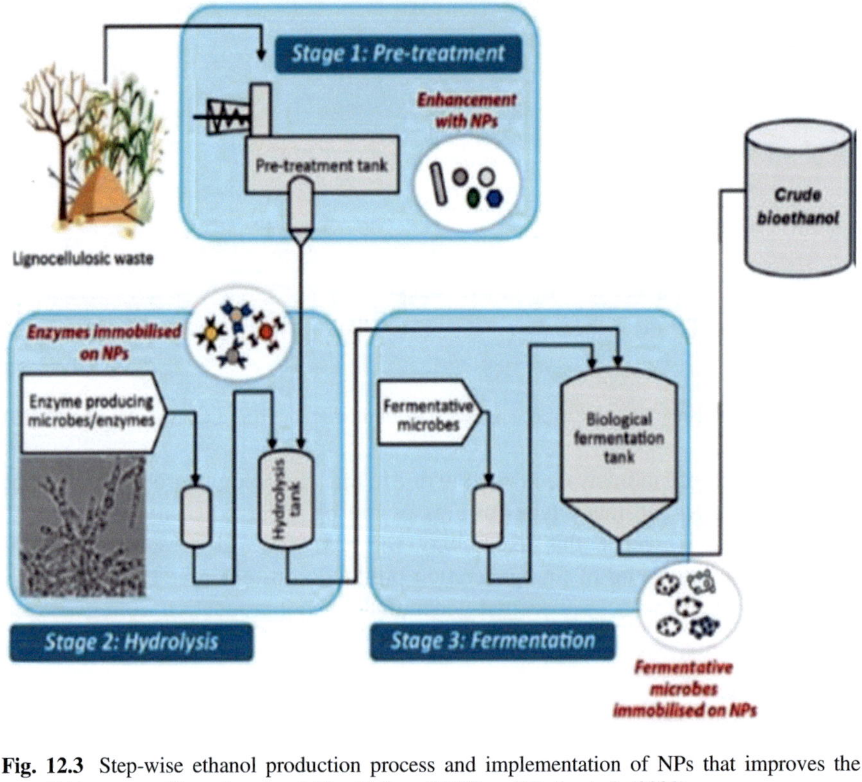

Fig. 12.3 Step-wise ethanol production process and implementation of NPs that improves the overall bioprocess (Awogbemi and Von Kallon (2022) and Dutta et al. (2023)

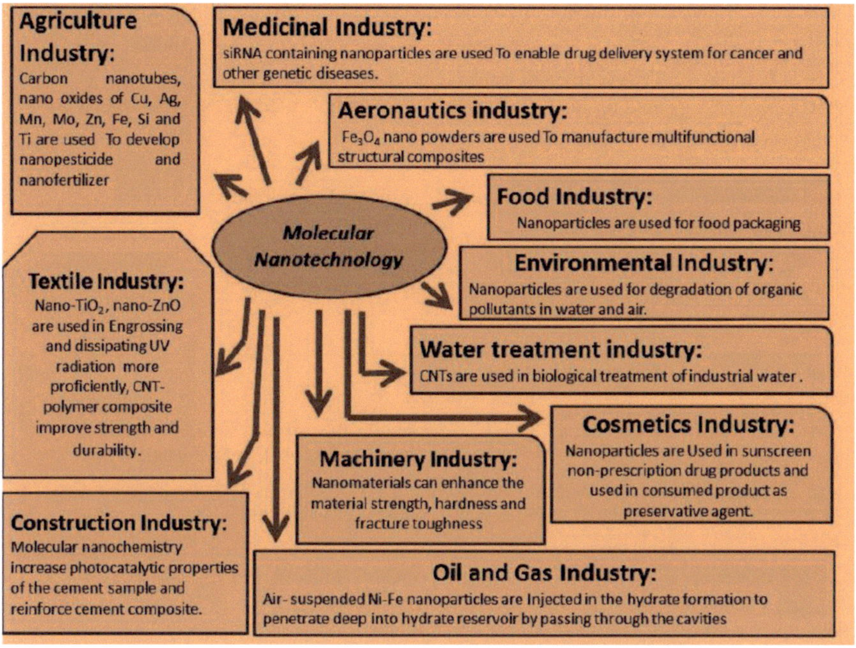

Fig. 12.4 Diverse use of nanobiotechnology in different industrial/medical/environmental purposes (Subhan et al. 2021)

fermentation of sugars (Fig. 12.4). *It has been also noticed that NPs acted as micronutrients and showed the significant impact on performance of Saccharomyces cerevisiae growth and alcohol production* (Table 12.1).

12.4 Challenges Before the Applications of Nanobiotechnology in the Process of Ethanol Production

Although implementation of nanobiotechnology in the process of bioethanol production is just at its budding stage and not commercialised yet, it needs to become perfect through serious scientific efforts. Nanobiotechnology possessed a huge potential to transform the entire process of ethanol production from LCB; rest of it is that this nanobiotechnology showed many other applications also (Fig. 12.4). In the case of ethanol production, implementation of NPs in whole process has significant challenges that need to be addressed carefully. One of the major concerns is innate toxicity of NPs; during the scale up of ethanol production, large quantity of NPs will be required but leakage of NPs with effluent is so easy. As a result, NPs can pollute the water reservoirs as well as may cause the terrestrial pollution of soil.

Table 12.1 Utilisation of different types of NPs for the improvement of fermentation process (Ndaba et al. 2025)

Type of biomass wastage	Type of microorganism used for the fermentation	Type of NPs used for the fermentation
Potato wastage	*Saccharomyces cerevisiae*	Fe_3O_4
Basal medium	*Clostridium ljungdahlii*	Silica and Cobalt based
Wastage of potato and black liquor	*S. cerevisiae*	CuO
Yeast extract, peptone and dextrose	*S. cerevisiae*	Hist. altered Fe_3O_4
Wastage of banana (peel)	*Pichia kudriavzevii*	NiO
Yeast extract peptone, dextrose	*S. cerevisiae*	Nanozymes such as FeCO-Prussian blue analogue, FeCo-N-C, FeN_3P-single-atom nanozymes
Microcrystalline cellulose	TT yeast	Silica NPs-cellulase
Peels of potato	*S. cerevisiae*	Nine metallic oxide NPs (Cu, Mn, Ni, Fe, Fe_3O_4, Fe–Ag, Zn, Co and Ag)
Potato peel waste	*S. cerevisiae*	NiO

Handling of the NPs is very crucial, especially in the case of metal-based NPs always be scared of inhaling or ingesting NPs that may cause severe toxicity to lungs and cause other health hazards. Consistent exposure to the NPs that may cause the DNA damage through reactive oxygen species (ROS) mechanism as a result chronic diseases may emerge in NPs handling individuals.

Another big challenge related to soil pollution by NPs is highly risky, especially to the fertile soil that supports the growth of edible crops. NPs decreased the fertility of soil and killed the friendly microflora of soil (Sun et al. 2022). Application of NPs at a large scale, especially for the deconstruction of LCB as a pretreatment process in the form of nanocatalysts, is highly cost intensive. To reduce the cost of NPs during scale-up, bioprocessing is all-time big challenge; pretreatment of LCB by NPs has only been tried in laboratory conditions (Cheah et al. 2020). In recent future, the cost of NPs may hinder the use of nanobiotechnology at large scale especially in the distilleries where the production of ethanol is in thousands of litres on a daily basis.

There is numerous scientific evidence present in literature that shows the nanobiotechnology contained huge practical potential to be used in the ethanol production process, especially in the case of the deconstruction of LCB and lateral stages as depicted in Fig. 12.3. But there should be need of balance such as research should be serious on reducing the toxicity of NPs and reducing the challenges of the cost of NPs, which is major problem for the developing countries (Yadav et al. 2022). Only through a comprehensive understanding and management of these challenges can nanotechnology be safely and effectively integrated into the biofuel sector (Mehejabin et al. 2024).

12.5 Development in India Through Biofuel Policies and Challenges

In the recent years India has showed the deep concern on bioethanol production to curb the dependency on oil. In current scenario it seems India is not crossing the limits to over-grow the biofuel related crops on its arable land; although India is the largest importer of vegetable oils in the world, due to feed its huge population. Even in the long run India needs to follow the strict policy of not to allow overgrow of biofuel related crops (Basavaraj et al. 2013; Chandel et al. 2017; Roy and Chandra 2019). India adopted the ethanol blended petrol program (EBPP) during first month of 2003, to avoid the over use of petrol and to bring the reduction on oil dependency. Off course ethanol blending with oil will increase the demand of ethanol more than 3.4 billion litters in coming years. In India primary source or raw material for ethanol production is molasses although production from grains is also increasing. If India needs to continue supplying huge ethanol stocks, then the food security of being a highly populous country may become threatened, especially when it requires growing ~736 million tons of sugar canes. To meet the molasses requirement, cultivation of sugarcane needs 10 million hectares of land. That seems like a tough situation that may cause the scarcity of land for the growth of food-based crops (Shinoj et al. 2011; Ray et al. 2012).

12.5.1 Employment Opportunities Related to Bioenergy/Bioethanol Sector

Numerous opportunities are produced by the distilleries for skilled and non-skilled workers in India. On the other hand, the sugar industry is the reason for providing the employment to millions of people all over India; approximately 6.0 million people depend on agricultural activities related to growing the sugar cane and transportation to sugar mills, while 0.5 million are skilled and semi-skilled industry workers. Additionally, this sector generates significant employment in ancillary and allied activities. In reality distilleries and sugar mills are the major industries in India. This industry plays the prominent role in the development of rural populations who live in villages that do not have any other sustainable work opportunities. The social lifestyle and socio-economic status of Indian villagers have excessive differences with people living in cities. This socio-economic difference may reduce if the rural population gets social security for work without any anxiety in the workplace (Chandel et al. 2017).

12.6 Conclusions

Demand of oil is increasing everywhere in the world; researchers are looking for the alternate energy sources. In this situation first generation of alcohol production is the easiest way by use of molasses/grains and yeast. But in this case use of land for the production of energy crops should be not exceeded than the land required for the production of food related crops. On the other hand, second-generation bioethanol is fantastic option that is still tedious to commercialise in developing countries due to its less production and high cost. While in the case of second-generation ethanol production, nanobiotechnology can play the major role in the future by judicious use of NPs at different stages of alcohol production. It is estimated that the number of distilleries and sugar mills will increase in future for the production of alcohol from the molasses/grains to meet the policy of blending the ethanol with gasoline. As the number of distilleries and sugar mills upsurges employment opportunities will also increase. Blending of ethanol to gasoline is beneficial and farsighted for increasing employment and reduction in the cost of gasoline.

References

Arora A, Nandal P, Singh J, Verma ML (2020) Nanobiotechnological advancements in lignocellulosic biomass pretreatment. Mater Sci Energ Technol 3:308–318

Arshad M, Abbas M, Iqbal M (2019) Ethanol production from molasses: environmental and socioeconomic prospects in Pakistan: feasibility and economic analysis. Environ Technol Innovation 14:100317

Awogbemi O, Von Kallon DV (2022) Valorization of agricultural wastes for biofuel applications. Heliyon 8(10)

Basavaraj G, Rao PP, Basu K, Reddy CR, Kumar AA, Rao PS, Reddy BVS (2013) Assessing viability of bio-ethanol production from sweet sorghum in India. Energy Policy 56:501–508

Chandel AK, Bhatia L, Garlapati VK, Roy L, Arora A (2017) Biofuel policy in Indian perspective: socioeconomic indicators and sustainable rural development. In: Sustainable biofuels development in India, pp 459–488

Cheah WY, Sankaran R, Show PL, Tg Ibrahim TNB, Chew KW, Culaba A, Chang JS (2020) Pretreatment methods for lignocellulosic biofuels production: current advances, challenges and future prospects. Biofuel Res J 7(1):1115–1127

Cherian E, Dharmendirakumar M, Baskar G (2015) Immobilization of cellulase onto MnO2 nanoparticles for bioethanol production by enhanced hydrolysis of agricultural waste. Chin J Catal 36(8):1223–1229

Devi A, Bajar S, Sihag P, Sheikh ZUD, Singh A, Kaur J et al (2023) A panoramic view of technological landscape for bioethanol production from various generations of feedstocks. Bioengineered 14(1):81–112

Dutta S, Saravanabhupathy S, Anusha, Rajak RC, Banerjee R, Dikshit PK et al (2023) Recent developments in lignocellulosic biofuel production with nanotechnological intervention: an emphasis on ethanol. Catalysts 13(11):1439

González-Leonardo M, Rowe F, Fresolone-Caparrós A (2022) Rural revival? The rise in internal migration to rural areas during the COVID-19 pandemic. Who moved and where? J Rural Stud 96:332–342

Kaur S, Khatri M, Arya SK, Singh G (2019) Stimulating effect of nanoparticles and salts on thermo and halo-tolerant cell-bonded laccase synthesis in Acinetobacter sp. UIETPU. Biocatal Agric Biotechnol 18:101085

Kazmi A, Sultana T, Ali A, Nijabat A, Li G, Hou H (2025) Innovations in bioethanol production: a comprehensive review of feedstock generations and technology advances. Energ Strat Rev 57: 101634

Mehejabin F, Musharrat A, Ahmed SF, Kabir Z, Khan TY, Saleel CA (2024) Sustainable biofuel production utilizing nanotechnology: challenges and potential solutions. GCB Bioenerg 16(10): e70001

Ndaba B, Rama H, Bingwa N, Roopnarain A (2025) Unravelling the role of nanoparticles during bioethanol production: a review on pretreatment, hydrolysis and fermentation. Fuel 396:135336

Ray SR, Amrita Goldar AG, Smita Miglani SM (2012) The ethanol blending policy in India

Roy MM, Chandra A (2019) Promoting biofuels: the case of ethanol blending initiative in India. Clean Techn Environ Policy 21:953–965

Sharma A, Singh G, Arya SK (2020) Biofuel from rice straw. J Clean Prod 277:124101

Shinoj P, Raju SS, Chand R, Kumar P, Msangi S (2011) Biofuels in India: future challenges

Shukla A, Kumar D, Girdhar M, Kumar A, Goyal A, Malik T, Mohan A (2023) Strategies of pretreatment of feedstocks for optimized bioethanol production: distinct and integrated approaches. Biotechnol Biofuels Bioprod 16(1):44

Singh G, Kumar S, Afreen S, Bhalla A, Khurana J, Chandel S et al (2023) Laccase mediated delignification of wasted and non-food agricultural biomass: recent developments and challenges. Int J Biol Macromol 235:123840

Singh G, Sahu S, Bharti S, Arya SK (2024) Significance of enzymes for the recycling of wasted non-food biomass to value added products: a sustainable stewardship towards the cleaner environment. Process Saf Environ Prot 190:395

Subhan MA, Choudhury KP, Neogi N (2021) Advances with molecular nanomaterials in industrial manufacturing applications. Nano 1(2):75–97

Sun C, Hu K, Mu D, Wang Z, Yu X (2022) The widespread use of nanomaterials: the effects on the function and diversity of environmental microbial communities. Microorganisms 10(10):2080

Yadav AK, Purushotham M, Gour NI, Gurnule GG, Choudhary VC, Yadav KR (2022) Brief review on nanotechnology as an effective tool for production of biofuels. Adv Sci Technol 117:3–8

Chapter 13
Future of Green Nanotechnology Applications in Bioenergy Area: Promises and Feasibility Analysis

Muhammad Anas, Waseem Ahmed Khattak, Erdogan E. Hakki, Javed Iqbal, Banzeer Ahsan Abbasi, and Akhtar Munir

Abstract This chapter presents a comprehensive overview of the integration of green nanotechnology in the bioenergy sector, emphasizing its potential to enhance efficiency, environmental performance, and long-term sustainability. Green nanotechnology, defined using environmentally benign materials and energy-efficient processes, has become increasingly relevant in addressing the limitations of conventional bioenergy systems. With rising global energy demand and the need to reduce greenhouse gas emissions, innovations at the nanoscale offer promising solutions for cleaner and more effective energy production from biomass. The chapter explores a range of applications where nanomaterials contribute significantly to bioenergy processes. These include nanocatalysts that accelerate biodiesel synthesis, nano-adsorbents that improve biomass pretreatment, and nanostructures that enhance microbial activity in anaerobic digestion and algal biofuel systems. Emphasis is placed on performance improvements such as higher bioconversion rates, increased gas yields, improved reaction kinetics, and the stability and reusability of nanomaterials. Furthermore, the environmental and energy impacts of nano-enabled bioenergy are evaluated using life cycle analysis (LCA) and carbon footprint comparisons. While the benefits are considerable, the chapter also addresses critical

M. Anas · E. E. Hakki
Department of Soil Science & Plant Nutrition, Selçuk University, Konya, Türkiye
e-mail: muhammed.anas@selcuk.edu.tr; eehakki@selcuk.edu.tr

W. A. Khattak
Department of Plant Sciences, Faculty of Biological Sciences, Quaid-i-Azam University Islamabad, Islamabad, Pakistan
e-mail: waseemkhattack@bs.qau.edu.pk

J. Iqbal (✉)
Department of Botany, Bacha Khan University, Khyber Pakhtunkhwa, Pakistan

B. A. Abbasi
Department of Botany, Rawalpindi Women University, Rawalpindi, Pakistan

A. Munir
Department of Chemistry, Quaid-i-Azam University, Islamabad, Pakistan
e-mail: akhtarmunir@qau.edu.pk

N. Srivastava (ed.), *Sustainable Nanostrategies for Biorefineries*, Clean Energy Production Technologies, https://doi.org/10.1007/978-981-95-5875-9_13

challenges, including nanotoxicity, material safety, regulatory uncertainty, and public perception. Ethical considerations and ecological risks are acknowledged as key concerns in the widespread deployment of nanomaterials. In conclusion, the chapter outlines a future innovation roadmap focused on biodegradable nanomaterials, scalable synthesis methods, smart sensor integration, and collaborative governance. It concludes that, when guided by sustainability principles and responsible innovation, the fusion of nanotechnology with bioenergy systems can become a cornerstone of the global transition toward a cleaner and more resilient energy future.

Keywords Green nanotechnology · Bioenergy · Nanocatalysts · Life cycle assessment · Sustainable energy systems

13.1 Introduction

13.1.1 Overview of Green Nanotechnology

The increasing demand for addressing climate change, environmental pollution, and energy security has initiated the developing concept of green nanotechnology for sustainable innovation (Naik et al. 2023). At the heart of green nanotechnology is the use of the knowledge and tools of nanoscience to create relevant processes and products that are efficient and environmentally friendly. Whereas conventional nanotechnology merely considers performance, typically leading to hidden environmentally costs, green nanotechnology focuses on sustainability throughout the production, use, and disposal of nanoparticles and nanoproducts. It encourages the use of nontoxic, renewable starting materials as well as processes with low energy consumption and low generation of byproducts. Recent developments have demonstrated the possibility of synthesizing NPs by biological routes, such as with plant extracts or microorganisms, without the use of toxic solvents and reagents (Anas et al. 2024a, b). These green composites are being used today in clean water systems, energy storage, pollution treatment, and renewable energy production. Green nanotechnology is the intersection of nanotechnology with environmental science such as green chemistry and environmental engineering (Khan 2019).

13.1.2 Relevance to the Bioenergy Sector

Bioenergy is a key component in the move towards renewable energy on a global scale (Gielen et al. 2019). This includes ethanol, biodiesel, biogas, and other forms of advanced motor vehicle fuels that are derived in whole or in part from biological feedstocks. Bioenergy systems have the potential to mitigate carbon emissions and enhance energy accessibility, but they are also restricted by practical constraints. Problems such as low conversion rate, low reaction rate, and impurity in the ultimate fuel product restrict the application of biofuel technology. In this regard, nanotechnology provides new routes for enhancement. By virtue of their high surface areas and reactivity, nanomaterials serve as excellent catalysts, adsorbents, and support materials in many bioenergy processes. For instance, metal oxide nanoparticles as catalysts have been known to increase the rate of biodiesel production, and carbon-based nanomaterials have enhanced the microbial activities in biogas systems. When environmental benign and resource-efficient synthetic methodologies are employed, these nanotechnologies may be regarded as "green" chemistry orienting towards renewable energy. This synergism of bioenergy and green nanotechnology is not just a performance story, it is a story targeting to the larger demand of clean, circular, and responsible industry (Naik et al. 2023).

13.1.3 Objectives of the Chapter

This chapter describes how green nanotechnology may significantly enhance the promise of bioenergy for the future, improving technical efficiency and providing a safer, cleaner, and greener alternative to traditional fossil energy release. The first objective is to give a straightforward overview on what green nanotechnology refers to, the principles, and materials. It subsequently shifts its attention to a more specific type of bioenergy application of these nanotechnologies including an increased yield and concentration of biofuels, enhancing microbial performance and the use of waste-to-energy systems. Case studies and examples of the latest research are included to draw attention to what can already be done and what the future holds. The last part of the chapter deals with the applied side, that is, the discussion how these technologies can be realized on large scale. Concerns about cost, their environmental safety, public perception, and policy readiness are also examined, together with areas for addressing the current situation. Our aim is to provide a balanced account, with optimism about the potential of green nanotechnology accompanied by a realistic appraisal of the work to make these solutions more widely available.

13.2 Green Nanotechnology: Concepts and Materials

13.2.1 Definition and Principles of Green Nanotechnology

Green nanotechnology is the development of clean technologies, which minimize potential environmental and human health risks, associated with the use of nanomaterials, and whose products and processes are relatively benign to the environment and human health (Anas et al. 2024a, b). It is concerned not only with improving performance, but primarily with minimizing adverse environmental effects over the entire life cycle from choice of a raw material (or other inputs) to the production process, to our use of the product, to the post-use fate of the product. It is

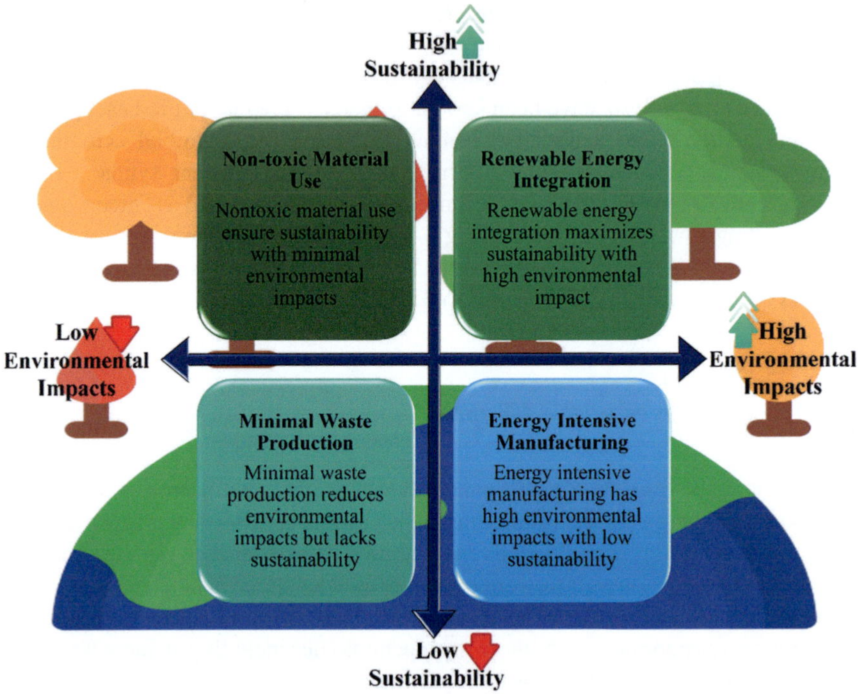

Fig. 13.1 Quadrant analysis illustrating the relationship between sustainability and environmental impacts for different production approaches—non-toxic material use, renewable energy integration, minimal waste production, and energy intensive manufacturing

based on two central concepts: that of mitigating risk in relation to the use of nanomaterials and that of using nanotechnology to develop cleaner, more sustainable processes in energy, manufacturing, agriculture, and environmental management. Key principles are based on the use of non-toxic species, renewable inputs where possible, low-energy or ambient synthesis, minimal/no waste, and integration into systems for resource circularity. Most importantly, it transcends performance to challenge whether a solution is feasible, safe, and sustainable into the distant future (Fig. 13.1).

13.2.2 Types of Green Nanomaterials

Green nanomaterials are characterized by either being synthesized by environmentally friendly methods or being nontoxic and biodegradable from the point of origin. Typical candidates are metal and metal oxide nanoparticles produced via biological methods, carbon nanomaterials graphene, carbon nanotubes and bio-based nanomaterials cellulose, starch, chitosan (Kamal et al. 2025). Each comes with a host of distinct benefits. For instance, silver and gold filaments produced by bioactive cells are characterized by active nanotopography and are used in nanoparticles because they show an excellent antimicrobial activity and only minimal toxicity after stabilization. Biochar nanoparticles prepared by biomass pyrolysis are especially convenient adsorbents for water and soil remediation (Liu et al. 2020). Indianapolis: It may sound futuristic, but *E. coli* bacteria could be the future of renewable energy. It also could help researchers at the Department of Energy's Oak Ridge National Laboratory clean up certain environmental pollutants. Such materials are chosen not just because of their functionality, but because they are developed to fit within ecosystem (Zari 2019).

13.2.3 Eco-Friendly Synthesis Techniques

Green nanotechnology relies on sustainable production methods. Nanoparticle fabrication typically uses high temperature, toxic solvents, and extremely energy-consuming methods. In contrast, green synthesis relies on biological entities or extracts, e.g., bacteria, fungi, algae, and plant leaves, for reducing metal ions and producing NPs under mild conditions (Anas et al. 2024a, b). In the process, hazardous chemicals are avoided, and it is possible to control particle size and shape using natural compounds such as flavonoids, tannins, or proteins. In another development, mechanochemical synthesis activity minimizes the use of solvents altogether. Sol-gel and hydrothermal approaches are also considered green routes providing that low temperature processes and low-waste productions are optimized (Gonçalves et al. 2021). The focus is to bring down ecological pressures along with maintaining usability. These advancements also enhance safety in production and use, as well as attenuate concerns pertaining to disposal and environmental emissions (Fig. 13.2) (Javaid et al. 2021).

13.3 Application of Nanotechnology in Bioenergy Systems

13.3.1 Nanocatalysts for Biofuel Production

Catalyst is an important factor in biofuel production, particularly to enhance the reaction rate and conversion (Bohlouli and Mahdavian 2021). At the nanoscale, catalysts have larger surface area and more catalytic sites; therefore, the reaction could be faster and more efficient. In the synthesis of biodiesel, nanocatalysts like nano zinc oxide, nano titanium dioxide, and nano calcium oxide nanoparticles have been employed to accelerate the transesterification reaction (Pandya et al. 2022). Several of these materials perform better than traditional catalysts, as they lower the energy requirements and the necessary reaction temperature. In addition, a majority of nanocatalysts are recyclable and stable in several cycles and in these cases help to

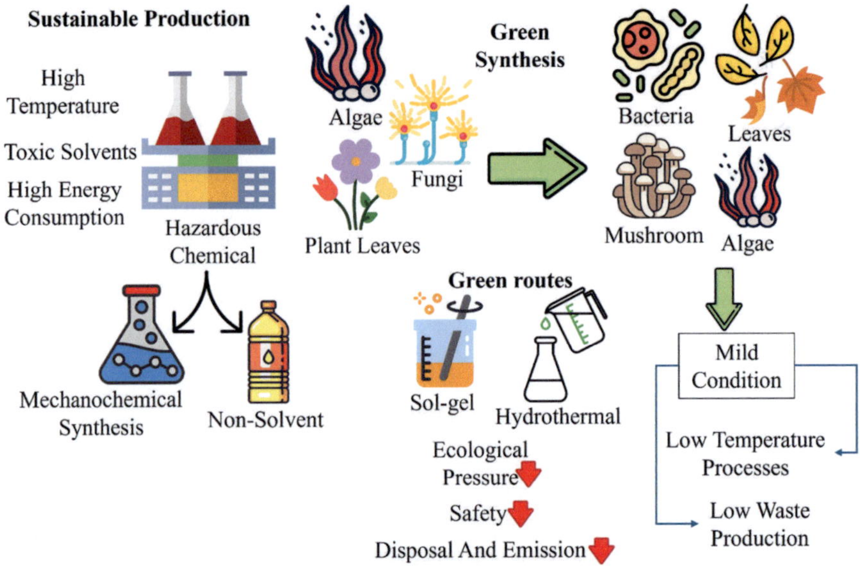

Fig. 13.2 Schematic representation of sustainable and green synthesis approaches. Traditional methods rely on high temperature, toxic solvents, and hazardous chemicals, contributing to ecological burden and energy consumption. In contrast, green synthesis utilizes biological sources such as algae, fungi, bacteria, mushrooms, and plant leaves under mild conditions through eco-friendly routes like sol-gel and hydrothermal processes. These methods offer low-temperature processing, improved safety, and reduced waste and emissions

lower the cost of the process. In bioethanol synthesis, nanoparticles have also been applied to improve cellulose enzymatic hydrolysis by making enzymes bind better or substrate be more accessible. Because nanocatalysts can speed up complex biochemical reactions without generating any toxic byproducts, they are particularly well-applicable in the clean energy sector (Abd El Rahman and Gepreel 2013).

13.3.2 Nano-adsorbents in Biomass Pretreatment

Biomass typically needs to be pretreated before it is made into fuel; this treatment melts tough structural materials such as lignin and cellulose. Old-school methods can be laborious and even chemical-laden. Nano adsorbents are a potential new and clean alternative to this process. Nanoparticles such as nano-silica, nano-clay, iron oxide nanoparticles, etc., can easily interact with the impurities because of their high surface area, and reactivity can help in the adsorption of impurities and removal from the system. In certain systems, they help delignify—loosen the lignocellulosic material so that it becomes easier for enzymes to work on cellulose and hemicellulose. These materials precision and efficiency can decrease both chemicals and time and increase the fermentable sugar concentration, which is used as a raw material for biofuel or planktonic strain culture (Rawoof et al. 2021).

13.3.3 Nanostructures in Algal Bioenergy Systems

Algae are considered a highly promising feedstock for third-generation biofuels because of their high lipid content and fast growth (Chowdhury et al. 2019). Nevertheless, the harvesting, lipid extraction, and processing challenges are still preventing their successful commercialization. In each of these steps, nanostructured materials are currently being used as enhancements. For instance, magnetic nanoparticles can simplify the process of algal biomass-water separation (Safarik et al. 2017). In the presence of well-suited functional groups, such particles bind algal cells and help in the collection process with the aid of magnetic force. Nanoporous membranes are also used in lipid recovery, with selective sieving and avoidance of organic solvents. In addition, nanoscale micronutrients and growth promoters can be supplemented to the algae cells to achieve high lipid production. These technologies enhance both energy-efficiency and enhance the utility's entire system economics and environmental performance (Fig. 13.3) (Vujanović et al. 2021).

13.3.4 Role of Nanomaterials in Anaerobic Digestion

Anaerobic digestion has become a popular process to convert organic waste to biogas, which contains mostly methane. While this method is successful, it is relatively slow and sensitive to operating disturbances. Nanomaterials are considered as a promising enhancement agent to treat anaerobic digestion systems with respect to the stabilization and production. For example, iron oxides and carbon derivatives can stimulate the enzymatic activity of microorganisms by acting as

Fig. 13.3 Enhancing biofuel production with nanotechnology

electron carriers in syntrophic microbial systems to accelerate the degradation of complex organics. There are other materials such as silver nanoparticles, which are antimicrobial agents, used in controlled amounts to suppress the growth of rubber-inhibitory organisms (Kandil 2024). Furthermore, use of nanostructured additives enhances the separation of the sludge and minimizes toxic by-product formation. In general, the strategic employment of nanomaterials in anaerobic system assists in enhancing (a) methane yields, (b) process time, and (c) strengthening the performance of the systems to fluctuate conditions (Rocha-Meneses et al. 2022).

13.4 Performance Enhancements Through Nanointegration

13.4.1 Efficiency Improvement in Bioconversion

The use of nanomaterials in bioenergy systems extends a novel horizon to accelerate the efficiency of bioconversion, an important performance figure of merit in the production of renewable fuel (Mumtaz et al. 2022). Bioconversion is the process of converting organic matter, such as lignocellulosic biomass or microalgal feedstock, into simpler compounds such as sugars, alcohols, or methane, using either a biological or a chemical process. It has historically suffered from many limitations such as recalcitrant biomass structure, enzyme inhibition, and non-ideal reaction conditions. Nanotechnology has a number of potential tools to address these issues, to facilitate faster, more specific, and energy-efficient biomass degradation (Ayub et al. 2025). One of the major roles for nanomaterials is that they can play in the enzymatic hydrolysis, a critical step in the conversion of polymeric complex biomass into fermentable sugars. This process is often accompanied by enzymes, such as cellulases and hemicelluloses, which however degrade rapidly, are costly, and require specific environmental conditions. When enzymes are loaded onto nanocarriers (e.g., magnetic nanoparticles, mesoporous silica, graphene oxide), they show superior thermal stability, activity, and reusability properties. The nanosize support offers a high active surface area and proper orientation and hence facilitates the enzyme-substrate interaction, subsequently diminishing enzyme leaching. In addition, immobilized enzymes can be recovered simply by a magnetic separation or filtration with a low operating cost (Bilal et al. 2018).

Nanoparticles can have also be beneficial for increasing pretreatment, necessary to break down the sturdy lignin-cellulose network of biomass. Through disrupting the lignin structure and enhancing biomass porosity, nanomaterials contribute to increased enzyme accessibility and sugar release, nano-sized CaO and ZnO particles can function as alkaline agents during delignification, employing lower amounts of chemicals and shorter times of treatment than conventional means. These upstream process enhancements directly lead to improved fermentation performance and yield downstream. In biologically mediated systems, such as anaerobic digestion and bioethanol and fermentation, nanomaterials can serve as a carrier of micronutrient and electron shuttle. For instance, iron oxide nanoparticles provide micronutrients for microbial metabolisms and enhanced electron flow between communities, hence resultant in more efficient metabolism. Overall, the incorporation of nanotechnology in bioconversion is not only enhancing the bioconversion to better conversion efficiencies but also leading to lower energy cost, less chemical feedstock, and greener process area.

13.4.2 Enhanced Gas Yields and Reaction Kinetics

One of the most evident advantages of nanointegration in bioenergy is the enhancement in gaseous production (mainly methane and hydrogen) and reaction kinetics. Gas production is a direct measure of the energy produced by a system, and thus, increased gas production enhances the energy return on investment of the system. Faster reaction kinetics also allow a reduction in retention time, in reactor volume, and in cost of operation. In anaerobic digestion, a common method for biogas recovery from organic waste, nanoscale materials contribute in microbial and physicochemical terms (Dikshit et al. 2023). Metallic nanoparticles such as Fe_3O_4, Ni, Co, and CuO have exhibited the potential to be used as traces of nutrient offer for the stimulation of the metabolic activity of methanogens (Jadhav et al. 2022). These materials mediate the syntrophic between microorganisms and the extracellular electron transfer that result in an enhanced decomposition of complex organics. This, in response, results in a higher methane yield rate and higher volumetric product productivity. In addition, nanomaterials serve buffering of the pH, which also hampers an accumulation of inhibitory compounds like ammonia or volatile fatty acids, which in turn contributes to the maintenance of the ideal digester environment. In thermochemical processes, such as gasification or pyrolysis, nanocatalysts increase the capability to convert biomass to syngas (comprising hydrogen, carbon monoxide, and methane) (Chidhambaram et al. 2023). Nanocatalysts consisting of cerium oxide, nickel alloys, or noble metals greatly mitigate tar formation (a common challenge in gasification) and enhance the carbon conversion efficiency. Because of their (high) surface area and specific active sites, they reduce the activation energy of reactions, thus making it possible to work with lower temperature and with a greater diversity of feedstocks. The outcome is boat gas with a higher calorific value and hence less gas processing required downstream.

Nanomaterials also enhance the hydrogen production by dark fermentation and photo fermentation, and microbial strains, in these ranging from photoautotrophic (Biswal et al. 2023), photoheterotrophic organisms (Biswal et al. 2023), to specific strains of microorganisms like bacteria, are used to convert carbohydrates into hydrogen gas. In such systems, nanoparticles such as molybdenum disulfide or titanium dioxide can serve as photocatalysts, enhancing hydrogen production by facilitating light absorption and promoting efficient charge separation. Furthermore, the nanoparticles could alleviate the accumulation of hydrogen-consuming bacteria resulting in higher hydrogen partial pressures. Reaction kinetics (i.e., the rate at which biomass is converted to biofuel or gas) could, for example, be improved with nanoscale additives that enhance substrate reactivity, mass transfer, and catalytic contacts. Utilizing carbon nanotubes, for example, in digesters can enhance the degree of organic matter dispersions and can facilitate for more rapid microbial access. The enhanced exchange of ions and electrons between the anode and cathode results in high current densities and can result in increased energy recovery for Bioelectrochemical systems such as Microbial Fuel Cells (MFCs). These gas yield and kinetics improvements, when combined, mean that assigned plant performance in addition to the overall environmental progress and commercial feasibility accomplish ability becomes better (Wu et al. 2021).

13.4.3 Stability and Reusability of Nanomaterials

The advantages of using nanomaterials in bioenergetics are clear, but long-term stability and reuse are important parameters for their economic and environmental costs. The high recycling cost of the expensive nanomaterials would be prohibitive, and the unstable or aging materials may destroy system reliability and ecological safety (Brar et al. 2022). For the sake of stability, scientists have produced surface-engineered nanomaterials that would remain undegraded when subject to harsh operating conditions. It has been found that, for instance, coating metal oxide nanoparticles with silica and embedding them within a polymer matrix, among other treatment approaches, can help eliminate oxidation, agglomeration and leaching. Also, other carbon-coated magnetic nanoparticles have a strong structural integrity and magnetic responsivity, being easy to recover and recycled in catalytic or microbial systems. A notable feature of any successful catalytic system is the ability to reuse the nanomaterial, which may provide the common or distinct catalytic species over several cycles. For example, nanocatalysts employed in biodiesel transesterification are frequently recoverable by simple filtration or magnetic separation and have shown low loss of activity upon repeated application (Tamjidi et al. 2021). It has been reported that some nano-catalysts maintain >90% of their initial activity even after five or more cycles. This not only has economic advantages in terms of material cost but also lessens the environmental impact of disposal of nanomaterials.

In enzymatic system, immobilized enzymes on the carriers like silica or carbon nanotube have a longer operational life than do free enzymes (Kumar et al. 2019). The immobilization not only limits the enzymes from shear forces and temperature changes but provides reusability of extreme importance for commercial applications. For MFCs and anaerobic reactors, nano-structured-based electrode materials show good stability in both aerobic and anaerobic environments, and long-term operation with low amount of maintenance is feasible. However, considerations of ecotoxicity and residual accumulation are still issues when the use of nanomaterials in open or semi-open systems is concerned. The leach of nanoparticles may have an impact on the diversity of the microbial community or may cause effluent contamination if not managed properly (Simonin and Richaume 2015). For this reason, efforts are now being made to develop nanomaterials that are biodegradable or bio-inspired (e.g., chitosan, cellulose, lignin) and that provide the same advantages in performance, but with reduced environmental hazard (Saraswat et al. 2023). Life cycle analysis is also receiving consideration with studies of the total environmental impact of nanoparticle manufacture, use, and disposal. The green synthesis of nanomaterials and the development of nanomaterials that can be recycled or safely degraded represent the main approaches of coupling performance with sustainability. The performance improvements achieved by nanointegration do not end after immediate increases in efficiency or optical output. The re-usability of the material and its long-lasting stability are factors that provide the right balance for further scaling these gains in a responsible manner. If nanomaterials are engineered for longevity and safety, it may be possible that they can provide sustainable, efficient, and recycle-friendly solutions to bioenergy systems from feedstock to process to end value products.

13.4.4 Life Cycle Analysis (LCA) of Nano-enabled Bioenergy

The integrated approach for environmental and energetics of the nanotechnology within the bioenergy systems can be considered in the life cycle analysis (Elsamadony et al. 2021). Certainly, life cycle analysis (LCA) provides an integrated approach to environmental impact assessment from cradle to grave (raw material acquisition through production, use, and disposal). For bioenergy systems in which nanomaterials are applied as catalysts, adsorbents, or supporting structures, both the nanomaterial and the bioenergy system should be considered across the entire life cycle. It includes the energy and material involved in nanoparticle production, emitted emissions during application, and the consequences of disposal or reuse. Initial LCA investigations of bioenergy enabled by nanotechnology have indicated promising benefits as well as novel challenges. Nanocatalysts for accelerated biodiesel production or enhanced enzymatic hydrolysis can lower energy consumption per unit of biofuel (Zhang et al. 2022). Nonetheless, if this kind of nanomaterials is strictly generated with energetic processes and/or toxic reactants, the "environmental" shoulder could be moved upstream the production process (Fig. 13.4). Thus, greener methods to synthesize nanoparticles, such as from plant extracts and at low-temperatures, can then help further enhance the sustainability perspective of the complete setup. Integrating recyclable or biodegradable nanomaterials can further close the loop and thereby minimize long-term ecological risks (Solomon et al. 2024). On the whole, the full LCA supports researchers and industry practitioners in going beyond individual performance scores and helps them to weigh the different trade-offs across the value chain. It encourages system analysis and optimization not only to maximize synergy but also to minimize hidden environmental

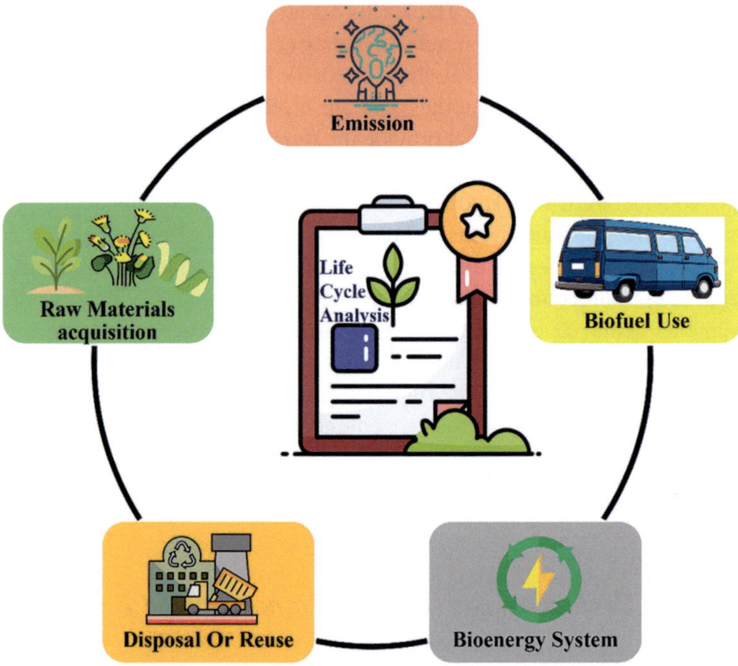

Fig. 13.4 Life cycle analysis (LCA) framework illustrating key stages in a bioenergy system. The cycle begins with raw material acquisition, followed by integration into a bioenergy system and biofuel utilization. Subsequent phases include emission evaluation and disposal or reuse of byproducts. LCA serves as a comprehensive tool to assess environmental impacts across each stage, promoting sustainable biofuel development

costs of the nano-assisted energy production, making nano-enabled bioenergy indeed enhance sustainability.

13.4.5 Carbon Footprint Comparison with Conventional Systems

This includes bioenergy's potential to reduce greenhouse gas (GHG) emissions when compared with fossil fuel systems (Staples et al. 2017). But once you add nanomaterials to the mix, the carbon equation looks markedly different in some keyways some good, others requiring caution. If used adequately, nanotechnology has the potential to reduce the carbon footprint of bioenergy through higher conversion efficiency, decreased process energy requirements, and higher yields with the use of less input material. For example, nanocatalysts that are used in biodiesel production can work at lower temperatures and shorter reaction times, reducing energy usage considerably. Similarly, gas separation in biogas upgrading can be achieved by nano-structured membranes, which can substitute much more energy consuming techniques as cryogenic distillation or chemical scrubbing (Aworanti et al. 2023). These enhancements result in direct CO_2 emissions reduction in the fuel processing section. But we have also to consider the copoints of the nanomaterials themselves. Energy and resources: Although there are much, we do not know about it, the process of mining, processing, and manufacturing nanoparticles, particularly those using rare or non-renewable materials, can be energy consuming. In this sense, the evaluation of the net carbon impact of nano-enabled systems requires careful reconciliation of on-site savings with off-site production emissions. Comparative reports of nano-enabled bioenergy and traditional fossil-based systems always present large emissions reductions, especially when considering that many times green synthesis and circular design approaches represent net negative emissions. For instance, substituting diesel with nano-optimized biogas can cut lifecycle CO_2 emissions by 60–80%, based on feedstock and process variables. Such benefits, however, require responsible materials selection and effective recovery or reuse of nanomaterials following use (Keshavan et al. 2024).

13.4.6 Waste Reduction and By-Product Utilization

Waste reduction and resource wastage are key issues in energy production and material production (Shah et al. 2021). One of the less-focused aspects of the contribution of nanotechnology on bioenergy systems is its role in waste minimizing and the conversion of the bi-products previously forced to deposal. Biomass pretreatment, for example, with nano-adsorbents, proves to be more selective in removal of inhibitory components and contaminants, thereby lessening the use of strong chemicals in pretreatment, and reduces the amount of sludge produced (Janani et al. 2022). Also, nano-structured catalysts and membranes lead to a more complete conversion reducing the level of residual biomass and un-reacted intermediates. Nanoparticles have also been found in anaerobic digestion systems to enhance the decomposition of complex substrates, to decrease the detention time, and to reduce the volume of non-digested waste. In addition, the presence of nanomaterial is frequently the source of new possibilities of by-product valorization. Nanocomposite pyrolysis-derived biochar is a versatile approach to make use of biochar which is the residue of pyrolysis laden with nanoparticles aimed for soil conditioning, pollutant adsorption, or carbon sequestration. The glycerol that can be produced, for example, during biodiesel production, may be upgraded to valuable chemicals by nanocatalysts with economic and environmental benefit (Esmaeili 2022). Residues of algae after oil extraction can also be a source of fertilizers, animal feed, or biochar, especially when combined with nanoscale additives in which nutrient retention or stability is increased. Adopting a waste-to-resource mindset at the design stage converts bioenergy systems from a linear conversion process to a more circular and restorative system. Nanotechnology is the supporting actor it isn't just about enhancements Lunarscape Increased recovery rates performance, but it is having a bonanza time because it can recover at higher levels, get a cleaner output, and actually have other revenue streams even out of "waste" streams (Pino et al. 2022).

13.5 Economic and Technical Feasibility

13.5.1 Cost-Benefit Analysis of Nano-bioenergy Integration

The incorporation of nanotechnology to bioenergy systems leads to significant technical benefits, but it must be balanced by its economic input. The cost-benefit is a useful tool to evaluate whether benefits (in terms of efficiency of yield and reduced environmental impact) FW (2012) by nanomaterials compensate the effort and costs of R&D and production (Sharma et al. 2021). Use of nano-catalysts, nano-adsorbents, or functional nanostructures has been proven to decrease the running cost of bioenergy plants by decreasing the energy input requirement, reducing the reaction time, and escalating the yield of conversion process. This translates into material savings in fuel production and treatment. Nevertheless, the anticipated cost on the nanomaterials is a matter of concern. High-purity nanoparticles, particularly of rare or engineered metals, can be costly if they are synthesized, and large-quantity availability is commonly restricted (Baig et al. 2021). For example, noble-metal nanoparticles (e.g., platinum based, or palladium based) are very good catalysts; however, their cost prohibits their use for mass production. Chinese market volatility has also been attributed to coronavirus, but as trends emerge in green synthesis (utilizing agricultural waste or plant extracts for nanoparticle synthesis) that can decrease costs (economic and environmental), one can hope the economic scenario will balance out. In systems where the nanomaterial can be recovered and reused (e.g., with magnetic catalysts or immobilized enzymes), the long-term economic argument is more convincing (Mariño et al. 2021). Repeated use across cycles decreases material cost and waste and increases overall return on investment. Moreover, with growing maturity of such technologies, it is anticipated that economies of scale, better manufacturing methods would lead to reduction in costs. Thus, although cost is a hurdle in the near term, the long-term economic perspective of nano-bioenergy integration is promising as well, especially if the benefits for environment and performance are considered in the valuation.

13.5.2 Industrial Scalability and Supply Chain Considerations

Whether a technology is new, and yet imperfect, the leap from the lab to the industrial deployment is a litmus test. For nano-bioenergy systems, scalability is limited by factors such as process compatibility, feedstock availability, material supply, and adaptation to existing systems. Although nanomaterials have demonstrated impressive performance in the laboratory, their performance in large and complex industrial settings remains to be established. Up scaling the nanotechnology leads directly to issues such as particle quality consistency, stability at operation conditions or even reactors, pumps, membranes, etc. Supply chain dependability is another consideration. A lot of the metals involved in the construction of nanomaterial, for instance, cobalt, nickel, and the rare earth elements, are affected by market forces, geopolitics, and availability. If supplies are tight or there are few sources of a material, spots of supply can also push up prices and make supplies scarce. This risk emphasizes the need to transition to biobased or locally sourced solutions (e.g., bio-based nanomaterials or waste-based nanostructures) that can be produced more sustainably and responsibility (Fritsche et al. 2020). Additionally, the development of industry is dependent on the availability of a technical workforce possessing an understanding of nanoscience and skills for operating in the bioenergy area. Training and empowerment is also a requirement to have a learning curve for the operation and maintenance. Regulatory compliance also must be taken into account by companies—particularly those in waste treatment or air. Equipment modifications might be necessary to handle nanoscale additives in a safe and effective manner, increasing up-front costs. Despite these barriers, a few pilot-scale systems have effectively implemented nanomaterials in the conversion of biodiesel plants, anaerobic digesters, and algal biofuel systems. These demonstrations underscore the scalability and versatility of nano-bioenergy technologies for industrial application, given careful system and reactor design.

13.5.3 Technology Readiness Levels (TRL) Assessment

Technology Readiness Levels (TRLs) provide a structured framework for assessing how close a technology is to full commercial deployment. In the case of nano-bioenergy systems, scalability presumes that process is compatible with existing infrastructure, that feedstock is available, that material can be supplied, and that it can be integrated with existing technologies. Whilst the laboratory-based performance of nanomaterials is impressive, their performance still needs to be proven on a large scale and in complex industrial scale systems. The nanotechnology to up-scales raised directly issues such as quality consistency of particles, stability at operation conditions or even at reactors, pumps, membranes, etc., and supply chain robustness. Many of the metals needed for the development of nanomaterial, such as cobalt, nickel, and the rare earth elements, are subject to market forces, geopolitics and supply. And when supplies are tight, or few suppliers offer a material, supply spots also can pinch prices and lead materials to become scarce. This risk illustrates the necessity to shift towards bio-based or local solutions (e.g., bio-based nanomaterials or waste-based nanostructures) that can be sustainably and socially acceptably manufactured. Further, industry also requires a technical workforce with skill in nanoscience and the ability to work in the field of bioenergy. It is also a matter of training and empowering—we need to have a learning curve in the operation and maintenance. Companies also need to consider regulatory compliance—especially those for waste treatment or air (Mazzi et al. 2024). For processing, costly modifications to equipment may be required to handle nanoscale additives in a safe and efficient manner, resulting in an increase in initial outlay. Despite these hurdles, some pilot scale systems have successfully deployed nanomaterials for conversion of biodiesel plants, anaerobic digesters, and algal biofuel systems. These results highlight the potential for scalable and adaptable nano-bioenergy technologies for industrial and other purposes, with appropriate system and reactor design.

13.6 Challenges and Risk Considerations

13.6.1 Nanotoxicity and Ecological Risks

Nanotechnology can bring great benefits to bioenergy systems, but the potential toxicity and environmental impact of nanomaterials is one of the most urgent issues. Because of their small size and high reactivity, nanoparticles can have unpredictable effects on biological tissues. Upon entering the environment (either intentionally or accidentally), some nanomaterials can last long periods of time, concentrate, or change into states hazardous for soil, water, and potentially human health. For instance, some metal oxide nanoparticles such as ZnO or silver NPs have already reported to cause oxidative stress in aquatic organisms, to impact microbial communities in soil and to disturb enzyme activities in plants and animals (Anas et al. 2025; Rajput et al. 2018). Nanoparticle release into digestates, effluents, or emissions In bioenergy processes, where nanomaterials are frequently employed for catalytic activation of feedstock or as microbial enhancers, there is clear potential for nanomaterials to be released into digestates, effluents, and emissions, which must be carefully monitored and (where required) managed. Open systems like anaerobic digesters or wastewater-fed algal ponds are especially at risk from uncontrolled spillage. In addition, long-term consequences are poorly understood. Most toxicity investigations are of teeny duration and performers in artificial conditions and thus ignore the full range of Factor X occurring in the environment. These are exactly the concerns that need to be addressed, but more effort must be paid to safer nanomaterial design—those that biodegrade, bind strongly to a substrate, or are made of biocompatible materials. The post-use lifecycle should be considered in the lifecycle assessment, and effective monitoring mechanisms should be introduced in operations to avoid environmental pollution. If nanotoxicology is ignored, the green facade of nano-enabled bioenergy could receive a blow from unexpected ecological impacts (Kumar et al. 2024).

13.6.2 Regulatory Barriers and Knowledge Gaps

The quick evolution of nanotechnology sometimes has progressed at a faster pace than the emergence of well-defined and consistent regulatory initiatives. In some countries, nanomaterials are not explicitly included in current environmental or industrial safety regulations. Definitions, taxonomies, and testing standards are inconsistent across locations, which increases uncertainty for manufacturers, investors, and policymakers. Such regulatory uncertainty may delay the adoption of nano-platform bioenergy technologies and create legal and practical risks for developers. For example, the European Union's REACH regulation mandates safety data for chemicals—and many nanomaterials are a bit iffy there, especially when they are in the form of a composite or embedded in a complex matrix. Different nanomaterials cannot be compared or judge by the safety with the lack of standard testing technology. This problem is particularly pronounced in developing countries, where regulatory bodies may not have the technical ability to assess new materials or to monitor handling. Another significant deficiency is the lack of longer-term environmental and health data. Much of the work to date on nanomaterials has been on immediate reactivity and acute toxicity, with few data available on chronic exposure, bioaccumulation and multi-generational effects (Ellis et al. 2020). Such lack of knowledge renders risk assessments speculative and policy responses tardy. Close of such gaps will depend on increased cooperation between academic researchers, industry, and regulators. Investment in independent risk research, public databases, and transparent reporting mechanisms also will be crucial to creating a regulatory environment that ensures public and environmental safety without choking off innovation.

13.6.3 Public Perception and Ethical Issues

Public perception is crucial in the take-up and success of new technologies. Nanotechnology has great promise, but it also can be met with concern or fear because it is new and has unknown risks. For bioenergy, which is already plagued with land use disputes, food security issues, and emission arguments, the use of nanomaterials just adds another debate for consumers, communities, and environmental groups to have about the technology (Behera and Varma 2019). A need for better information and public involvement has led to misunderstandings about nanotechnology. For example, even technologies that are well studied will struggle to gain acceptance if fears about “invisible particles” entering the food or water supply, or the notion that nanoparticles are inherently dangerous, create resistance. There are also ethical concerns about the transparency of application of nano materials in energy or agricultural systems. And there is an equity aspect to this as well. Rewards from new technology often flow to richer areas or to particular industries, while the environmental and health risks can fall disproportionately on the vulnerable, particularly residents of communities near biomass facilities or waste treatment plants (Mittlefehldt 2018). Left unchecked, this has the potential to widen social divides and undermine public trust. To get through these obstacles, the communication must be open and honest. Developers need to work with communities early in the planning process, share data on safety, and address concerns in a transparent manner. Ethical considerations that focus on fairness, environmental justice, as well as informed consent can facilitate the ethical development of nano-bioenergy solutions.

13.7 Future Directions and Innovation Roadmap

Nanotechnology and bioenergy are emerging as most promising field of clean and efficient next generation energy technologies. Yet, achieving this potential in practice will require a visionary approach that goes beyond technical progress and considers issues of safety, scalability, and societal readiness. The prospect of nano-bioenergy will be on smart, adaptive, and sustainable systems that can adapt to the complex-environmental-economic challenge. One of the key innovation directions is the trend toward bio-mimetic and bio-degradable nanomaterials. Conventional nanomaterials are very practical, but they are usually a source of environmental problems and toxic effects. Moreover, the future evolution of nanostructures should be directed toward natural and renewable sources, like cellulose, chitin or lignin, as precursors. These can be designed to have catalytic, adsorptive, or structural function and to safely disintegrate at the end of its lifetime. To be one step closer to these principles will allow the nano-bioenergy systems to meet the circular economy, instead of posing concerns about long-term ecological contamination. The utilization of smart sensing/automation in nano-bioenergy platforms is yet another potential frontier. Nanosensors that are measuring the pH, temperature, gas composition, or the microbial activity allow that the systems are becoming self-optimizing. This can increase process robustness, and yields, and decrease manual intervention. These technologies combined with machine learning algorithms and remote monitoring explain how, further down the line, decentralized, modular bioenergy units are conceivable for rural and off-grid applications. From a technology development standpoint, the roadmap must focus on the scalable syntheses of the materials. Many high-performance nanomaterials are still fabricated by laboratory procedures not commercially viable at industrial level. It will be necessary to mainstream the use of cost-effective, energy-saving, and reproducible high-throughput production technologies, particularly those relying on green chemistry.

Such technology includes mechanochemistry, microbial synthesis, and low-temperature catalysis with reduction of waste and energy consumption.

Interagency cooperation will be crucial, as will intersectoral cooperation. The carrier will be more economical and practical by involving, for example, "the energy (sector), agriculture, materials, and environmental policy sectors in reaching the solution." Public-private partnerships and research consortia can assist in the translation of academically inspired discovery into commercially viable products as well as promoting the dissemination of knowledge and the establishment of common practices. Finally, further policy support and citizen engagement are both key components of the roadmap. The government also has a role in creating incentives for green innovation, funding high-risk research and establishing benchmarks for safety and sustainability for the use of nanomaterials. Also important is going to be open, transparent communication with the public to engender their trust, correct misunderstandings, and guarantee that the innovation moves forward in a way that respects our ethical and societal values. Designing not-but-also-advanced but safe, fair, or/and sustainable systems for bioenergy is a grand challenge of green nanotechnology. Through appropriate innovation, regulation, and collaboration, nanobioenergy could be a key enabler of a low-carbon, secure energy future.

13.8 Conclusion

The incorporation of green nanotechnology in bioenergy is seen as a significant step forward in the development of new energy systems that are more sustainable and efficient and environmentally acceptable. In the context of the growing world energy demands as well as the need to mitigate carbon emissions, the convergence of nanoscale technologies and bio-based energy systems may bring about technical and ecological opportunities. In this chapter, we have been discussing the nascent impacts of nanomaterials in the form of catalyst, adsorbent, membrane, and structural support in all phases involved in the generation of biofuel and biogas. These clear indicators from improving bioconversion and gas yields to enhancing system stability further highlight the potential for nanotechnology to refine and improve the performance of conventional systems in bioenergy. But as always, with innovation, comes responsibility. The concerns about nanotoxicity, environmental persistence, and lack of regulation cannot be dismissed lightly. A successful future for nanobioenergy will not be measured only in terms of technical performance, but rather by its commitment to safety, ethics, and inclusivity. In all of this, public trust, policy clarity, and equitable access will be crucial to ensuring these technologies benefit more than a few sectors or regions but society at large. Moving forward, the community should give priority to scalable green synthesis approaches, focusing on the development of biodegradable and bio-based nanomaterials and the promotion of interdisciplinary cooperation. Pathways to the mainstream will depend also on government, industry, research, and community involvement. Through such a linkage between scientific progress and societal necessities, values, and participation, nano-enabled bioenergy can become a pillar of an innovative clean energy infrastructure of the future. Green nanotechnology is not just a means of improving bioenergy, it is a revolutionary way of looking at energy, and if handled responsibly, could completely change the way energy is generated, consumed, and priced in a world with finite resources.

References

Abd El Rahman AM, Gepreel MA (2013) Nanotechnology applications in water treatment: future avenues and challenges: a review. Sci Technol Applic Alexandria, Egypt

Anas M, Falak A, Khan A, Khattak WA, Nisa SG, Aslam Q, Khan KA, Saleem MH, Fahad S (2024a) Therapeutic potential and agricultural benefits of curcumin: a comprehensive review of health and sustainability applications. J Umm Al-Qura Univ Appl Sci:1–16

Anas M, Khattak WA, Fahad S, Alrawiq N, Alrawiq HS, Abdelsalam NR, Jaremko M, Quraishi UM (2024b) Mitigating heavy metal pollution in agriculture: a multi-omics and nanotechnology approach to safeguard global wheat production. J Hazardous Mater Adv 17:100584

Anas M, Falak A, Hassan S, Khattak WA, Saleem MH, Khan KA, Khalid A, Fahad S (2025) Microbial interactions and bacterial responses to metal stress in plants: mechanisms, adaptations, and applications for sustainable agriculture. J Crop Health 77(1):36

Aworanti OA, Ajani AO, Agbede OO, Agarry SE, Ogunkunle O, Laseinde OT, Kalam MA, Fattah IMR (2023) Enhancing and upgrading biogas and biomethane production in anaerobic digestion: a comprehensive review. Front Energy Res 11:1170133

Ayub A, Wani AK, Chopra C, Sharma DK, Amin O, Wani AW, Singh A, Manzoor S, Singh R (2025) Advancing dye degradation: integrating microbial metabolism, Photocatalysis, and nanotechnology for eco-friendly solutions. Bacteria 4(1):15

Baig N, Kammakakam I, Falath W (2021) Nanomaterials: a review of synthesis methods, properties, recent progress, and challenges. Mater Adv 2(6):1821–1871

Behera BK, Varma A (2019) Bioenergy for sustainability and security. Springer

Bilal M, Zhao Y, Rasheed T, Iqbal HM (2018) Magnetic nanoparticles as versatile carriers for enzymes immobilization: a review. Int J Biol Macromol 120:2530–2544

Biswal T, Shadangi KP, Sarangi PK (2023) Application of nanotechnology in the production of biohydrogen: a review. Chem Eng Technol 46(2):218–233

Bohlouli A, Mahdavian L (2021) Catalysts used in biodiesel production: a review. Biofuels 12(8): 885–898

Brar KK, Magdouli S, Othmani A, Ghanei J, Narisetty V, Sindhu R, Binod P, Pugazhendhi A, Awasthi MK, Pandey A (2022) Green route for recycling of low-cost waste resources for the biosynthesis of nanoparticles (NPs) and nanomaterials (NMs)-a review. Environ Res 207: 112202

Chidhambaram N, Kay SJJ, Priyadharshini S, Meenakshi R, Sakthivel P, Dhanbalan S, Shanavas S, Kamaraj S-K, Thirumurugan A (2023) Magnetic nanomaterials as catalysts for syngas production and conversion. Catalysts 13(2):440

Chowdhury H, Loganathan B, Mustary I, Alam F, Mobin SM (2019) Algae for biofuels: the third generation of feedstock. In: Second and third generation of feedstocks. Elsevier, pp 323–344

Dikshit PK, Padhi SK, Pattanaik L, Khan A, Ranjan A, Sadhu S (2023) A critical review on nanotechnological advancement in biogas production from organic waste. Biomass Convers Biorefinery 15:26737–26759

Ellis L-JA, Valsami-Jones E, Lynch I (2020) Exposure medium and particle ageing moderate the toxicological effects of nanomaterials to Daphnia magna over multiple generations: a case for standard test review? Environ Sci Nano 7(4):1136–1149

Elsamadony M, Elreedy A, Mostafa A, Fujii M, Gescher J, Shakeri Yekta S, Schnurer A, Gaillard J-F, Pant D (2021) Perspectives on potential applications of nanometal derivatives in gaseous bioenergy pathways: mechanisms, life cycle, and toxicity. ACS Sustain Chem Eng 9(29): 9563–9589

Esmaeili H (2022) A critical review on the economic aspects and life cycle assessment of biodiesel production using heterogeneous nanocatalysts. Fuel Process Technol 230:107224

Fritsche U, Brunori G, Chiaramonti D, Galanakis C, Hellweg S, Matthews R, Panoutsou C (2020) Future transitions for the bioeconomy towards sustainable development and a climate-neutral economy—knowledge synthesis final report. Publications Office of the European Union, Luxembourg, 10, 667966

Gielen D, Boshell F, Saygin D, Bazilian MD, Wagner N, Gorini R (2019) The role of renewable energy in the global energy transformation. Energ Strat Rev 24:38–50

Gonçalves RA, Toledo RP, Joshi N, Berengue OM (2021) Green synthesis and applications of ZnO and TiO2 nanostructures. Molecules 26(8):2236

Jadhav P, Khalid ZB, Zularisam A, Krishnan S, Nasrullah M (2022) The role of iron-based nanoparticles (Fe-NPs) on methanogenesis in anaerobic digestion (AD) performance. Environ Res 204:112043

Janani R, Gurunathan B, Sivakumar K, Varjani S, Ngo HH, Gnansounou E (2022) Advancements in heavy metals removal from effluents employing nano-adsorbents: way towards cleaner production. Environ Res 203:111815

Javaid M, Haleem A, Singh RP, Suman R, Rab S (2021) Role of additive manufacturing applications towards environmental sustainability. Adv Industr Eng Polymer Res 4(4):312–322

Kamal A, Yang J, Batool M, Ara U, Khattak WA, Touhami D, Aljowaie RM, Elshikh MS (2025) Green synthesis of zinc oxide nanoparticles, characterization and its applications in inducing disease resistance against spot blotch disease in wheat by enhancing its physiological and biochemical parameters. J Crop Health 77(1):1–12

Kandil H (2024) Significant role of novel additives in improving the antibacterial properties of rubber composite: a comprehensive review. J Vinyl Additive Technol 30(6):1372–1395

Keshavan S, Petri-Fink A, Rothen-Rutishauser B (2024) Understanding the benefits and risks of sustainable nanomaterials in a research environment. Chimia 78(6):397–402

Khan SH (2019) Green nanotechnology for the environment and sustainable development. In: Green materials for wastewater treatment. Springer, pp 13–46

Kumar A, Park GD, Patel SK, Kondaveeti S, Otari S, Anwar MZ, Kalia VC, Singh Y, Kim SC, Cho B-K (2019) SiO2 microparticles with carbon nanotube-derived mesopores as an efficient support for enzyme immobilization. Chem Eng J 359:1252–1264

Kumar A, Jayeoye TJ, Mohite P, Singh S, Rajput T, Munde S, Eze FN, Chidrawar VR, Puri A, Prajapati BG (2024) Sustainable and consumer-centric nanotechnology-based materials: an update on the multifaceted applications, risks and tremendous opportunities. Nano-Struct Nano-Objects 38:101148

Liu X, Yang L, Zhao H, Wang W (2020) Pyrolytic production of zerovalent iron nanoparticles supported on rice husk-derived biochar: simple, in situ synthesis and use for remediation of Cr (VI)-polluted soils. Sci Total Environ 708:134479

Mariño MA, Fulaz S, Tasic L (2021) Magnetic nanomaterials as biocatalyst carriers for biomass processing: immobilization strategies, reusability, and applications. Magnetochemistry 7(10): 133

Mazzi A, Barbiero C, Miserocchi F, Nisato F, Tassinato G, Cerchier P (2024) Life cycle assessment of Al-Cu-Ag-Si recycling process from photovoltaic waste. Resources, Conservation and Recycling 211:107885. https://doi.org/10.1016/j.resconrec.2024.107885

Mittlefehldt S (2018) Wood waste and race: the industrialization of biomass energy technologies and environmental justice. Technol Cult 59(4):875–898

Mumtaz M, Baqar Z, Hussain N, Bilal M, Azam HMH, Iqbal HM (2022) Application of nanomaterials for enhanced production of biodiesel, biooil, biogas, bioethanol, and biohydrogen via lignocellulosic biomass transformation. Fuel 315:122840

Naik SU, Chistie SM, Mishra RK (2023) Recent advances of bio-nanotechnology potential in the scaling up of sustainable green energy production. In: Bionanotechnology towards green energy, pp 1–32

Pandya HN, Parikh SP, Shah M (2022) Comprehensive review on application of various nanoparticles for the production of biodiesel. Energy Sour Part A Recov Utilizat Environ Effects 44(1):1945–1958

Pino P, Salmeri A, Hugo A, Hume S (2022) Waste management for lunar resources activities: toward a circular lunar economy. New Space 10(3):274–283

Rajput VD, Minkina TM, Behal A, Sushkova SN, Mandzhieva S, Singh R, Gorovtsov A, Tsitsuashvili VS, Purvis WO, Ghazaryan KA (2018) Effects of zinc-oxide nanoparticles on soil, plants, animals and soil organisms: a review. Environ Nanotechnol Monitor Manag 9:76–84

Rawoof SAA, Kumar PS, Vo D-VN, Devaraj K, Mani Y, Devaraj T, Subramanian S (2021) Production of optically pure lactic acid by microbial fermentation: a review. Environ Chem Lett 19:539–556

Rocha-Meneses L, Hari A, Inayat A, Shanableh A, Abdallah M, Ghenai C, Shanmugam S, Kikas T (2022) Application of nanomaterials in anaerobic digestion processes: a new strategy towards sustainable methane production. Biochem Eng J 188:108694

Safarik I, Pospiskova K, Baldikova E, Safarikova M (2017) Magnetic particles for microalgae separation and biotechnology. In: Food bioactives: extraction and biotechnology applications. Springer, pp 153–169

Saraswat P, Singh S, Prasad M, Misra R, Rajput VD, Ranjan R (2023) Applications of bio-based nanomaterials in environment and agriculture: a review on recent progresses. Hybrid Adv 4: 100097

Shah AV, Srivastava VK, Mohanty SS, Varjani S (2021) Municipal solid waste as a sustainable resource for energy production: state-of-the-art review. J Environ Chem Eng 9(4):105717

Sharma A, Mittal K, Arora D, Ganti SS (2021) A comprehensive review on strategies for new drug discovery and enhanced productivity in research and development: recent advancements and future prospectives. Mini-Rev Org Chem 18(3):361–382

Simonin M, Richaume A (2015) Impact of engineered nanoparticles on the activity, abundance, and diversity of soil microbial communities: a review. Environ Sci Pollut Res 22:13710–13723

Solomon NO, Simpa P, Adenekan OA, Obasi SC (2024) Sustainable nanomaterials' role in green supply chains and environmental sustainability. Eng Sci Technol J 5(5):1678–1694

Staples MD, Malina R, Barrett SR (2017) The limits of bioenergy for mitigating global life-cycle greenhouse gas emissions from fossil fuels. Nat Energy 2(2):1–8

Tamjidi S, Esmaeili H, Moghadas BK (2021) Performance of functionalized magnetic nanocatalysts and feedstocks on biodiesel production: a review study. J Clean Prod 305:127200

Vujanović M, Wang Q, Mohsen M, Duić N, Yan J (2021) Recent progress in sustainable energy-efficient technologies and environmental impacts on energy systems. Appl Energy 283:116280

Wu D, Peng X, Li L, Yang P, Peng Y, Liu H, Wang X (2021) Commercial biogas plants: review on operational parameters and guide for performance optimization. Fuel 303:121282

Zari MP (2019) Ecosystem services impacts as part of building materials selection criteria. Mater Today Sustain 3:100010

Zhang Y, Duan L, Esmaeili H (2022) A review on biodiesel production using various heterogeneous nanocatalysts: operation mechanisms and performances. Biomass Bioenergy 158:106356

MIX
Papier aus verantwortungsvollen Quellen
Paper from responsible sources
FSC® C105338

If you have any concerns about our products,
you can contact us on
ProductSafety@springernature.com

In case Publisher is established outside the EU,
the EU authorized representative is:
Springer Nature Customer Service Center GmbH
Europaplatz 3, 69115 Heidelberg, Germany

Printed by Libri Plureos GmbH
in Hamburg, Germany